D0209332

UNIFIED SEPARATION SCIENCE

UNIFIED SEPARATION SCIENCE

J. CALVIN GIDDINGS
University of Utah

A Wiley-Interscience Publication
JOHN WILEY & SONS, INC.
New York • Chichester • Brisbane • Toronto • Singapore

In recognition of the importance of preserving what has been written, it is a policy of John Wiley & Sons, Inc. to have books of enduring value published in the United States printed on acid-free paper, and we exert our best efforts to that end.

Copyright © 1991 by John Wiley & Sons, Inc.

All rights reserved. Published simultaneously in Canada.

Reproduction or translation of any part of this work
beyond that permitted by Section 107 or 108 of the
1976 United States Copyright Act without the permission
of the copyright owner is unlawful. Requests for
permission or further information should be addressed to
the Permissions Department, John Wiley & Sons, Inc.

Library of Congress Cataloging in Publication Data:
Giddings, J. Calvin (John Calvin), 1930–
   Unified separation science / J. Calvin Giddings.
     p.   cm.
   "A Wiley-Interscience publication."
   Includes bibliographical references and indexes.
   ISBN 0-471-52089-6
   1. Separation (Technology) 2. Chromatographic analysis.
I. Title.
QD63.S4G53   1991
543'.089--dc20
                                90-38139
                                   CIP

Printed in the United States of America

10 9 8 7 6 5 4 3 2 1

For information about our audio products, write us at:
Newbridge Book Clubs, 3000 Cindel Drive, Delran, NJ 08370

To mother nature who, through separation, has created beauty and bounty from formless matter.

PREFACE

The origin of this book dates back to 1969 when I first taught the course *The Principles of Chemical Separations* at the University of Utah. Lacking suitable materials, I started writing notes and issuing them to students. The notes expanded with each course offering. These quasiperiodic bursts of writing, extending over 21 years, have led to the present book.

The object of my teaching and writing from the beginning was to develop separation science as a unified discipline rather than as a farflung collection of unrelated techniques. Thus in an early preface to my notes, I wrote:

> *At the core of this approach, mass transport phenomena underlying all separation processes are developed in a simple physical-mathematical form. The rules of transport become the rules defining the potential and limits of separations. Thus we obtain a common theoretical denominator for a heretofore scattered methodology. The basic theory is applied to many of the principal techniques in such a way that limits and optima are obtained and alternate methods compared.*

A milestone was reached in 1982 when my evolving material was published at the invitation of Prof. Eli Grushka as a 100-page chapter in the *Treatise of Analytical Chemistry*. The building and refining continued. Now, in early 1990, after the sixteenth course offering (one at Cayetano Heredia University in Peru in 1974 and one at the University of Washington in 1984), the manuscript has been temporarily closed and sent to the publisher.

I use the word "temporarily" because the book is still not complete. There are dozens of changes and additions I would make given time. Once one starts forging links between different methods based on common theoretical principles, and once global optima begin to spill out, the possibilities for further integration descend as a flood. The task of unifying the diverse human endeavor of separations (which other writers have approached in different ways) is indeed formidable. Keeping the text current with the explosive growth of many branches of the discipline is equally daunting. I admit to many shortcomings in these areas. (I hope readers will be generous in their suggestions to improve future editions.) However I have been encouraged to publish the book because as techniques proliferate, the need for a more unified treatment becomes increasingly acute.

This book, as an outgrowth of its history, is written foremost as a textbook for a graduate level course in analytical separations. However I have always kept in mind the needs of lifelong students (like myself) to gain a deeper understanding and a broader perspective of the separations field. I have introduced much new material in pursuit of these goals.

As a textbook, this book aligns itself with the philosophy that educating students takes priority over training them. Students, as emerging scientists, are on the threshold of a lifelong career. Throughout chemistry, mathematics, and physics, they are taught subjects that will, to our best knowledge, endure. We have an equal obligation throughout analytical chemistry and separation science, specifically in the discipline of analytical separations.

For a lifelong education, it is more important to understand how it works than how to do it; basics rather than recipes. Procedures will change and evolve and old recipes are soon obsolete, but the underlying mechanisms of separation will be around for a long time.

Even now, changes in methodology are rampant. New techniques in chromatography are rapidly emerging and field-flow fractionation is beginning to have practical impact. Most revolutionary of all, capillary zone electrophoresis, unknown a decade ago, has assumed immense analytical significance. Yet the structural core of this book, particularly that developed in Chapters 3 and 5, has needed no substantial revision to accommodate the revolution in electrophoresis and other techniques. The "Treatise" chapter of 1982 contained much the same core and a similar (but abbreviated) treatment of electrophoresis based on that core.

As a monograph for working scientists, the same fundamental emphasis applies. Descriptions of apparatus are brief; recipes are absent. The book will be of most interest to those seeking a long-range perspective. I hasten to add, however, that the book should have immediate as well as long-term value for those wishing to think beyond recipes. The numerous discussions of optimization, the comparison of diverse methods, and the evaluation of factors enhancing and limiting separation power should be quickly profitable for the inquisitive.

I have tried to provide a flavor of research frontiers, unsolved problems, and imperfect methodology for students of all ages. I have included more references than is common for a textbook for this purpose. I have drawn heavily on my own research experience and publications in bringing this material together.

In terms of organization, the text has two main parts. The first six chapters constitute generic background material applicable to a wide range of separation methods. This part includes the theoretical foundations of separations, which are rooted in transport, flow, and equilibrium phenomena. It incorporates concepts that are broadly relevant to separations: diffusion, capillary and packed bed flow, viscous phenomena, Gaussian zone formation, random walk processes, criteria of band broadening and resolution, steady-state zones, the statistics of overlapping peaks, two-dimensional separations, and so on.

Chapter 7 is a transition chapter in which methods are compared and classified. The multifaceted role of flow in separations is dissected.

In the final five chapters, the general principles of the first part are utilized to better understand families of techniques and specific methods. The coverage highlights electrophoretic and sedimentation techniques, field-flow fractionation, and chromatography. Future editions are expected to cover extraction and membrane methods in more detail.

The last three chapters are devoted solely to chromatography. Such coverage (three of twelve chapters) may appear scanty to those who focus on chromatography in their separations course or work environment. However the coverage of chromatographic principles becomes much broader when we consider the many topics relevant to chromatography treated in the early chapters. These topics include phase equilibrium (Chapter 2), general transport theory (in 3), flow and viscous phenomena (4), the formation and properties of Gaussian zones, diffusion, zone broadening, concepts of plate height, resolution, and peak capacity (5), two-dimensional separation and the nature and statistics of peak overlap (6), the role of flow (7), and, more specifically, the introduction to chromatographic peak migration and broadening in Chapter 9. These topics are needed as a foundation for the study of chromatography, but here they serve also as the basis for electrophoresis, solvent extraction, membrane separations, field-flow fractionation, sedimentation, and other methods.

How well this book strikes a balance between theoretical and descriptive elements remains for future generations to say. For better or worse, scientific development is trending in a more formal, more mathematical direction. I have tried to use the main advantages of this formalism, which is to present a broad and general treatment applicable to many techniques. This obviously saves time in coming to grips with our exploding scientific knowledge. At the same time I have tried to develop physical insights (sadly missing from many mathematical treatments) and to keep lines open with intuition, the wellspring of creativity. Thus I have emphasized the physical and chemical basis of mathematical developments by using examples, by discussing the roles of key mathematical terms, by pointing out analogs, and lastly by writing exercises to help the student (young and old) reduce mathematics to numbers.

Acknowledging those who have assisted in this 21 year old project is doomed to failure by the antiquity of its beginnings and the ebb and flow of helpful students and colleagues too numerous to mention. I apologize to those left out. Early on, I benefited from the feedback of Drs. Marcus Myers, Karin Caldwell, Eli Grushka, Frank Yang, and Roy Keller. (I have learned much from the lucid writing style of Keller.) Drs. Rick Hartwick and Vicky McGuffin were kind enough to "field test" some of the text material in their courses. Dr. Joe Davis simulated several of the complicated peak overlap profiles used as figures. Dr. Steve Williams has also developed illustrative material and has critically read the text. More recent help in evaluating exercises, checking answers, and proofing text material has been provided by Mr. Jiang Yong and by Drs. Bhajendra Barman, Kim Ratanathanawongs, Maria Anna Benincasa, and Laya Kesner. My illustrator, Mr. Alexis Kelner, who is both a scientist and a longtime friend, worked closely with me in developing meaningful figures.

Last but not least, Ms. Julie Westwood has skillfully and patiently typed and organized multiple revisions of the manuscript, including this final one. The patience of my family, especially Leslie, in enduring the torture of writing a book, is gratefully acknowledged.

J. CALVIN GIDDINGS

Salt Lake City, Utah
March 1990

CONTENTS

MATHEMATICAL SYMBOLS

d	number of doublet peaks
d_c	capillary (tube) diameter
d_p	mean particle diameter
d'	distance between flow channels
D	diffusion coefficient
D_g	diffusion coefficient in gas
D_m	diffusion coefficient in mobile phase
D_s	diffusion coefficient in stationary phase
D_n	effective diffusion coefficient for nonequilibrium
D_T	total effective diffusion coefficient
e	proton charge
E	excess solute accumulating in polarization layer
E	electrical field strength
E	internal energy
E_η	activation energy for viscous flow
f	friction coefficient for one mole
f_1	term correcting plate height for gas compression
f'	friction coefficient for one particle
F	force per mole of particles
F'	Force exerted on single particle
$\mathscr{F}$	Faraday, or charge per mole of protons
F_1	driving force for motion
F_2	dissipative force
F_η	viscous drag force
$F(t)$	$\text{const}/t^{1/2}$
g	acceleration due to gravity
G	Gibbs free energy
G	acceleration, $\omega^2 r$
$G(t)$	$1/4Dt$
h	reciprocal of electrical double layer thickness
h	reduced plate height, H/d_p
H	plate height
H_c	mobile phase plate height accounting for coupling
H_D	diffusion component of H_c
H_f	flow component of H_c
H_g	plate height originating in gas phase
H_m	mobile phase nonequilibrium term
H_n	nonequilibrium contribution to plate height
H_s	stationary phase nonequilibrium term

$\hat{H}$	apparent plate height, $L\tau^2/t_r^2$
H	enthalpy
$\bar{H}^0$	partial molar enthalpy under standard state conditions
$\dot{H}$	heat input per unit volume and time
i	electrical current
I	gas compression constant, $(p_i^2 - p_0^2)/2L$
j	James–Martin pressure-gradient correction factor
J	flux density
J_0	constant value of flux density
k_d	first-order rate constant for desorption
k'	capacity factor, $(1 - R)/R$
K_0	specific permeability
K	partition distribution or coefficient
ℓ	mean layer thickness, $D/|W|$
l	step length of random walk
L	length of separation path: column or tube length
L_{kj}	coefficient in Onsager's equation of reciprocity
$\bar{L}$	mean external length of molecule or colloid
m	mass
m	number of solute components
$\bar{m}$	statistical estimate of component number
M	molecular weight
n	number of random steps
n	number of moles
n	exponent in Eq. 6.14, $w = -ay^n$
$\dot{n}$	number of random steps taken per unit time
n_c	peak capacity
N	number of theoretical plates
$\mathcal{N}$	Avogadro's number
p	number of apparent peaks
p	pressure
p_c	critical pressure
p_i	inlet pressure
p_i	partial pressure of component i
p_o	outlet pressure
P	compression ratio, p_i/p_o
$\mathcal{P}$	potential energy
pI	isoelectric point on pH scale
q	coefficient in plate height term; configuration factor

q	increment of added heat
Q	flowrate through single tube
q	flowrate through unit area
r	radius of curvature of meniscus
r	radial coordinate
r_c	capillary (tube) radius
R	retention ratio; zone velocity/mean fluid velocity
R'	equilibrium fraction in mobile phase
Re	Reynold's number
R_f	distance migrated by zone/distance to liquid front
R_s	resolution
$R_s^{\ddagger}$	critical resolution required of separation
$\mathscr{R}$	gas constant
s	surface area per unit pore volume
s	number of single component peaks
s	sedimentation coefficient
s	ionic strength
S	entropy
$\bar{s}^0$	partial molar entropy under standard state conditions
S	velocity correlation distance in packed bed
t	time; separation time
t	number of triplet peaks
t_d	mean desorption time
t_{ds}	stationary phase component of t_d
t_D	diffusion time
t_{eq}	equilibrium time; relaxation time for equilibration
t_r	retention time
t_s	time spent in stationary phase
t^0	emergence time of nonretained peak
T	temperature
T_b	boiling temperature
T_c	critical temperature
U	drift velocity of solute due to external fields, etc.
$\bar{U}$	the sum of U and diffusion velocity
v	flow velocity
v_c	flow velocity used for reduced velocity scale; $v_c = D_m/d_p$
v_i	inlet flow velocity
v_o	outlet flow velocity
$\langle v \rangle$	mean cross-sectional fluid flow velocity

$\tilde{v}$	time average $\langle v \rangle$
$\bar{v}$	distance average $\langle v \rangle$
$\bar{v}$	partial specific volume of solute
v_f	velocity of liquid front in capillary flow
V	volume
V	voltage or voltage drop
$\mathcal{V}$	mean velocity of component zone
$\bar{V}$	molar volume
V_m	volume of mobile phase in column
V_r	retention volume
V_s	volume of stationary phase in column
V^0	void or dead volume of channel or column
w	channel thickness
w	increment of work done on system
w	zone width
W	overall component velocity, $W = U + v$
$\bar{W}$	average displacement velocity
W_{err}	erratic component of W
x	distance along transport or separation path
x_0	distance between peaks yielding resolution $R_s^{\ddagger}$
X	distance to center of gravity of zone
X	gradients or forces
X	specified interval along axis x
X_f	distance to liquid front in TLC or PC
y	distance relative to zone center; lateral coordinate
$\langle y^n \rangle$	nth moment about the mean
z	net effective charge in proton units
z_{ion}	total ionic charge on species

GREEK ALPHABET

α	thermal diffusion factor
α	saturation factor, $\bar{m}/n_c$
α'	separation factor
α, β	designations for two regions, open systems, or contacted phases
γ	coefficient of thermal expansion
γ	activity coefficient
γ	surface tension

δ	solubility parameter
Δ	general symbol for the difference in two like quantities
ε	porosity
η	viscosity
η_b	viscosity at boiling point
θ	dispersion coefficient, D_T/D
θ	contact angle
κ	electrical conductivity
λ	retention parameter in FFF, ℓ/w
λ	density of components along axis x
λ, λ_i	eddy diffusion coefficients
μ	electrophoretic mobility, U/E
μ	chemical potential
μ^{ext}	component of μ due to external forces
μ^{int}	contribution to μ from internal chemical environment
μ^0	standard state μ reflecting intermolecular forces
μ^*	overall chemical potential, $\mu^0 + \mu^{\text{ext}}$
ν	reduced velocity, $d_p v / D_m$
ρ	solvent, carrier, or dense gas density
ρ_s	solute (sample) density
σ	standard deviation (length units)
σ^2	variance (length units)
σ_T^2	total variance due to multiple processes
σ_{12}	collision diameter of solute-carrier pair in gas phase
τ	transient time for acceleration to steady motion
τ	standard deviation in time units
$\tau_{1/2}$	half-width at half-height in time units
ϕ	flow resistance parameter
ω	rotation rate in radians per second
ω, ω_i	constants in mobile phase plate height
ω_j	mobile phase constants ($j = \alpha, \beta, \gamma$)

ABBREVIATIONS

SEPARATION TECHNIQUES

CCD	countercurrent distribution
CE	capillary electrophoresis
CZE	capillary zone electrophoresis
ElFFF	electrical FFF
FlFFF	flow FFF
FFF	field-flow fractionation
GC	gas chromatography
GE	gel electrophoresis
GFC	gel filtration chromatography
GLC	gas liquid chromatography
GPC	gel permeation chromatography
GSC	gas solid chromatography
HIC	hydrophobic interaction chromatography
HPLC	high performance liquid chromatography
IEC	ion exchange chromatography
IEF	isoelectric focusing
IEF × GE	a 2D combination of IEF and GE
IMS	ion mobility spectrometry
LC	liquid chromatography
LLC	liquid liquid chromatography
LSC	liquid solid chromatography

MCE	micellar capillary electrochromatography
PAGE	polyacrylamide gel electrophoresis
PC	paper chromatography
PFGE	pulsed field gel electrophoresis
RPLC	reversed-phase liquid chromatography
SdFFF	sedimentation FFF
SEC	size exclusion chromatography
SFC	supercritical fluid chromatography
ThFFF	thermal FFF
TLC	thin layer chromatography

OTHER ABBREVIATIONS

1D	one-dimensional
2D	two-dimensional
c	continuous μ^* profile
CBP	chemically-bonded phase
cd	continuous/discontinuous μ^* profile
Cy	cyclical field mode of FFF
Ch	chromatographic hybrid mode of FFF
d	discontinuous μ^* profile
Dl	dielectrical field (for FFF)
em	electromagnetic
EMG	exponentially modified Gaussian
F	flow system
F(=)	system with flow and μ^* gradient parallel
F(+)	system with flow and μ^* gradient perpendicular
Hy	hyperlayer mode of FFF
Mg	magnetic field (for FFF)
Nl	normal mode of FFF
S	static (nonflow) system
SF	supercritical fluid
St	steric mode of FFF
Sy	secondary equilibrium mode of FFF

UNIFIED SEPARATION SCIENCE

1

INTRODUCTION

Separation is as old as the earth. From the cloud of dust and gases that assembled as protoearth, heavier elements sank inward and condensed to form our planet. Remaining clouds of hydrogen and helium were blown away (separated) by the sun's radiation. The hot, formless core of early earth then began to fractionate further. Leaving dense elements like nickel and iron within, the lighter elements floated to the surface and crystallized into the different minerals that make up the crust of the earth. The gas-forming elements emerged from volcanoes and formed the atmosphere. Water condensed out as oceans. Thus the basic layercake of earth—core, mantle, crust, hydrosphere, and atmosphere—is structured according to the driving forces of separation acting on a grand scale.

The emergence of life brought—indeed depended upon—a kaleidoscopic variety of separation processes. Mechanisms evolved to sort out and thus organize cellular components. Carbon and oxygen were split apart by photosynthesis, leading to a carbon-rich biosphere and an oxygen-rich atmosphere. Trace metals were selectively retained to participate in biological functions.

The dabblings of homo sapiens into separations are very recent indeed when viewed in the context of these primordial separations. But even here the beginnings are lost in human antiquity. Crude food processing, such as milling, metallurgical processes to obtain metals, the extraction of dyes, flavors, and medicines, and the concentration of many materials by evaporation are certainly very ancient on a human time scale.

The separation of complex materials to obtain analytical information rather than usable products is still more recent. The selective precipitation of metal species has been used for metal determinations for a few hundred years. Complex organic separative analysis goes back to the invention of chromatography at the turn of the century. Chromatographic, electrophoretic, and other separation methods are now used in many thousands of laboratories worldwide to unravel countless complex biological, environmental, and industrial samples. The intricate makeup of our world is in much clearer focus than would have been possible without these multifaceted analytical separation techniques.

It is the objective of this book to describe the underlying science of separation systems, to describe the fundamental relationship of one method to another, to evaluate some of their relative strengths and weaknesses, and to broadly discuss the promising directions as well as the limits encountered in seeking a maximum realization of the goals of separation. We focus particularly on multicomponent separation systems of analytical importance such as chromatography, electrophoresis, field-flow fractionation, sedimentation, and their variants. We show how each of these methodologies relate to basic separative principles and thus to each other. We omit coverage of most methods used for large-scale industrial purposes (see below) and of less common separation systems.

The present book contains a great deal of new and uniquely organized material. Many of the details have not been published before, although some of the central concepts on displacement, transport, and method classification have appeared in several published articles [1–4].

In the remainder of this chapter we present some concepts which help us understand the nature and purpose of separations in the larger context of modern science and technology.

1.1 BASIC REPRESENTATIONS OF SEPARATIONS

A complete separation of a mixture of chemical constituents can be represented by (1)

$$(a + b + c + d + \cdots) \rightarrow (a) + (b) + (c) + (d) + \cdots \qquad (1.1)$$

where the parentheses represent different regions of space and the letters a, b, c, d, ... represent the individual constituents occupying those regions. According to this scheme, a group of components, originally intermixed, are forced into different spatial locations by the process of separation.

A partial separation might be of the form

$$(a + b + c + d + \cdots) \rightarrow (a) + (b + c + d + \cdots) \qquad (1.2)$$

or

$$(a + b + c + \cdots) \rightarrow (a + b) + (b + a) + \cdots \qquad (1.3)$$

which, in the first case, represents the separation of one solute, *a*, from all the others, and in the second case the enrichment of *a* with respect to *b* (*a* + *b*) in one region, and of *b* over *a* (*b* + *a*) in another, and so on.

These schemes, at whatever level of separation and enrichment, demonstrate an essential feature of all separations: components must be transported and redistributed in space in order to realize the goals of the separative operation. One must, therefore, have a close look at transport processes if one is to have a root understanding of separations. However, the transport basics do not stand alone: auxiliary considerations are the equilibrium conditions which serve as a driving force for transport, the macroscopic, microscopic, and molecular structure of the system through which transport is occurring, the details of flow, the mechanics of sample handling, and so on. These features should not be thought of as passive parameters rigidly governing separations, but as interacting elements with wide flexibility that are to be manipulated toward separation goals. This manipulation is best achieved through an understanding of the basic phenomena involved in separations, the potential range of controllable parameters, and the physical and chemical limitations that ultimately impose themselves on the achievement of separation.

1.2 ANALYTICAL AND PREPARATIVE SEPARATION

The general goals of separation can be divided into two broad categories: preparative and analytical [5].

In preparative work, the purified (or partly purified) components are generally used as drugs, fuels, metals, chemical feedstock for synthesis, or products for industrial or human consumption. There is ordinarily a fixed value to a unit weight of the purified product. The total value of product from a given separation system, of course, is roughly proportional to the amount of material isolated. Therefore, preparative separations tend to be continuous in operation and large in scale. Some of them are plant scale— enormous industrial processes such as distillation, extraction, and smelting, typically producing many tons of separated products per day (see Figure 1.1).

Preparative techniques used by industry at a reduced scale of production include adsorption, crystallization, zone refining, filtration, and electrolysis. The techniques of gas and liquid chromatography provide products of relatively high purity but have resisted spectacular scaleup; they are used in preparative work at a lower level of productivity (usually grams to kilograms/day) with columns 0.1 m or more in diameter (see Figure 1.2).

Small scale preparative work is important for recovering valuable products like recombinant proteins from complex mixtures. Because of its high resolving power, preparative scale liquid chromatography (LC) has moved rapidly into the biotechnology industry where it is used for purifying such

Figure 1.1. Crude fractionation tower at the Phillips Oil Refinery in Woods Cross, Utah. In this crude fractionator, 25,000 barrels per day of incoming crude oil is divided into six fractions by distillation. (Photo by Alexis Kelner.)

exotic components. While large quantities of biologically derived materials may be processed through LC columns, specific proteins of great value and corresponding scarcity (e.g., insulin and human growth hormone) may be recovered only at milligram levels. The diminished scale of recovery in such preparative separations may resemble that of analytical methods, but the objectives are uniquely preparative.

Preparative methods, both large- and small-scale, are used widely for the simple purpose of removing low-level contaminants from products of value (e.g., sulfur from coal). There is major emphasis now on developing better methods to remove pollutants from air and water.

Analytical goals, by contrast to those of preparative separation, are based upon the information generated in the process of separation, or realized by

Figure 1.2. Preparative scale liquid chromatography unit with 6 ft (1.8 m) long column (in center) having an inner diameter of 6 in (15 cm). Up to 2 L of sample solution can be injected and processed in a single 90-min run. (Courtesy Thomas J. Filipi and Whatman Chemical Separation Division.)

subsequent measurements made possible by separation. In a strict analytical sense, separation methods are simply tools to multiply the quantity of detailed information available about complex mixtures and to enhance the quality of that information. Most complex mixtures, especially those of biochemical, environmental, and fossil origin, contain so many components with common backbone and functional groups that it is impossible to identify or quantify most of them by chemical or spectroscopic methods as long as they remain in the mixture. Therefore, separation can be thought of as a means of removing components that interfere with measurement [6–9].

If the separation is good enough and components of high purity emerge, the method of measurement need not be selective at all because there will be no interference (10, 11). The simplest measuring devices (e.g., most common chromatographic detectors) are those that respond to a single physicochemical property (e.g., light absorption at a given wavelength). The response of such devices over the coordinate of separation reveals the presence and the relative amount of components in different regions. Therefore, with proper calibration, these devices can quantify the isolated components. However, the identification of each member in a train of separated components requires further information—either the correlation of molecular structure with component position or independent measurements by methods such as mass spectroscopy.

The sequence, position, and distribution of separated components contain a good deal of information on the mixture. If properly measured and interpreted, this can serve many analytical goals without further tests. The quality of this information naturally improves as the system is better understood, characterized, and controlled. Informational content is greatest when, through theory and/or calibration, one can identify zones or peaks located at defined positions in the sequence with specific molecular species. At that point, using a suitable sensor (detector), both qualitative and quantitative analyses follow. One can, at the same time, often measure certain physicochemical constants for the components, such as partition coefficients and diffusion constants.

With incomplete resolution or inexact calibration, the separated fractions may require additional analytical steps to characterize them. This is particularly true of highly complex mixtures of natural origin. The additional information may be sought by optical spectroscopy, mass spectroscopy, biochemical assay, and so on. It may also be sought through additional separation steps (see Chapter 6).

In gathering the above information, sample size by itself has no merit. Nevertheless, the goals of analysis require that there be enough of each component to detect and possibly submit to subsequent analytical procedures. Fortunately, modern detection and analysis techniques require only minute quantities of material, often at the microgram, nanogram, or picogram levels. Hence it is now uncommon to work with large amounts of

Figure 1.3. Thin fused silica capillary tubes similar to (and frequently smaller than) those shown here are often used for the analytical-scale separation of complex mixtures by chromatography and electrophoresis. The inside diameters of these capillaries are only 220 μm (lower two) and 460 μm (upper). The diameters of the smaller tubes are 680 times less than that of the LC column shown in Figure 1.2. The cross-sectional area, roughly proportional to separative capacity, is over 400,000 times less. (Photo by Alexis Kelner.)

material in analytical work. The dimensions of the separation system have shrunken accordingly, often to hair-thin capillary size tubes (see Figure 1.3). However, there are cases where the component of interest exists at the ultratrace level and its isolation in sufficient quantities for analysis may require kilograms of the starting material. Then the separation system must be scaled up, an alternate system must be used, or repetitive runs must be employed.

The coverage of separation techniques in many important textbooks in chemical analysis is largely limited to chromatography (e.g., refs. 9, 12, 13). While chromatography is of central importance in analysis, the omission of modern electrophoretic (both gel and capillary) and field-flow fractionation techniques leaves a large void in the description of separative capabilities, particularly in the biochemical and macromolecular realm.

1.3 LITERATURE OF ANALYTICAL SEPARATIONS

The literature dealing with analytical separations is immense and widely scattered. Much of it appears in journals emphasizing individual methods such as chromatography, for example, the *Journal of Chromatography*, or subclasses of those methods, the *Journal of Liquid Chromatography*, the *Journal of High Resolution Chromatography*, and so on. Other articles appear in the general analytical literature, particularly in *Analytical Chemistry*. Both specialized and integrative papers on separations, preparative as well as analytical, appear in *Separation Science and Technology* and in *Separation and Purification Methods*. Important journals emphasizing preparative methods of separation include the *Journal of Membrane Science* and *Solvent Extraction and Ion Exchange*. Dozens of other journals cover specialized aspects of the field.

Books, like journals, are generally limited in scope, as, for example, Brown and Hartwick's *High Performance Liquid Chromatography* [14], or the monograph series (now 30 in number) *Advances in Chromatography* (e.g., [15]). Some are engineering and preparative oriented, such as King's *Separation Processes* [16] or the *Handbook of Separation Process Technology* [17]. An exception is the excellent (but somewhat outdated) book by Morris and Morris, *Separation Methods in Biochemistry*, which details the theory and practice of many analytical separation methods used in biochemical work [18]. A few textbooks also cover analytical separations rather broadly, the most complete being Karger, Snyder, and Horvath's *An Introduction to Separation Science*, published in 1973 [19]. Many other excellent books are widely used.

The listings of most science libraries are replete with material from the specialized literature on separations, but the discipline of separations, when considered as a whole science, is rather poorly developed. Some of the best sources are those noted above and some to be cited later in this book.

1.4 SEPARATION CRITERIA

Numerous indices have evolved to characterize separation power and to evaluate and compare the outcome of separative operations. To be useful, the index employed should match the structure of the separation system. For many systems, particularly those employed for preparative purposes, there are two (or more) discrete product regions or streams. Components are generally distributed homogeneously within a region, but the levels may be enriched or depleted relative to other regions. A simple example is liquid–liquid extraction where components typically partition between an aqueous phase and an organic phase. Polar components tend to partition into the aqueous phase and nonpolar components into the organic phase. Thus polar and nonpolar components can be separated as they accumulate in their respective phase regions.

The separation of two components, say i and j, is readily achievable if (1) the concentration c_i^β of i in its preferred phase or region β well exceeds its concentration c_i^α in phase α, $c_i^\beta/c_i^\alpha >> 1$, and (2) component j partitions preferentially in α so that $c_j^\beta/c_j^\alpha << 1$. Effective separation requires that the ratio α' of those two quantities be large: $\alpha' >> 1$. The ratio α' can be written in the two forms

$$\alpha' = \frac{c_i^\beta/c_i^\alpha}{c_j^\beta/c_j^\alpha} = \frac{c_i^\beta/c_j^\beta}{c_i^\alpha/c_j^\alpha} \tag{1.4}$$

the latter expression emphasizing the relative concentration of the two components (e.g., c_i^β/c_j^β) within a phase, compared for the two phases.

Quantity α' is known as the *separation factor* [16]. (Usually, to get α', concentrations are expressed as mole fractions.) The parameter α' is a measure of the extent of segregation of components between two phases (as described for liquid-liquid extraction above), regions (e.g., above and below a membrane filter), or product streams (as in distillation). For a single separation step or stage, α' must be large for effective separation; for certain multistage processes, a small enrichment (e.g., $\alpha' = 1.1$) in one stage can often be compounded by linking stages in series to achieve high enrichment levels in the final product [20]. In chromatography, very low enrichment levels between mobile and stationary phases are amplified by flow, often yielding complete component separation (see Chapters 7 and 12).

The separation factor is a useful index describing the differential accumulation of two components in two well-defined regions. It is applicable whether the components are dilute or concentrated. However, it is specific to each component pair; by differing in value from pair to pair it provides minimal information about the global capabilities of the separation step.

Analytical separations generally evolve over a continuum of space rather than over space divided into a few discrete elements. In chromatography and electrophoresis, components separate into individual concentration pulses or zones distributed along a continuous axis. The separation factor does not describe the separation (although in chromatography the α' for mobile/stationary phase partitioning influences separation) because no two regions of the space distributed along the main separation axis can be identified as more critical to separation than any other two regions. The most useful criterion for the effectiveness of separation of two components in this case is the *resolution*, which is the ratio of the distance between the centers of the two component zones to the average zone width (see Chapter 5 for details).

Resolution, like the separation factor, differs for each specific component pair and therefore fails as a global criterion of separation. For analytical separations, more universal criteria have evolved, such as plate height, number of plates, rate of generation of plates, and peak capacity (Chapter 5). While these indices differ somewhat from one component to another, they effectively establish a "ballpark" figure of merit for different systems and different conditions of operation.

1.5   CONCENTRATION LEVELS

While separation processes must generate selective increases and decreases in the mole fraction of components, the absolute concentration levels differ enormously. Separation at such different levels of purity may be referred to by specific terms: enrichment, concentration, and purification [21, 22]. Rony has proposed that enrichment be used when the mole fraction of the desired component remains under 0.1, concentration when it stays below 0.9, and purification when the mole fraction is above 0.9 [22]. These levels, while arbitrary, help illustrate the variability of concentration levels employed in preparative separations.

Analytical separations generally deal with highly dilute solutions. Important components are sometimes found in the parts per billion range or lower, which is sufficient to produce the desired information with highly sensitive detection.

1.6   THERMODYNAMIC LIMITATIONS

It seems enigmatic that we often struggle so hard to achieve desired separations when the basic concept of moving one component away from another is inherently so simple. Much of the difficulty arises because separation flies in the face of the second law of thermodynamics. Entropy is gained in mixing, not in separation. Therefore it is the process of mixing that occurs spontaneously. To combat this and achieve separation, one must apply and manipulate external work and heat and allow dilution in a thermodynamically consistent way. While in this book we will not present the specific techniques of separation as if they were thermodynamic machines, it should be kept in mind that all separation processes must be thermodynamically consistent [16, 23, 24].

The inherent difficulties of separation can be largely explained in terms of two closely related processes. The second law of thermodynamics tells us that both processes are accompanied by an increase in entropy, making them spontaneous. Both act to hinder separation. First we have

$$(a) + (b) + (c) + (d) \xrightarrow{\text{Mixing at constant volume}} (a + b + c + d) \qquad (1.5)$$

which represents mixing in a fixed total volume, a process diametrically opposite to separation. Second is any process of dilution

$$(a) \xrightarrow{\text{Dilution}} (\, a \,) \qquad (1.6)$$

which occurs by the diffusion or dispersion of narrow component zones into adjoining space, eventually throughout the space available to them, often into the space occupied by other components.

The two processes represented by Eqs. 1.5 and 1.6 are related because mixing is simply the dilution of components over the space they mutually occupy. Thermodynamically, each component in Eq. 1.5 is behaving much like the single component of Eq. 1.6, it being unimportant entropy-wise whether dilution occurs into other components as in Eq. 1.5 or into a background of solvent as in Eq. 1.6.

The entropy of dilution of a component of an ideal gas or liquid increases with the logarithm of the volume available for dilution. Specifically, the entropy gain of n moles of a component due to dilution is

$$\Delta S = n\mathcal{R} \ln \frac{V(\text{final})}{V(\text{initial})} \qquad (1.7)$$

where $\mathcal{R}$ is the gas constant, $V(\text{final})$ is the final volume of occupancy of the component, and $V(\text{initial})$ is the starting volume.

While dilution is favored by an entropy gain and its opposite, concentration, is opposed by an entropy loss, there is nothing in thermodynamics to oppose the separative (differential) transport of components from a shared region into different regions providing they are not simultaneously concentrated [23]. However, any space through which transport can be used to structure the separation is also space through which dispersive (entropic) transport can act to counteract the separation.

Separation is the art and science of maximizing separative transport relative to dispersive transport. In pursuit of this goal, work of recent decades has increasingly amplified the former. However, without a concomitant effort to reduce the latter—to suppress the universal thermodynamic inclination toward dispersion and randomization—modern analytical tech-

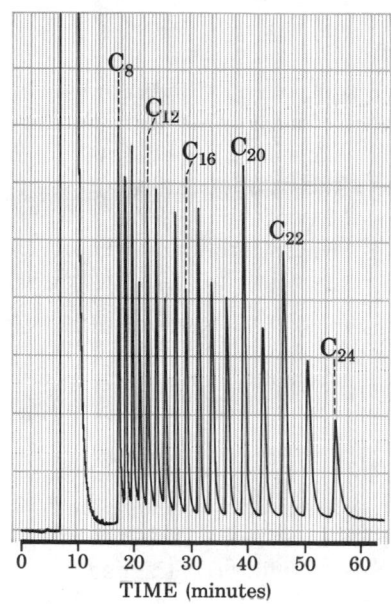

Figure 1.4. High-resolution separation of free fatty acids of indicated carbon numbers by supercritical fluid chromatography using a compressed CO_2 mobile phase at 170°C. Column is 50 cm long fused silica capillary tube of 250 μm internal diameter. (Courtesy of Frank J. Yang.)

niques would yield only blurred images of their sharp separative structure. The basis of the capability for the simultaneous isolation of over 100 components by several existing chromatographic and electrophoretic techniques can be understood by inspection of the separation profiles: the peaks and bands, within which the different components are isolated, are remarkably narrow (see example in Figure 1.4). Clearly, the control of dispersive transport has had a major role in the development of these high resolution systems.

In this book the mechanisms of both separative and dispersive transport are discussed, with an emphasis on their role in analytical separation processes.

1.7 ORIGIN OF SEPARATION IN DISPLACEMENT

Inasmuch as all chemical separations require the movement of components from a region commonly shared to (or toward) individual regions, processes of *differential displacement* must underlie the entire methodology. Thus, by its most fundamental nature, a separation process must entail a selective movement through space of one component with respect to another (1, 2). As a result, the science of separations is rooted foremost in transport phenomena.

While the displacement of components must be selective—different for each species—we need to focus first on the underlying phenomenon: the mechanism for getting a single component transported from one location to another. All such displacements fall into one of two broad classes (1, 2). These are described below.

Bulk Flow Displacement

We can displace the medium that contains the component in bulk, in which case the component follows the gross movement of mass. This displacement can be achieved in two ways: by direct mechanical means or by flow.

Direct mechanical transfer simply means that the component, alone or in solvent, is physically moved by relocating it or the container holding it. This occurs, for example, in many fraction collectors where fractions accumulate in small tubes that are periodically shifted in position to allow for new collections.

Flow displacements are more common. In this case, the entrained component is simply swept along in a fluid stream. This type of displacement is far more significant in separation methodology than is direct mechanical transfer.

The word "flow" implies fluid moving through (or across) a rigid framework or conduit (a container, tube, or packed bed) and not being carried with it as in the case of mechanical transfer. Flow is an integral part of many separation techniques, including chromatography, field-flow fractionation, ultrafiltration, and elutriation. The flow process is not itself selective, but it enables one to multiply by many times the benefits of separations attempted without flow. This point is explained in Chapter 7.

Some subtleties and complications arise in the details of bulk (flow) displacements, especially in the case of complex flow, but on the whole these displacements reduce to the simple matter that if an element of fluid medium is displaced by a certain amount, entrained components are displaced identically in direction and distance.

Relative Displacement

Relative displacement entails the motion of a component through its surrounding medium. This can be realized through thermal (Brownian) motion alone, and in that form it is described by diffusion equations (Sections 3.4 and 3.6). However, relative displacements in separations are usually impelled by some force or gradient. For instance, if the component species is charged, the application of an electrical field will drive it through its surroundings. Sedimentation forces generated in a centrifuge also cause relative displacement. Unbalanced intermolecular forces at an interface will drive components to or across the interface. Other forces exist and are discussed later. In general these forces fall into two categories: (1) external forces, resulting from external gradients or fields (such as electrical); and (2) internal molecular forces, arising in intermolecular interactions with the surrounding medium and subject to abrupt changes at interfaces.

Of the two classes of displacement, only the relative displacement is selective. While the two may, and often are, used in combination, bulk displacement cannot be used alone because of its nonselectivity. All separation methods must rely on relative displacement of some form, alone or in conjunction with bulk displacement (see Chapter 7).

Relative displacement differs for each of the various forces that can be applied; its magnitude depends on how different components respond to these forces. Details are provided in Chapter 3. The origin and nature of bulk (flow) displacement will be described in Chapter 4.

1.8 METHODS OF SEPARATION

The basic techniques of separation, according to most tabulations, number about 20, but many variations exist. The variations themselves are realized through a vast and highly diversified array of instruments and procedures described in the scattered literature. For the most part, the basic operating principles of the subtechniques can be understood in terms of certain parent or core techniques. However, because of multiple elements of similarity and differences among method, it is nearly impossible to define a set of techniques that stand uniquely at the core of separations, the others being derivative. Complications arise, for example, in choosing the features of separation systems that are to be compared (see Chapter 7). Even the naming of separation methods reflects this complexity, with different names chosen to describe different features of the separation.

Often, basic methods are described by and named after the underlying forces or phenomena that give rise to fractionation: extraction, adsorption, crystallization, precipitation, ion exchange, diffusion, thermal diffusion, sedimentation, centrifugation, and so on. At other times the name implies a distinct form of operation: chromatography, distillation, zone melting, filtration, dialysis, thermogravitational separation, elutriation, field-flow fractionation, countercurrent distribution, parametric pumping, electrostatic precipitation, and so on. In some cases, a given form of operation, chromatography, can employ any one of a number of forces (adsorption, ion exchange, extraction). Altogether, there are numerous and subtle cross ties in the relationships that exist between the above methods.

Clearly, the naming of separation techniques has proceeded along historical lines rather than logical lines. The word "chromatography" literally means "color writing" and arose from its original application to colored plant pigments. The word is inappropriate as a description of today's technology, but it is ingrained in scientific usage. The diversified origin of individual separation methods had led to a nomenclature that is overlapping and confusing when one tries to look at the field as a whole.

Our emphasis in this book is not so much on the subtleties that distinguish one historical name from another, or one result from another. The basic classification of methods described in Chapter 7, for example, is founded on the arrangements of the fundamental forces and flows that give rise to separation and ultimately determine their efficacy. It is the belief of this author that this approach is ultimately the most lasting one; it will remain relevant despite the ebb and flow in the popularity of individual separation methods. It is also the most useful one to the student and practitioner faced with difficult separation problems and having to choose among a myriad of historical techniques that have not generally been related and compared.

REFERENCES

[1] J. C. Giddings, in I. M. Kolthoff and P. J. Elving, Eds., *Treatise on Analytical Chemistry*, Part I, Wiley, New York, 1982, Chapter 3.

[2] J. C. Giddings, *Sep. Sci. Technol.*, **13**, 3 (1978).

[3] J. C. Giddings, *Sep. Sci. Technol.*, **14**, 875 (1979).

[4] J. C. Giddings, *Anal. Chem.*, **53**, 945A (1981).

[5] L. B. Rogers, in S. Siggia, Ed., *Survey of Analytical Chemistry*, McGraw-Hill, New York, 1968, Chapter 11.

[6] W. E. Harris and B. Kartochvil, *Chemical Separations and Measurements*, W. B. Saunders, Philadelphia, PA, 1974, Chapter 10.

[7] L. B. Rogers, in I. M. Kolthoff and P. J. Elving, Eds., *Treatise on Analytical Chemistry*, Part I, Vol. 2, Wiley, New York, 1961, Chapter 22.

[8] D. C. Harris, *Quantitative Chemical Analysis*, 2nd ed., W. H. Freeman, New York, 1987, Chapter 23.

[9] H. H. Willard, L. L. Merritt, Jr., J. A. Dean, and F. A. Settle, Jr., *Instrumental Methods of Analysis*, 7th ed., Wadsworth, Belmont, CA, 1988, Chapter 1.

[10] J. A. Bernard and R. Chayen, *Modern Methods of Chemical Analysis*, McGraw-Hill, New York, 1965, Chapter 6.

[11] D. J. Pietrzyk and C. W. Frank, *Analytical Chemistry*, Academic, New York, 1974, Chapter 22.

[12] D. A. Skoog, D. M. West, and F. J. Holler, *Fundamentals of Analytical Chemistry*, 5th ed., W. B. Saunders, New York, 1988, Chapters 23–25.

[13] H. A. Strobel and W. R. Heineman, *Chemical Instrumentation: A Systematic Approach*, 3rd ed., Wiley, New York, 1989, Chapters 24–26.

[14] P. R. Brown and R. A. Hartwick, Eds., *High Performance Liquid Chromatography*, Wiley, New York, 1989.

[15] J. C. Giddings, E. Grushka, and P. R. Brown, Eds., *Advances in Chromatography*, Vol. 30, Marcel Dekker, New York, 1989.

[16] C. J. King, *Separation Processes*, McGraw-Hill, New York, 1971.

[17] R. W. Rousseau, Ed., *Handbook of Separation Process Technology*, Wiley, New York, 1987.

[18] C. J. O. R. Morris and P. Morris, *Separation Methods in Biochemistry*, 2nd ed., Pitman, London, 1976.

[19] B. L. Karger, L. R. Snyder, and C. Horvath, *An Introduction to Separation Science*, Wiley, New York, 1973.

[20] H. R. C. Pratt, *Countercurrent Separation Processes*, Elsevier, Amsterdam, 1967.

[21] *International Encylcopedia of Chemical Sciences*, Van Nostrand, Princeton, NJ, 1964.

[22] P. R. Rony, *Chem. Eng. Prog., Symp. Ser.*, **68**, 89 (1972).

[23] G. H. Stewart, *J. Chromatogr. Sci.*, **14**, 69 (1976).

[24] G. H. Stewart, *Sep. Sci. Technol.*, **13**, 201 (1978).

EXERCISES

(Note that the number of star symbols (*), varying from one to three, provides a rough scale of difficulty.)

1.1(*) Use a parenthetical expression like those shown in Eqs. 1.1–1.3 to represent the separation of five components (a, b, c, d, e) into three regions, one containing all of (and nothing more than) components a and d and a second all of (and only) c and e.

1.2(**) For a binary solution in equilibrium with its vapor in accordance with Raoult's law, show that the vapor/liquid separation factor is given by the ratio of pure component vapor pressures.

1.3(*) The preparative scale column of Figure 1.2 (length 6 ft, inside diameter 6 in) has been used to purify tributylphenol (TBP) using 2-L injections of a 10% (by volume) solution of TBP in a methanol/water mobile phase flowing at 1.8 L/min. The separation is complete in 90 min. How many liters of TBP can be processed by this column per 24-h day? How much mobile phase is consumed?

1.4(**) Calculate the entropy change, in calories/degree, that accompanies the separation of four components from one another in an ideal solution containing one mole of each. When separated, each component occupies one-quarter of the original volume. Deduce from your results whether or not the separation is thermodynamically spontaneous.

1.5(*) The distribution coefficient K (see Chapter 2) is the equilibrium concentration ratio c^β/c^α of a component distributed between phase β and phase α. At 25°C, the K for I_2 partitioned between $CCl_4(\beta)$ and $H_2O(\alpha)$ is 90. For NH_3, $K = 0.0045$. What is the separation factor α' when I_2 and NH_3 are partitioned at equilibrium between these two phases?

1.6(*) Ionic components a and b, after being driven through a capillary electrophoresis column by an electrical field for 10 min, form bands (zones) whose centers are located at 8.714 and 9.035 cm from the inlet, respectively. Their widths are 0.350 and 0.442 cm, respectively. Calculate the resolution of the two bands.

2

EQUILIBRIUM: DRIVING FORCE FOR SEPARATIVE DISPLACEMENT

All isolated systems move, rapidly or slowly, by one path or another, toward equilibrium. In fact essentially all motion stems from the universal drift to eventual equilibrium. Therefore, if we wish to obtain a certain displacement of a component through some medium, we must generally establish equilibrium conditions that favor the desired displacement. Clearly, a knowledge of the equilibrium state is indispensable to the study of the displacements leading to separation.

In many separation processes (chromatography, countercurrent distribution, field-flow fractionation, extraction, etc.), the transport of components, in one dimension at least, occurs almost to the point of reaching equilibrium. Thus equilibrium concentrations often constitute a good approximation to the actual distribution of components bound within such systems. Equilibrium concepts are especially crucial in these cases in predicting separation behavior and efficacy.

2.1 MECHANICAL VERSUS MOLECULAR EQUILIBRIUM

We can identify two important classes of equilibria:

(a) Mechanical—defines the resting place of macroscopic bodies.
(b) Molecular—defines the spatial distribution of molecules and colloids at equilibrium.

Of the two, (a) is more simple. With macroscopic bodies, it is unnecessary to worry about thermal (Brownian) motion, which greatly complicates equilibrium in molecular systems. This is equivalent to stating that entropy is unimportant. This is not to say that entropy terms are diminished for large bodies, but only that energy changes for displacements in macroscopic systems are enormous compared to those for molecules, and the swollen energy terms completely dominate the small entropy terms, which do not inherently depend on particle size.

Without entropy considerations, equilibrium along any given coordinate x is found very simply as that location where the body assumes a minimum potential energy $\mathscr{P}$; the body will eventually come to rest at that exact point. Thus, mechanical equilibrium is subject to the simple criterion

$$\frac{d\mathscr{P}}{dx} = 0 \quad \text{or} \quad d\mathscr{P} = 0 \tag{2.1}$$

which is equivalent to saying that there are no unbalanced forces on the body.

Systems out of equilibrium—generally in the process of moving toward equilibrium—are characterized by $(d\mathscr{P}/dx) \neq 0$. A rock tumbling down a mountainside and a positive test charge moving toward the region of lowest electrical potential are both manifestations of the tendency toward simple mechanical equilibrium.

Molecular equilibrium, by contrast, is complicated by entropy. Entropy, being a measure of randomness, reflects the tendency of molecules to scatter, to diffuse, to assume different energy states, to occupy different phases and positions. It becomes impossible to follow individual molecules through all these conditions, so we resort to describing statistical distributions of molecules, which for our purposes simply become concentration profiles. The molecular statistics are described in detail by the science of statistical mechanics. However, if we need only to describe the concentration profiles at equilibrium, we can invoke the science of thermodynamics.

We discuss below some of the arguments of thermodynamics that bear on common separation systems. We are particularly interested in the thermodynamics of equilibrium between phases and equilibrium in external fields, for these two forms of equilibrium underlie the primary driving forces in most separation systems. A basic working knowledge of thermodynamics is assumed. Many excellent books and generally simple monographs on this subject are available for review purposes (1–4). In the treatment below, we seek the simplest and most direct route to the relevant thermodynamics of separation systems, leaving rigor and completeness to the monographs on thermodynamics.

2.2 MOLECULAR EQUILIBRIUM IN CLOSED SYSTEMS

A *closed system* is one with boundaries across which no matter may pass, either in or out, but one in which other changes may occur, including expansion, contraction, internal diffusion, chemical reaction, heating, and cooling. The first law of thermodynamics gives the following expression for the internal energy increment dE for a closed system undergoing any such change

$$dE = q + w \qquad (2.2)$$

where q is the increment of added heat (if any) and w is the increment of work done on the system. If we assume for the moment that only pressure-volume work is involved, then $w = -p\,dV$, the negative sign arising because positive work is done on the system only when there is contraction, that is, when dV is negative. For q we write the second law statement for entropy S as the inequality: $dS \geq q/T$, or $T\,dS \geq q$. With w and q written in the above forms, Eq. 2.2 becomes

$$dE \leq T\,dS - p\,dV \qquad (2.3)$$

an equation which contains the restraints of both the first and the second law of thermodynamics. We hold this equation briefly for reference.

By definition, the *Gibbs free energy* relates to enthalpy H and entropy S by

$$G = H - TS = E + pV - TS \qquad (2.4)$$

from which direct differentiation yields

$$dG = dE + p\,dV + V\,dp - T\,dS - S\,dT \qquad (2.5)$$

The substitution of Eq. 2.3 for the dE in Eq. 2.5 yields

$$dG \leq -S\,dT + V\,dp \qquad (2.6)$$

Therefore, all natural processes occurring at constant T and p must have

$$dG \leq 0 \qquad (2.7)$$

while for any change at equilibrium

$$dG = 0 \qquad (2.8)$$

In other words, equilibrium at constant T and p is characterized by a minimum in G. This is analogous to mechanical equilibrium, Eq. 2.1, except that G is the master parameter governing equilibrium instead of $\mathscr{P}$.

For example, if a small volume of ice is melted in a closed container at $0°C$ and 1 atm pressure, we find by thermodynamic calculations that $dG = 0$, representing ice-water equilibrium, which is reversible. At 10°C, we have $dG < 0$, representing the spontaneous, irreversible melting of ice above 0°C, its melting (equilibrium) point. Spontaneous processes such as diffusion, of course, are likewise accompanied by $dG < 0$.

2.3 EQUILIBRIUM IN OPEN SYSTEMS

An *open system* is one which can undergo all the changes allowed for a closed system and in addition it can lose and gain matter across its boundaries. An open system might be one phase in an extraction system, or it might be a small-volume element in an electrophoretic channel. Such systems, which allow for the transport of matter both in and out, are key elements in the description of separation processes.

In open systems, we must modify the expression describing dG at equilibrium in closed systems, namely

$$dG = -S \, dT + V \, dp \qquad (2.9)$$

to account for small amounts of free energy G taken in and out of the system by the matter crossing its boundaries. For example, if dn_i moles of component i enter the system, and there are no changes in T and p and no other components j crossing in or out, G will change by a small increment proportional to dn_i

$$dG = \left(\frac{\partial G}{\partial n_i}\right)_{T,p,n_j} dn_i \qquad (2.10)$$

The magnitude of the increment depends, as the above equation shows, on the rate of change of G with respect to n_i, providing the other factors are held constant. This magnitude is of such importance in equilibrium studies that the rate of change, or partial derivative, is given a special symbol

$$\mu_i = \left(\frac{\partial G}{\partial n_i}\right)_{T,p,n_j} \qquad (2.11)$$

Quantity μ_i is called the *chemical potential*. It is, essentially, the amount of "G" brought into a system per mole of added constituent i at constant T and p. Dimensionally, it is simply energy per mole.

If we substitute μ_i for the partial derivative $\partial G/\partial n_i$ of Eq. 2.10, we find that dG for the addition of dn_i moles of i to the system is simply

$$dG = \mu_i \, dn_i \qquad (2.12)$$

By an identical argument, the addition of dn_j molecules of j yields $dG =$

$\mu_j\,dn_j$. For the transfer of any arbitrary number of components, dG is the sum of the individual contributions

$$dG = \sum \mu_i\,dn_i \tag{2.13}$$

If T and p are now allowed to vary, we must add the increment that describes their contribution, which is identical to that for a closed system, Eq. 2.9. All of these contributions together give a general equilibrium expression for an open system subject to any combination of incremental changes

$$dG = -S\,dT + V\,dp + \sum \mu_i\,dn_i \tag{2.14}$$

where the summation extends over all components entering or leaving the system. For those leaving, of course, dn_i is negative.

We now examine two open systems—for example, two immiscible phases—which may exchange matter between them. These systems are labeled α and β, as shown in Figure 2.1. Taken together, α plus β constitute a closed system, for which equilibrium is defined by $dG = 0$ for any change at constant T and p. Therefore, for the transfer of a small amount of component i from α to β at equilibrium

$$dG = dG^\beta + dG^\alpha = (\mu_i^\beta - \mu_i^\alpha)\,dn_i = 0 \tag{2.15}$$

This equation shows that equilibrium (specified by $dG = 0$) between interconnected regions at constant T and p is defined by

$$\mu_i^\beta = \mu_i^\alpha \tag{2.16}$$

.This equality applies, for example, to solute distributed between two or more phases of an extraction, adsorption, or chromatographic system.

The μ_i value for solute i in a given phase depends on two factors: the intrinsic thermodynamic "affinity" of the solute to the phase and the dilution of the solute. The latter affects μ_i through entropy (i.e., the entropy of dilution). In analytical separations, solute concentrations are generally

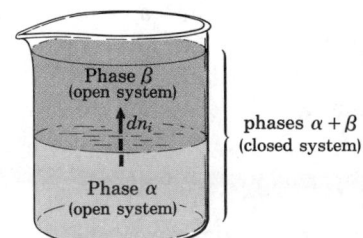

Figure 2.1. Transfer of dn_i moles of component i from phase α to phase β.

quite low, although they can vary considerably from case to case. We therefore describe μ_i in a form appropriate for the study of dilute solutions. Specifically, we account for *affinity* and *dilution* factors in highly dilute solutions by the two respective terms on the right side of the equation

$$\mu_i = \mu_i^0 + \mathcal{R} T \ln c_i \tag{2.17}$$

Where $\mathcal{R}$ is the gas constant, T is the temperature, c_i is the concentration of component i (usually in moles or mass per unit volume), and μ_i^0 is a reference value of the chemical potential called the *standard-state chemical potential*. The latter is the chemical potential in a hypothetical standard state with component i at unit concentration but with each molecule of i surrounded by solvent as found at infinite dilution. (We note that Eq. 2.17 can be written in many forms involving different concentration units and different definitions of the standard state. A common approach is to express concentration as mole fraction and to let the pure solute represent the standard state. The convention chosen above serves present purposes best.)

The value of μ_i^0 in Eq. 2.17 depends strongly on the energy and to some degree on the entropy of solute-solvent interactions; it thus undergoes a change in value when solute i moves from phase to phase. The value of μ_i^0 is generally lowest in phases where intermolecular interactions are strongest (i.e., where solute-solvent affinity is greatest).

As stated above, the term $\mathcal{R} T \ln c_i$ represents a contribution to chemical potential stemming from the entropy associated with the dilution and concentration of components (see Section 1.6). Dilution and concentration processes are fundamental to separations, so it is clear that this term lies at the roots of separative equilibria.

If Eq. 2.17 is substituted into Eq. 2.16 and the resulting expression solved for the ratio of the two concentrations c_i^α and c_i^β, we get the equilibrium expression

$$\left(\frac{c_i^\beta}{c_i^\alpha}\right)_{eq} = \exp\left(\frac{-\Delta\mu_i^0}{\mathcal{R} T}\right) \tag{2.18}$$

where $\Delta\mu_i^0 = \mu_i^{0\beta} - \mu_i^{0\alpha}$. This equilibrium ratio, $(c_i^\beta/c_i^\alpha)_{eq}$, is simply the *partition* coefficient or *distribution coefficient K*, giving

$$K = \exp\left(\frac{-\Delta\mu_i^0}{\mathcal{R} T}\right) \tag{2.19}$$

An interpretation is in order. Since the μ_i in Eq. 2.17 must be equal for solute i in two partitioning phases (Eq. 2.16), the phase with the lowest μ_i^0 must compensate with a larger $\mathcal{R} T \ln c_i$ term, that is, a higher concentration. Thus, if $\Delta\mu_i^0$ is positive, it means that μ_i^0 in phase α (termed $\mu_i^{0\alpha}$) is lower than that in phase β, and the concentration in α is therefore higher than in phase β. Indeed, Eq. 2.18 shows that a positive $\Delta\mu_i^0$ yields $c_i^\beta/c_i^\alpha < 1$.

Essentially all separative equilibria, therefore, involve adjustments in concentration to such relative levels that the general equilibrium criterion $\mu_i^\beta = \mu_i^\alpha$ is satisfied. These adjustments differ from component to component because different intermolecular interactions give different $\Delta\mu^0$ values, thus giving the unequal concentration distributions that ultimately yield separation.

For concentrations sufficiently high that the infinite dilution approximation breaks down, Eq. 2.17 must be modified by the incorporation of an *activity coefficient* γ, which compensates for additions to the chemical potential from such nonidealities as solute-solute interactions. We have

$$\mu_i = \mu_i^0 + \mathcal{R}T \ln \gamma_i c_i \tag{2.20}$$

where $\gamma_i \to 1$ as $c_i \to 0$. When γ_i departs from unity, the equilibrium ratio of Eq. 2.18 should be expressed as $(c_i^\beta \gamma_i^\beta / c_i^\alpha \gamma_i^\alpha)_{eq}$. In this case, calculation of the concentration ratio c_i^β / c_i^α requires that the γs be known.

For ideal gas mixtures, which may be used to approximate the mobile phase of gas chromatography, Eq. 2.17 can be expressed as

$$\mu_i = \mu_i^0 + \mathcal{R}T \ln p_i \tag{2.21}$$

where p_i is the partial pressure of component i. In developing the subsequent equilibrium expression, Eq. 2.18, p_i replaces c_i for any gaseous phases involved.

Further details on the thermodynamics of solution equilibrium can be found in books on thermodynamics and solutions (5–11).

2.4   EQUILIBRIUM IN EXTERNAL FIELDS

External fields are applied widely in separation systems. The most common fields used are based on electrical and sedimentation (both centrifugal and gravitational) forces. Gradients in solvent composition and temperature maintained by actions external to the system may also be considered as external fields defined in their broadest context (see Chapter 8). All of these fields are capable of changing the equilibrium distribution of chemical components. Furthermore, they may be selective, affecting one component differently from another, a basic requirement for separation.

In general, a field can be assumed to impart to a molecule or particle a certain potential energy, which varies with position. This potential energy becomes an additive component to the Gibbs free energy G since, after all, G is simply a special form of energy. (There are rigorous ways of proving this, such as using the thermodynamic fact that ΔG is equal to the reversible work done on a system, which in turn equals the increase in potential energy for simple transfer processes in an external field.) The potential energy for one mole becomes, therefore, an additive contribution to chemical potential, which can be designated as μ_i^{ext} for component i. We distinguish this term from the chemical potential having internal origins, appearing in the last section as μ_i, by labeling the latter μ_i^{int}, where the superscript "int" refers to the internal chemical environment.

With this change and the addition of μ_i^{ext}, Eq. 2.14 becomes

$$dG = -S\,dT + V\,dp + \sum (\mu_i^{int} + \mu_i^{ext})\,dn_i \tag{2.22}$$

At constant T and p, this becomes

$$dG = \sum (\mu_i^{int} + \mu_i^{ext})\,dn_i \tag{2.23}$$

As before, we transfer dn_i moles of component i from region α to region β at equilibrium. We have

$$dG = 0 = (\mu_i^{int,\beta} + \mu_i^{ext,\beta} - \mu_i^{int,\alpha} - \mu_i^{ext,\alpha})\,dn_i$$
$$= (\Delta\mu_i^{ext} + \Delta\mu_i^{int})\,dn_i \tag{2.24}$$

If we substitute Eq. 2.17 for each μ_i^{int} as follows

$$\mu_i^{int} = \mu_i^0 + \mathcal{R}T \ln c_i \tag{2.25}$$

we get from Eq. 2.24

$$-(\Delta\mu_i^{ext} + \Delta\mu_i^0) = \mathcal{R}T \ln \left(\frac{c_i^\beta}{c_i^\alpha}\right)_{eq} \tag{2.26}$$

or rearranged

$$K = \left(\frac{c_i^\beta}{c_i^\alpha}\right)_{eq} = \exp\left(\frac{-\Delta\mu_i^0 - \Delta\mu_i^{ext}}{\mathcal{R}T}\right) \tag{2.27}$$

an equation much like Eq. 2.18, except for the addition of the external field term $\Delta\mu_i^{ext}$. The terms are

$$\Delta\mu_i^0 = \mu_i^{0\beta} - \mu_i^{0\alpha} \tag{2.28}$$
$$\Delta\mu_i^{ext} = \mu_i^{ext,\beta} - \mu_i^{ext,\alpha} \tag{2.29}$$

Equation 2.27 is general for dilute solutes: it accounts for the intermolecular forces of phase transfer in the term μ_i^0 and the potential energy of external fields in μ_i^{ext}. The two enter equally in a mathematical sense. Either may be zero, the former in one-phase systems and the latter in field-free systems. However, in a physical sense, $\Delta\mu_i^0$ and $\Delta\mu_i^{ext}$ are dissimilar in that

$\Delta \mu_i^0$ changes abruptly at phase boundaries and $\Delta \mu_i^{ext}$ changes continuously in space. This difference is crucial to the type of separations achieved in multiphase systems versus external field systems, as we shall see later.

A more complete discussion of thermodynamic equilibrium in an external field can be found in the book by Guggenheim [9] or treatments by other authors[10, 11].

2.5   FACTORS CONTROLLING EQUILIBRIUM BETWEEN PHASES

The thermodynamics outlined above has the advantage of being universally applicable to all dilute systems and the associated disadvantage of lacking specifics. Equation 2.18, for example, relates the equilibrium concentration ratio of component i between two phases to $\Delta \mu_i^0$, but it fails to specify how $\Delta \mu_i^0$ is to be calculated. Thermodynamics is mute concerning actual values of parameters like $\Delta \mu_i^0$. Indeed, on closer examination, it is found that $\Delta \mu_i^0$ cannot be related rigorously by any means to molecular parameters.

Statistical mechanics, the science that should yield parameters like $\Delta \mu_i^0$, is hampered by the "multibody" complexity of molecular interactions in condensed phases and by the failure of quantum mechanics to provide accurate interaction potentials between molecules. Because pure theory is impractical, progress in understanding and describing molecular equilibrium between phases requires a combination of careful experimental measurements and correlations by means of empirical equations and approximate theories. The most comprehensive approximate theory available for describing the distribution of solute between phases—including liquids, gases, supercritical fluids, surfaces, and bonded surface phases—is based on a lattice model developed by Martire and co-workers [12, 13].

The first step in understanding the nature of $\Delta \mu_i^0$ is the recognition that it is composed of two unlike terms: an enthalpy term and an entropy term. This follows if we recognize that μ_i^0 is the gain in the Gibbs free energy per mole of added i under standard conditions, and furthermore that each G or μ_i value is composed of both enthalpy (H) and entropy (TS) terms as shown in Eq. 2.4. Since $\Delta \mu_i^0$ is the difference in two μ_i^0 values, one for each of two phases, $\Delta \mu_i^{00}$ can be expressed by

$$\Delta \mu_i^0 = \Delta \bar{H}_i^0 - T \Delta \bar{S}_i^0 \tag{2.30}$$

where the symbols with bars, $\bar{H}^0$ and $\bar{S}^0$, represent the partial molar enthalpy and entropy under standard conditions, respectively. (The entropy $\bar{S}^0$, since it is defined for standard conditions, has nothing to do with the entropy of dilution, accounted for in the last term of Eq. 2.17. Instead, $\bar{S}^0$ reflects the randomness of the immediate molecular environment of a species.)

For most separation systems involving a partitioning of components between phases, $|\Delta \bar{H}^0| \gg |T\Delta \bar{S}^0|$; consequently the equilibrium is largely ruled by enthalpy (or energy) factors. Specifically, the increment $\Delta \bar{H}_i^0$ is most often controlled by intermolecular interactions between component i and the two phases it occupies. When intermolecular attractions are stronger (entailing a lower energy and thus enthalpy) in phase β than phase α, $\Delta \bar{H}_i^0 (= \bar{H}_i^{0\beta} - \bar{H}_i^{0\alpha})$ is negative, $\Delta \mu_i^0$ tends to be negative, and the distribution constant favors phase β, as specified by $K > 1$. This agrees fully with chemical intuition: the stronger a component is attracted to a phase, the more of it will partition into that phase.

Intermolecular Forces

Intermolecular attractive forces fall into several categories. In each, the magnitude of the forces depends upon the properties of the two molecules (e.g., a solute and a solvent molecule) subject to attraction [14, 15].

If both molecules are polar, then opposite charges on the two are drawn together by electrostatic forces. These forces are called *orientation forces* because the two molecules must be oriented such that the opposite charges are close enough for attraction. These forces are of major importance in polar substances.

In highly polar liquids such as water, there is a still stronger intermolecular force between the positive hydrogen and the negative oxygen atoms because of partial bond formation. These linkages are called *hydrogen bonds*. These bonds, because of their high strength, play a prominent role in determining the structure and properties of liquids such as H_2O, NH_3, and HF, and in attracting and dissolving certain highly polar solutes capable of hydrogen bonding with the molecules of these liquids.

When one molecule is polar and the other nonpolar, the polar molecule induces a dipole in the nonpolar one. The two dipoles are again attracted by electrostatic forces, in this case designated as *induction forces*. These forces are considerably weaker than orientation forces but stronger than the attractive forces between nonpolar molecules.

The finite but weak attraction between nonpolar species is due to *dispersion forces* or *London forces*. These forces originate in the fleeting dipoles created by electron motion and in their transient attraction to the resulting dipoles induced in neighboring molecules.

The above discussion makes it clear that the attractive interaction between two polar molecules (especially hydrogen bonded molecules) is stronger than that between other pairs of molecules. The attraction between polar and nonpolar molecules is considerably weaker, and the affinity between two nonpolar molecules is weakest of all. The actual strength of interaction, of course, depends upon the degree of polarity of the participating molecules and upon their potential for hydrogen bond formation, polarizability, size, and so on.

The strong attraction of water (or other hydrogen bonded) molecules to one another leads to the strong association of any given water molecule with a surrounding shell of H_2O. Any molecule or ion dissolving in water must disrupt this shell, which disruption is energetically favorable (or at least acceptable) as long as the intruding species has strong linkages with water. However, intruding nonpolar species disrupt the hydrogen bonds between adjacent water molecules and offer much weaker ties in return. Therefore the water, seeking its most energetically favorable state, tends to expel the solute so that the hydrogen bonds can be reformed. This effective "repulsion" of nonpolar species by water is termed the *hydrophobic effect*. The repelled solute generally associates with other nonpolar species that are similarly repelled, or with nonpolar surfaces. We note that entropy as well as energy changes play a strong role in governing the magnitude of this important effect [16, 17].

The tendency of polar substances to expel nonpolar molecules and to accept other polar species in order to maintain a strongly interactive environment is quite universal, not limited solely to water and the hydrophobic effect. The general tendency for substances of unlike polarity to lack miscibility and those of like polarity to dissolve freely in one another is summarized by the well-known maxim: "like dissolves like."

From the above we recognize that solubility and partitioning phenomena are quite generally governed by intermolecular interactions, influenced strongly by polarity. The picture is complicated for larger molecules containing different groups with different polarity and interactive characteristics. This complication can be dealt with by assuming that each group k of the molecule is associated with its own unique change $\Delta\mu_k^0$ in chemical potential, independent of the presence of other groups, when transferred between phases. Thus the total $\Delta\mu_i^0$ for species i can often be roughly approximated as the sum of its parts

$$\Delta\mu_i^0 = \sum \Delta\mu_k^0 \tag{2.31}$$

This relationship is attributed to Martin [18]. Its utility is discussed in ref. [15].

Supercritical Fluids

Consider now a gas-phase solute molecule (e.g., naphthalene) entering a liquid of like polarity. In the gas, the solute has very little interaction with other molecules, but on entering the liquid it is immediately surrounded by host molecules to which it is attracted by various intermolecular forces. These attractive interactions lower the enthalpy and chemical potential of the solute, thus generally favoring its transfer from gas to liquid.

If now the gas phase is compressed so that molecules in the gas are crowded closely together, the near proximity of solute and surrounding gas molecules leads to attractive interactions between the solute and the gas.

These interactions lower enthalpy and chemical potential just like interactions with liquids. The higher the compression and thus the higher the gas density, the stronger the interactions. Thus gases, adequately compressed and densified, become solvents. The compression must be carried out above the critical temperature T_c of the gas to avoid condensation to a liquid. Generally the pressure p must exceed the critical pressure p_c to gain adequate compression. Any gas for which $T > T_c$ and $p > p_c$ is referred to as a supercritical fluid (SF), or more simply as a dense gas [19].

SFs are unique—quite unlike liquids—because their solvent power is highly sensitive to pressure changes and can be varied over wide limits by pressure swings. Thus most solutes, once dissolved, can be rapidly and cleanly precipitated from SF solutions by dropping the pressure to one atmosphere, where little solvent power remains. Such flexible control of solubility, combined with improved transport characteristics relative to liquids, has made SFs popular for solvent extraction and chromatography [19–21].

2.6 MOLECULAR INTERACTION AND POLARITY SCALES

In view of the fact that the theoretical basis of $\Delta\mu^0$ is so complex, many efforts have been made to develop simple scales or series to describe partitioning and solubility properties. By finding the relative positions of solutes and solvents on these scales, solubility can often be roughly predicted. By comparing the scale values of a solute and of two solvents, the partitioning of the solute between the two solvents can be sometimes approximated.

In Section 2.5 we observed that the strength of intermolecular interactions is related to the polarity of the interacting species, with polarizability a secondary factor. A scale of polarities might therefore be useful. An obvious polarity scale is immediately available: the scale of molecular dipole moments. Values of dipole moments for many molecules can be found in the literature, making such a scale easy to assemble. However, like most simple scales, a scale of dipole moments fails to account for many aspects of intermolecular interactions. For example, while hydrogen bonding occurs between polar molecules, many polar pairs do not hydrogen bond. A scale of dipole moments does not by itself identify hydrogen bonding. The strength of hydrogen bonds is consequently far greater than predicted by a simple correlation with dipole moments. Another shortcoming of the dipole scale is that the dipole-dipole interaction of small dipolar molecules is generally greater than that of larger molecules having the same dipole moments. In addition, dipole moments account for neither polarizability effects nor the polar interactions between molecules (like para-dichlorobenzene) having local polarity but no net dipole moment.

No single scale can account for the entire spectrum of interactions between molecules. However some scales, despite limitations, serve to correlate a fairly broad range of solvent effects more successfully than a scale of dipole moments. Numerous scales and indices have been derived to describe polarity and chromatographic retention. Prominent among them are the *eluotropic series* of Snyder describing the solvent strength of liquid mobile phases [15], the Rohrschneider *phase constants* describing the polarity of GC stationary phases [22], and the $E_T(30)$ *solvatochromic scale* of solvent polarity developed by Dimroth et al. [23]. However, the best known scale of solvent power, used widely throughout science, is the *solubility parameter scale* of Hildebrand [14].

The Solubility Parameter

The solubility parameter δ of a substance is defined by

$$\delta = \left(\frac{\Delta E_v}{V}\right)^{1/2} \tag{2.32}$$

where ΔE_v is the energy necessary to vaporize (i.e., pull completely apart) the molecules in volume V. The ratio $\Delta E_v / V$ is the *cohesive energy density* of the substance. Since cohesive (intermolecular) forces between molecules increase with polarity, δ values correlate strongly with polarity. This is evident from the compilation of δ values shown in Table 2.1.

For so-called *regular solutions*, in which small solute molecules disperse randomly among like-size solvent molecules, $\Delta \bar{S}^0$ can be taken as zero [14]. The enthalpy change $\Delta \bar{H}^0_{i\beta}$ for transferring (mixing) pure solute of molar volume $\bar{V}_i$ into solvent β at high dilution is approximately [14]

$$\Delta \bar{H}^0_{i\beta} = \bar{V}_i(\delta_i - \delta_\beta)^2 \tag{2.33}$$

providing both solute and solvent consist of nonpolar molecules interacting via London dispersion forces. Similarly, for diluting solute i in phase α

$$\Delta \bar{H}^0_{i\alpha} = \bar{V}_i(\delta_i - \delta_\alpha)^2 \tag{2.34}$$

Transferring a mole of solute from α to β entails an enthalpy change equal to the difference

$$\Delta \bar{H}^0_i = \Delta \bar{H}^0_{i\beta} - \Delta \bar{H}^0_{i\alpha} = \bar{V}_i[(\delta_i - \delta_\beta)^2 - (\delta_i - \delta_\alpha)^2] \tag{2.35}$$

Since $\Delta \bar{S}^0_i = 0$, Eq. 2.30 gives

$$\Delta \mu^0_i = \bar{V}_i[(\delta_i - \delta_\beta)^2 - (\delta_i - \delta_\alpha)^2] \tag{2.36}$$

or equivalently

TABLE 2.1 Values of Solubility Parameter δ for Assorted Liquids (Temperature is 298 K unless otherwise stated)

Liquid	δ (cal/cm^3)$^{1/2}$
Water	23.4
Ammonia	16.3
Methanol	14.5
Ethanol	12.7
Nitromethane	12.6
Dimethyl sulfoxide	12.0
Bromoform	10.5
Nitrobenzene	10.0
Carbon disulfide	10.0
Chloroform	9.3
Benzene	9.2
Toluene	8.9
Carbon dioxide (223 K)	8.9
Ethyl benzene	8.8
Carbon tetrachloride	8.6
Cyclohexane	8.2
n-Octane	7.5
Methane (90 K)	7.4
n-Heptane	7.4
Ethyl ether	7.4
n-Hexane	7.3
Oxygen (90 K)	7.2
n-Pentane	7.1
Isopentane	6.8
Argon (90 K)	6.8

$$\Delta\mu_i^0 = \bar{V}_i(\delta_\beta - \delta_\alpha)(\delta_\beta + \delta_\alpha - 2\delta_i) \qquad (2.37)$$

Thus the distribution coefficient (from Eq. 2.19) is described by

$$\ln K = \frac{-\bar{V}_i}{\mathcal{R}T}(\delta_\beta - \delta_\alpha)(\delta_\beta + \delta_\alpha - 2\delta_i) \qquad (2.38)$$

Equations of this form were first developed by Hildebrand and Scatchard (see ref. 14).

The above equations, although highly approximate, are uniquely useful because they relate the thermodynamic terms that arise from interactions of solute and solvent molecules (e.g., solute i and solvent β in Eq. 2.33) to the properties (i.e., the δ values) of the pure solutes and solvents. Remarkably, the equations provide some guidance on solvent selection applicable not only to the small nonpolar molecules intended by theory but also to species of moderate polarity and size [14].

Equations 2.33 and 2.34 show that the heat of mixing of solute and solvent approaches zero whenever the δ values of the two substances are close; such substances mix readily. When the δ values diverge, the enthalpy of the mixing process increases, which in turn increases the free energy (or chemical potential) change and eventually limits free mixing (i.e., limits solubility). This reasoning gives semiquantitative meaning to the earlier discussed maxim "like dissolves like", where likeness is now measured by the closeness of δ values.

When solute i partitions between two phases or solvents, it dissolves preferentially in the one having the closest δ value to its own. Thus if $\delta_i = \delta_\beta \neq \delta_\alpha$, the $\Delta\mu_i^0$ (for $\alpha \to \beta$ solute transfer) from Eq. 2.36 is negative, driving solute into β. Conversely, when $\delta_i = \delta_\alpha \neq \delta_\beta$, $\Delta\mu_i^0 > 0$, driving solute into α. Thus a compilation of δs (Table 2.1) can help predict partitioning and aid in the choice of suitable phases to maximize selectivity, particularly the mobile and stationary phases of chromatography [15, 24].

Solubility Parameters of Supercritical Fluids

While solubility parameter concepts and equations are designed for liquid solvents, they have been shown to be applicable to dense gases or SFs [25, 26]. The solubility parameter δ for a dense gas can be estimated by [25]

$$\delta = 1.25 P_c^{1/2}(\rho/\rho_{\text{liq}}) \tag{2.39}$$

where ρ/ρ_{liq} is the ratio of dense gas density to liquid density and P_c is the critical pressure. More succinctly, δ can be related to the value δ_{liq} of the solubility parameter of the condensed parent liquid by the approximation [26]

$$\delta = \delta_{\text{liq}}(\rho/\rho_{\text{liq}}) \tag{2.40}$$

Equation 2.40 shows that the solvent power (as measured by δ) of an SF is governed by two factors (25). First, it is regulated by the liquid value δ_{liq} (see Table 2.1), reflecting the polarity and other "chemical" properties of molecules of the parent liquid. Highly polar substances like ammonia would thus yield high δs. By contrast, small relatively nonpolarizable species like helium have negligible solvent effects under any circumstances.

The problem in choosing a polar SF for solubilizing polar solutes is that both the boiling point and the critical point are elevated by the polarity; since supercritical operation requires $T > T_c$, high temperatures are mandated in such cases. Thus for ammonia, the critical temperature, $T_c = 132°C$, must be exceeded for practical operation. A widely used compromise between polar substances with high values of T_c and nonpolar substances with low values of δ_{liq} is carbon dioxide, for which $T_c = 31°C$ and $\delta_{\text{liq}} = 8.9$. Other factors involved in choosing an SF phase are elaborated by Schoenmakers et al. [19].

The second factor governing the solvent power of an SF is the "state" of compression of the gas, specifically the gas density. The critical role of density (as opposed, for example, to pressure) in SF solubility was first discussed at some length by this author and his colleagues [25], who showed that pressures of ~1000 atmospheres are needed to gain liquid-like solvent power. Density effects have recently been discussed by a number of authors [12, 19–21].

2.7 ENTROPY EFFECTS IN PHASE DISTRIBUTION: POROUS MEDIA

Finite entropy changes can often be expected when a component species moves from one phase to another. These entropy changes (reflected in $\Delta \bar{S}^0$) relate to the way the solute molecule fits into the liquid structure of the two respective phases and the associated reorientation and repositioning of the liquid molecules. In most cases the structural changes accompanying the arrival of a solute molecule are similar in different phases and the increment in entropy is relatively small. In the case of the hydrophobic effect, the presence of the nonpolar intruder induces a semirigid structure in the surrounding water molecules, leading to a significant reduction in entropy. In this case, the change in entropy plays a major role in influencing phase distribution.

There is another large and important class of phase distribution phenomena in which the distribution constant is profoundly influenced by entropy. Specifically, entropy plays a significant role whenever one of the phases has a porous structure, providing the mean pore diameter is of the same order of magnitude as the diameter of the partitioning species. Porous structures used in the separations field include various polymer gels, virtually all membranes, and chromatographic packing used for size exclusion chromatography. The partitioning species involved are generally of macromolecular or colloidal size: proteins, synthetic polymers, polysaccharides, viruses, inorganic colloids, and many others.

When a large molecule approaches a porous structure having pores in its own size range, entry into the pores can only be realized if the molecule is properly positioned and oriented to slip into the openings. Once within the porous network, the molecule is severely limited in the positions and orientations it can assume; if it moves outside of its restricted range, it will be repelled by the walls (or other fixed elements) of the porous structure. Because of the constraints imposed by the pore walls, the motions of contained molecules are severely restricted. This loss of freedom in molecular motion is associated with a corresponding loss of entropy, a factor thermodynamically unfavorable to the partitioning of the species into the pores [27]. In this way, pores (such as those in a membrane) can partially (in some cases almost totally) exclude molecules technically capable of fitting within the porous network. For example, a linear polymer molecule could snake its way into a long thin pore only a few angstroms in diameter, but in so doing the polymer would lose the normal conformational entropy associated with its bends and twists in space. The unfavorable entropy change leads to a rejection from the pores.

The distribution coefficient K for species i partitioning between a porous matrix and bulk solution is

$$K = \frac{c_i(pores)}{c_i(bulk)}$$

where $c_i(pores)$ is the amount (moles, grams, etc.) of i per unit volume of pore space (not counting the volume of the solid matrix) and $c_i(bulk)$ is the amount per unit volume of bulk solution. Because of the entropic constraints noted above we find that $c_i(pores) < c_i(bulk)$, and thus $K < 1$. To see more specifically how this inequality arises, we consider spherical molecules or colloids of radius a partitioning into a simplified pore: a cylindrical capillary tube of diameter d_c and length L. The partitioning process is illustrated in Figure 2.2.

Because of finite radius a, the centers of the spheres can get no closer than distance a from the cylinder wall. Thus the sphere centers are excluded from an annular volume extending from radial position $\frac{1}{2}d_c - a$ to $d_c/2$ in the cylinder. If the component species were point objects rather than bulky spheres, they could freely occupy the entire pore volume and (in the absence of attractive or repulsive forces near the wall) K would equal unity. Spheres, by contrast, are excluded from the annular region that point objects readily occupy. Excluded sphere positions are shown as open dashed circles in Figure 2.2. Since these excluded positions are forbidden due to overlap with the wall, the number of allowed spheres is reduced in number to the few spheres (shaded circles) falling in the accessible volume closer to the cylinder axis.

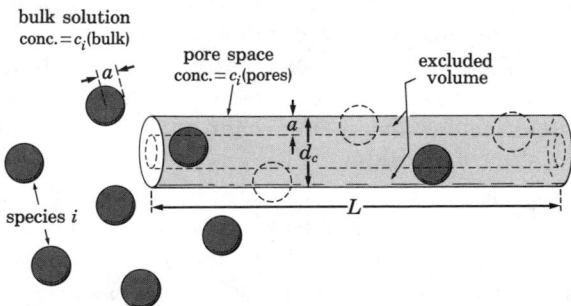

Figure 2.2. Partitioning of spherical bodies between a bulk solution and an idealized cylindrical pore. Spheres that are excluded from the pore volume because they would overlap with the wall are shown by open circles.

The exclusion process described above occurs because the "effective" volume of occupancy of the cylinder is reduced below its true volume. (A reduction in entropy naturally accompanies the shrinkage in effective volume.) The distribution coefficient (in the absence of disturbing forces) is simply the volume ratio

$$K = \frac{\text{accessible volume}}{\text{true volume}} = \frac{\pi(\frac{1}{2}d_c - a)^2 L}{\pi(\frac{1}{2}d_c)^2 L} \tag{2.42}$$

which reduces to [26]

$$K = \frac{(d_c - 2a)^2}{d_c^2} = \left(1 - \frac{2a}{d_c}\right)^2 \tag{2.43}$$

(This expression is valid for $2a \leq d_c$; $K = 0$ for $2a > d_c$.)

If d_c is replaced by $4/s$, where s is the wall area of the capillary per unit volume of pore space, we get

$$K = \left(1 - \frac{sa}{2}\right)^2$$

This expression is more widely applicable than Eq. 2.43 because few pores are true cylinders and d_c, but not s, loses meaning for noncylindrical pore geometries. Equation 2.44 can consequently be used as an approximation for other pore shapes and even for more complex pore space. For example, Eq. 2.44 proves to be exactly applicable to long pores of square cross section [27]; Eq. 2.43 cannot be applied without arbitrarily defining an apparent pore diameter to replace d_c. For any given pore geometry, s^{-1} is proportional to mean pore size.

We emphasize that the exclusion of partitioning objects from pore space occurs because the overlap of the objects with the pore walls is forbidden. It is rather natural to relate this phenomenon to the wall area subject to overlap, specifically area s per unit pore volume. (We add that the wall area due to surface roughness with a scale much smaller than that of the partitioning object should not be counted.)

For objects more complicated than spheres (e.g., rods or flexible chains), exclusion can be triggered not only by proximity to a section of pore wall (as above) but also by any inappropriate orientations and conformations that lead to overlap. This is illustrated by Figure 2.3. The distribution coefficient K for such complex bodies can no longer be considered as a simple volume ratio as expressed in Eq. 2.42. Instead, K becomes a ratio of volumes in multidimensional configuration space in which all possible positions, orientations, and conformations must be considered [27].

The theory of partitioning for bodies of complex geometry in convoluted pore spaces with different-sized passages is obviously quite complicated, but

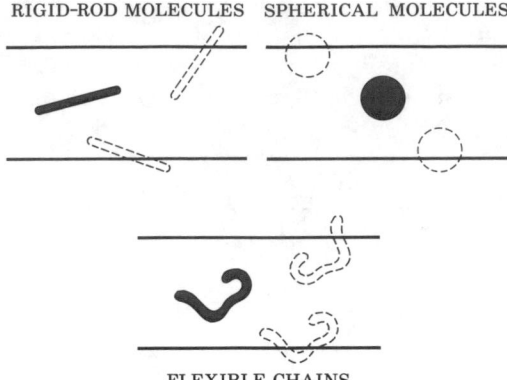

RIGID-ROD MOLECULES   SPHERICAL MOLECULES

FLEXIBLE CHAINS

Figure 2.3. Illustration of allowed and forbidden (overlapping) configurations of objects of different shapes in idealized pores. (Based on illustration in ref. 27.)

it is nonetheless important for describing many separation processes in real-world porous media. While no suitable all-inclusive theory exists (except in the form of complex integrals over ill-defined configuration space), a simple partitioning equation surprisingly emerges from theory for the rather general case in which rigid bodies of any arbitrary size and shape, no matter how complex, partition into an isotropic pore space defined by randomly spaced and oriented intersecting planes (the random-plane model of pore space). The equation is [27]

$$K = e^{-s\bar{L}/2} \tag{2.45}$$

where $\bar{L}$ is the *mean external length* (or *mean projection length*), which equals the maximum length of the object measured along any arbitrary axis, averaged over all possible orientations of the body in space. For spheres, $\bar{L}$ equals the diameter $2a$ and K becomes

$$K = e^{-sa} \tag{2.46}$$

which approaches the same limit as Eq. 2.44 for small values of sa, corresponding to large pores and/or small partitioning objects.

As the pore "diameter" increases in size (s decreases) relative to molecular or colloidal dimensions, less restrictions are imposed on the motions of contained species. Thus the exclusion effect gradually subsides as the pore size increases and consequently $K \rightarrow 1$. For the separation of two molecules of different size, it is important to pick a pore diameter that will substantially exclude one species but not another. Pore size selection is thus of utmost importance in membrane science and in choosing a support for size exclusion chromatography (SEC). Aspects of pore size optimization in SEC based on the above partitioning theory have been developed [28].

Experimental evidence on whether $\bar{L}$ or other molecular parameters (Stokes radius, viscosity radius, radius of gyration, the product of intrinsic viscosity and molecular weight, etc.) govern partitioning in SEC supports has been summarized by Dubin [29]. He concludes that none of these parameters perfectly correlates with SEC partitioning when a wide variety of macromolecules, of both rigid and flexible structure, are used as test probes. This may result from the complex uncharacterized nature of the pore space occupying the porous supports commonly utilized.

REFERENCES

[1] H. F. Franzen and B. C. Gerstein, *Rudimentary Chemical Thermodynamics*, Heath, Lexington, MA, 1971.

[2] J. R. Goates and J. B. Ott, *Chemical Thermodynamics*, Harcourt, New York, 1971.

[3] J. H. Knox, *Molecular Thermodynamics*, Wiley-Interscience, London, 1971.

[4] G. C. Pimental and R. D. Spratley, *Understanding Chemical Thermodynamics*, Holden-Day, San Francisco, CA, 1969.

[5] D. Buckingham, *The Laws and Applications of Thermodynamics*, Pergamon, Oxford, 1964.

[6] S. Glasstone, *Thermodynamics for Chemists*, Nostrand, New York, 1947.

[7] J. H. Hildebrand, J. M. Prausnitz, and R. L. Scott, *Regular and Related Solutions*, Van Nostrand, New York, 1970.

[8] A. G. Williamson, *An Introduction to Non-Electrolyte Solutions*, Wiley, New York, 1967.

[9] E. A. Guggenheim, *Thermodynamics*, 6th ed., North-Holland, Amsterdam, 1977, Chapters 9–11.

[10] J. C. Kirkwood and I. Oppenheim, *Chemical Thermodynamics*, McGraw-Hill, New York, 1961.

[11] F. H. MacDougall, *Thermodynamics and Chemistry*, Wiley, New York, 1939, Chapter XVI.

[12] D. E. Martire and R. E. Boehm, *J. Phys. Chem.*, **91**, 2433 (1987).

[13] D. E. Martire, *J. Chromatogr.*, **452**, 17 (1988).

[14] J. H. Hildebrand and R. L. Scott, *The Solubility of Nonelectrolytes*, 3rd ed., Dover, New York, 1964.

[15] B. L. Karger, L. R. Snyder and C. Horvath, *An Introduction to Separation Science*, Wiley, New York, 1973.

[16] C. J. van Oss, R. J. Good, and M. K. Chaudhury, *J. Colloid Interface Sci.*, **111**, 378 (1986).

[17] C. J. van Oss, R. J. Good, and M. K. Chaudhury, *Sep. Sci. Technol.*, **22**, 1 (1987).

[18] A. J. P. Martin, *Biochem. Soc. Symp.*, **3**, 4 (1949).

[19] P. J. Schoenmakers and L. G. M. Uunk, in J. C. Giddings, E. Grushka, and P. B. Brown, Eds., *Advances in Chromatography*, Vol. 30, Marcel Dekker, New York, 1989, Chapter 1.

[20] J. W. King, *J. Chromatogr. Sci.*, **27**, 355 (1989).

[21] M. L. Lee and K. E. Markides, *Science*, **235**, 115 (1987).

[22] L. Rohrschneider, in J. C. Giddings and R. A. Keller, Eds., *Advances in Chromatography*, Vol. 4, Marcel Dekker, New York, 1967, Chapter 7.

[23] K. Dimroth, C. Reichardt, T. Siepmann, and F. Bohlmann, *Justus Liebigs Ann. Chem.*, **661**, 1 (1963).

[24] P. J. Schoenmakers, H. A. H. Billiet, and L. deGalan, *Chromatographia*, **15**, 205 (1982).

[25] J. C. Giddings, M. N. Myers, L. McLaren, and R. A. Keller, *Science*, **162**, 67 (1968).

[26] J. C. Giddings, M. N. Myers, and J. W. King, *J. Chromatogr. Sci.*, **7**, 276 (1969).

[27] J. C. Giddings, E. Kucera, C. P. Russell, and M. N. Myers, *J. Phys. Chem.*, **72**, 4397 (1968).

[28] J. C. Giddings, *Anal. Chem.*, **40**, 2143 (1968).

[29] P. L. Dubin, in J. C. Giddings, E. Grushka and P. R. Brown, Eds., *Advances in Chemistry*, Vol. 31, Marcel Dekker, New York, in press.

EXERCISES

2.1(*) Sufficiently diluted, ethyl acetate has an equilibrium concentration at 20°C in an isobutanol-rich phase that is 7.20 times higher than that in a water-rich phase (that is, the distribution coefficient K for isobutanol/H_2O is 7.20). What is the value of $\Delta\mu^0$ for the transfer of ethyl acetate from water to isobutanol solutions?

2.2(**) The concentration of a colloidal suspension of 1.00 μm diameter polystyrene latex spheres in water at a distance of 5 μm above the bottom of a beaker under gravitational equilibrium at 298 K is 1.00×10^8 spheres/cm^3. Calculate the concentration at an elevation of 10 μm above the bottom wall. The densities of polystyrene and water at 298 K are 1.050 and 0.997 g/mL, respectively.

2.3(**) Prove that for spherical objects of radius a partitioned into long uniform pores of square cross section, the expression for the distribution coefficient is identical to Eq. 2.44.

2.4(*) Show that Eqs. 2.44 and 2.46 approach the same limiting form when the product $sa \rightarrow 0$. What is the limiting expression?

2.5(*) For thin rods of length l, it can be shown that $\bar{L} = l/2$ [27]. Estimate K for fibrinogen, which can be approximated as a thin rod of length 700 Å, partitioning into a porous solid with $s = 0.012/$Å. What does K change to if all pore dimensions are exactly doubled in size? Assume the applicability of the random-plane model of pore space.

3

SEPARATIVE TRANSPORT

We have observed that all separations require some form of displacement or transport. An element of that transport must be selective, which as we saw earlier requires the relative motion of solute components through the solvent or carrier medium in which they are contained. We describe the basics of such transport in this chapter. The other major form of transport—bulk or flow transport—will be detailed in Chapter 4.

3.1 TRANSPORT: DRIVEN BY EQUILIBRIUM

Essentially all transport is of such a nature that equilibrium is approached. Therefore, equilibrium dictates the direction of transport. Equilibrium considerations, however, do not detail the nature and the rate of the displacements toward equilibrium. These matters are extremely important. For one thing, the speed of separation is related to the speed of the elementary displacement steps leading toward equilibrium. For example, it is known that the motion of macromolecules through liquids is inherently slow. Accordingly, the speed of separation of macromolecules by chromatography or other methods is almost always slower than the speed of separation of small molecules.

In another vein, we must recognize that separation (e.g., electrophoresis) is often achieved on the path to equilibrium. Here an important question unanswered by equilibrium thermodynamics concerns the nature of the concentration profiles that evolve in the drift toward equilibrium. One must determine if they are suitable for separation and, if not, how they can be made so.

The issues raised above lie in the realm of transport phenomena. Below we develop elements of this discipline sufficient to address these issues for a majority of analytical separations.

3.2 IRREVERSIBLE THERMODYNAMICS

Displacements toward equilibrium are irreversible or, more descriptively, one way only. An elegant discipline describing these displacements is *irreversible thermodynamics*, sometimes called *nonequilibrium thermodynamics*. The four fundamental postulates of irreversible thermodynamics are (1):

1. Entropy S increases continuously for irreversible processes in isolated systems.
2. S, T, and other thermodynamic (equilibrium) functions can be defined and used even though equilibrium does not exist.
3. Fluxes are proportional to gradients or so-called "generalized forces." For example, diffusional flux is proportional to the concentration gradient dc/dx, not to $(dc/dx)^2$. Electrical current is proportional to voltage, and so on. This linearity is expressed by

$$J_k = \sum L_{kj} X_j \qquad (3.1)$$

in which fluxes J are related to gradients or generalized forces X by transport coefficients L_{kj}.
4. Onsager's reciprocity relationships are valid. According to Onsager the coefficients are related by

$$L_{kj} = L_{jk} \qquad (3.2)$$

Although we shall not directly use these four postulates of irreversible thermodynamics as a foundation to our study of molecular transport in separations, a number of important principles are illuminated here. For instance, postulate 2 permits us to use—and this is in no way obvious—equilibrium parameters such as entropy and temperature in descriptions of systems where no equilibrium exists. The importance of this is evident when we ask ourselves how we would describe a system if these parameters were not available. Postulate 3 demonstrates that in the range of our typical experiences, the fluxes of matter or of heat are proportional to the gradients or forces that drive them. However, there are exceptions; nonlinear terms enter if the forces become intense enough.

Equation 3.1 implies that we get all kinds of fluxes from any one of the separate gradients or forces that exist. Reciprocity, Eq. 3.2, is a concise statement of symmetry among the coefficients describing this. While we shall not use reciprocity in a quantitative sense, we should keep the symmetry in mind. Consider, for instance, *thermal diffusion*. Here, a temperature gradient, primarily responsible for heat flow, causes also a flux of mass, sometimes used for separation in thermogravitational columns or in field-flow fractionation. Postulate 4 makes us aware that the complementary effect also exists: a gradient in concentration, beyond its primary function of driving mass diffusion, will give rise to a flux of heat. This phenomenon is called the *diffusion thermoeffect* or the *Dufour effect*. Other such symmetrical effects complement one another throughout the study of transport phenomena. Some of these influence separation processes and should therefore be recognized by the student of separations. Further details can be found in books on the subject [1–6].

While the formalism of irreversible thermodynamics provides an elegant framework for describing molecular displacements, it provides too little substance and too much conceptual difficulty to justify its development here. For instance, it provides no values, not even estimates, for various transport coefficients such as the diffusion coefficient. Cussler has noted the disappointment of scientists in several disciplines with the subject [7]. It is the author's opinion that a clearer understanding of the transport processes and interrelationships that underlie separations can be obtained from a mechanical-statistical approach. This is developed in the subsequent sections.

3.3 MECHANICAL VERSUS MOLECULAR TRANSPORT

Mechanical motion—the motion of macroscopic bodies—is described by simple Newtonian physics. The study of mechanical motion is a logical precursor to the study of molecular displacements. It is useful to highlight the principal similarities and differences here.

We note first that the final goals of the descriptions of mechanical and molecular motions are distinctly different. In the theory of mechanical motion, we wish to follow the path of individual objects (like a space probe or baseball), while for molecular motion we are rarely interested in individual particles (molecules or colloids) but rather in the changes in the spatial distribution of large numbers of such particles.

Despite the difference in goals, single molecules respond to forces much like macroscopic bodies. The mathematics, too, is analogous, as shown by the two equations

$$\text{mechanical driving force} = -\frac{d\mathcal{P}}{dx} \tag{3.3}$$

$$\text{molecular driving force} = -\frac{d\mu}{dx} \tag{3.4}$$

in which $\mathcal{P}$ is potential energy and μ is chemical potential. Both $\mathcal{P}$ and μ have the dimensions of energy, which emphasizes the similarity.

Quantities $\mathcal{P}$ and μ differ only in the greater comprehensiveness of the latter. Chemical potential μ includes $\mathcal{P}$ in an implicit manner in the term for external forces μ^{ext}. However, μ encompasses even more than $\mathcal{P}$ for it also includes the effects of phase distribution forces in μ^0 and the entropy influence of the term $\mathcal{R}T \ln c$ as shown in Eq. 2.17. The entropy term is dwarfed by $\mathcal{P}$ and is thus negligible for macroscopic bodies, but it has a major influence in governing the transport of molecules and colloids. It is responsible for the diffusion of these microscopic particles and makes it necessary to describe molecular transport in terms of distributions (i.e., concentration distributions) as just noted.

While μ is more comprehensive than $\mathcal{P}$, its use is limited in one sense: Eq. 3.4 is strictly applicable only at constant temperature. This limitation can be traced back to the thermodynamic nature of μ, Section 2.3.

We note that the entropic contribution to $-d\mu/dx$ is not strictly a force in the mechanical sense of the word. It arises out of the incessant random molecular collisions that produce Brownian motion. Yet, for mathematical purposes, it acts like a force. The use of $-d\mu/dx$ as an effective force has been justified in the cited monographs on irreversible thermodynamics and by other authors [8, 9].

In the case of strictly mechanical motion (whose molecular analog we shall shortly show), the fundamental equation leading eventually to the description of an object's trajectory is the equation of motion. This begins with Newton's fundamental equation: force $(F') = $ mass $(m) \times$ acceleration (d^2x/dt^2), or

$$m \frac{d^2x}{dt^2} = F' \tag{3.5}$$

This differential equation can be integrated to show the object's position x at different times t, providing F' (and any variations in time and space) is known.

Force F' typically has several components. First is the driving force of Eq. 3.3

$$F'_1 = -\frac{d\mathcal{P}}{dx} \tag{3.6}$$

This is generally a so-called *conservative force* and depends only on the position of the body. Second is the frictional force, which always opposes the motion of the body through its medium. This includes sliding friction if the body is slipping across a solid surface. Most relevant is viscous friction due to the drag forces of air, water, or other fluids that retard the motion of any body passing through them. For low velocities, this force is directly proportional to velocity dx/dt; it acts in such a direction that it opposes the existing motion as indicated by the negative sign in the equation

$$F'_2 = -f' \frac{dx}{dt} \tag{3.7}$$

This force, being one of a class of dissipative or frictional forces, is termed a *nonconservative force*, as are all forces that depend on velocity or time.

The constant f' of Eq. 3.7 is known as the *friction coefficient* and, as used here, applies to a single object or particle. It is a fundamental parameter reflecting the magnitude of drag forces through fluids. It depends on the dimensions of the body in motion as well as upon the viscosity of the medium through which it moves (Chapter 4).

The substitution of the sum of F'_2 and F'_2 from the last two equations for F' in Eq. 3.5 yields the basic *equation of motion*

$$m \frac{d^2x}{dt^2} = -\frac{d\mathscr{P}}{dx} - f' \frac{dx}{dt} \tag{3.8}$$

Integration of this equation yields $x(t)$, that is it provides particle coordinate x as a function of time t, and it thus fully describes particle motion along axis x. The same mathematical form, of course, applies to the other spatial coordinates y and t. These equations, upon integration, describe completely the trajectory of a macroscopic body through space.

Now we must examine the differences that arise with molecules.

3.4 MOLECULAR TRANSPORT

Molecules and colloidal bodies are subject to all the forces that control the motion of macroscopic bodies. This includes the frictional force $-f' \, dx/dt$. In addition, they are subject to displacement by violent collisional forces from neighboring molecules, which cause erratic fluctuations in motion. These forces, as noted, are effectively accounted for by using the gradient in chemical potential rather than in potential energy as the driving force, as suggested by Eq. 3.4. Since chemical potential refers to a mole of solute, we write the equation of motion for Avogadro's number of molecules. Equation 3.8 for macroscopic bodies therefore assumes molecular relevance in the form

$$M \frac{\overline{d^2x}}{dt^2} = -\frac{d\mu}{dx} - f \frac{\overline{dx}}{dt} \tag{3.9}$$

where M is the molecular weight, the mass of the one mole of component material assumed to be in transport. Quantity $-d\mu/dx$ is the effective force per mole, and friction coefficient f must now be considered as a molar drag constant

$$f = \mathcal{N}f' \qquad (3.10)$$

where f' is the friction coefficient for a single particle or molecule and $\mathcal{N}$ is Avogadro's number. The bars in Eq. 3.9 represent the averaging of molecular motions for the entire mole of substance.

While Eq. 3.9 will serve our mathematical purposes, useful insights into molecular motions and processes such as diffusion can be gained by examining more closely the equation of motion, Eq. 3.8, as applied to molecules. In particular, we focus below on the effects of erratic molecular impacts.

For molecules and colloids the forces due to molecular impacts are large, but they are short in duration and random in direction, as can be surmised from the observation of Brownian motion. These forces can be described by a rapidly fluctuating term $A(t)$ in the equation of motion. With this "Brownian" term added to the other forces on the right, Eq. 3.8 assumes the form

$$m\,\frac{d^2x}{dt^2} = -\frac{d\mathscr{P}}{dx} - f'\,\frac{dx}{dt} + A(t) \qquad (3.11)$$

which is a form of the so-called *Langevin equation* [6]. The treatment of this expression (and equivalent equations) in a rigorous fashion requires considerably more mathematical sophistication than does the theory of mechanical transport. As a consequence, it has attracted the attention of mathematicians and mathematically inclined scientists [10–12]. In the following, we limit ourselves to a discussion of the implications of random force $A(t)$.

The observation of Brownian motion, which arises as a consequence of $A(t)$, shows that the fluctuating forces of molecular collisions constitute a randomizing influence, dispersing and leveling concentrations as a result of the random displacements of individual particles. However, entropy is also a randomizing parameter and it is not unreasonable to imagine that they are born in the same roots. With this assumed connection, the gradient in chemical potential, which contains entropy in the $\mathscr{R}T \ln c$ term (among others), can be employed in place of $A(t)$. It can also replace $-d\mathscr{P}/dx$ because chemical potential, as we saw, includes the potential energy originating in external fields.

One more consideration enters. Chemical potential is a thermodynamic parameter applicable to large collections of molecules, not to single molecules. Equation 3.11 applies to single molecules. Since we are more interested in collections of molecules, we can adapt Eq. 3.11 to express the average motion of a large assemblage, say one mole, of molecules. With this and the use of $-d\mu/dx$ in place of $A(t)$ and $-d\mathscr{P}/dt$, we arrive again at Eq. 3.9.

In regards to Eq. 3.9, it turns out that the frictional drag on <u>a mole</u> of molecules is so relatively great that the average acceleration d^2x/dt^2 is effectively zero under most circumstances, except under vacuum. (For example, it requires a force equivalent to the weight of 25,000 (metric) tons to drag a mole of typical solute through a solvent at only 0.1 mm/s at 300 K.) The reason for the high friction is that abundant molecular collisions inhibit the movement of molecules subjected to external forces. Thus when an electrical field is suddenly applied to an ionic solution at time $t = 0$, the ions accelerate from rest during the brief period of a few molecular collisions, after which the frictional resistance to motion becomes so high that no further acceleration can occur. Thereafter the ions follow a more or less steady-state movement, without overall acceleration. The velocity-time curve for the average motion of a large number of ions would appear roughly as shown in Figure 3.1. It can be shown that the transient time τ, in which acceleration is important, is given by

$$\tau = \frac{M}{f} \tag{3.12}$$

This is typically 10^{-12} s or less, a time totally negligible compared to the duration of most transport processes.

For all essential purposes then, we can ignore the average molecular acceleration term in Eq. 3.9. We consequently arrive at an expression for the average velocity (which we designate as $\bar{U}$) in the following form

$$\left(\frac{\overline{dx}}{dt}\right) = \bar{U} = -\frac{1}{f}\frac{d\mu}{dx} \tag{3.13}$$

Equation 3.13 fulfills our basic objective in describing average molecular motion. However, inasmuch as we cannot ordinarily measure directly the average velocity of molecules or ions undergoing transport, it is advantageous to transform $\bar{U}$ into the flux density J, a parameter that is more directly observable. The flux density of a component is the number of moles carried through a unit area in unit time. It is related to that component's mean molecular velocity $\bar{U}$ and concentration c by

$$J = \bar{U}c \tag{3.14}$$

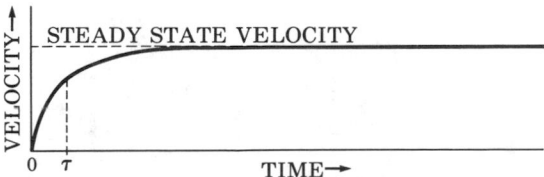

Figure 3.1. Mean velocity of solute molecules after application of a field at time zero. The transient time t is $\sim 10^{-12}$ s, which means that for all practical purposes the steady-state velocity is reached instantaneously.

Equation 3.14 is a widely used expression that can be simply derived. If we imagine the uniform displacement of a component at velocity $\bar{U}$ along the axis of a tube of unit cross section, the specific component material swept through a fixed cross section in unit time (1 sec) will be that part of the component within the distance $\bar{U}$ of that fixed cross section, since $\bar{U}$ is, by definition, the distance that the component moves in unit time (see Figure 3.2).

Since the cross-sectional area is unity, the volume swept out is simply $\bar{U}$. The amount of component swept across the fixed cross section in this unit time, J, is simply the product of concentration and volume or $c\bar{U}$, which gives Eq. 3.14.

If we substitute $\bar{U}$ from Eq. 3.13 into Eq. 3.14, we obtain

$$J = -\frac{c}{f}\frac{d\mu}{dx} \tag{3.15}$$

According to our discussion of chemical potential in Chapter 2, the general form of μ is

$$\mu = \mu^{\text{ext}} + \mu^0 + \mathscr{R}T \ln c \tag{3.16}$$

The substitution of this into Eq. 3.15 yields

$$J = -\frac{c}{f}\left(\frac{d\mu^{\text{ext}}}{dx} + \frac{d\mu^0}{dx}\right) - \frac{\mathscr{R}T}{f}\frac{dc}{dx} \tag{3.17}$$

in which J, c, and f are all expressed in terms of moles of the component of interest.

According to Eq. 3.17, all fluxes for relative transport are inversely proportional to friction coefficient f. Since relative transport rates—ultimately responsible for selectivity in separations—are keyed to f, *the friction coefficient is fundamental in establishing a time scale for virtually all selective transport and thus all separations.*

We look in detail at factors affecting f in Section 4.13. For example, for a spherical particle of radius a suspended in a liquid we can use Stokes law, $f = 6\pi\eta a$, where η is the liquid viscosity. We shall show more completely

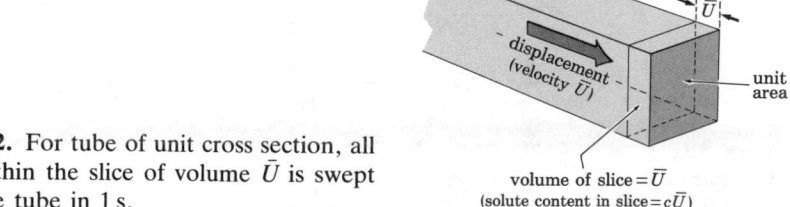

Figure 3.2. For tube of unit cross section, all solute within the slice of volume $\bar{U}$ is swept out of the tube in 1 s.

how f depends on molecular dimensions and solvent properties and how one can, as a practical matter, reduce its magnitude in order to hasten the separations that it governs.

Equation 3.17 can be written in a condensed form

$$J = Uc - D\frac{dc}{dx} \tag{3.18}$$

where U designates the mean displacement velocity caused by external fields and chemical (internal) effects in combination, but in the absence of concentration gradients. It is expressed by

$$U = -\frac{1}{f}\left(\frac{d\mu^{\text{ext}}}{dx} + \frac{d\mu^0}{dx}\right) \tag{3.19}$$

Of equal interest is the second term on the right of Eq. 3.18. This comes from the far-right term, $\mathcal{R}T\ln c$, of Eq. 3.16 and thus represents the contribution of dilution entropy to J. This term is proportional to the concentration gradient $-dc/dx$. This dependence on concentration gradient is universally characteristic of diffusion. We thus establish that diffusion has its origin in the drive of entropy to level out all concentrations and mix all components.

In the absence of external fields and internal forces, in which case $U = 0$, we are left with

$$J = -\frac{\mathcal{R}T}{f}\frac{dc}{dx} = -D\frac{dc}{dx} \tag{3.20}$$

The latter form is the basic equation of diffusion generally identified as Fick's first law, formulated in 1855 [13]. Fick's first law, of course, can be deduced from the postulates of irreversible thermodynamics (Section 3.2), in which fluxes are linearly related to gradients. It is historically an experimental law, justified by countless laboratory measurements. The convergence of all these approaches to the same basic law gives us confidence in the correctness of that law. However, the approach used here gives us something more.

The present approach, by giving the first equality of Eq. 3.20, $J = -(\mathcal{R}T/f)(dc/dx)$, yields not only Fick's first law but provides an expression for diffusion coefficient D in terms of the basic parameters $\mathcal{R}$, T, and f. Thus, comparison of the last two parts of Eq. 3.20 yields

$$D = \frac{\mathcal{R}T}{f} \tag{3.21}$$

This fundamental expression for D is the *Nernst–Einstein equation*. It shows still another dimension to the importance of friction coefficient f, which is now seen to control diffusion as well as nearly all other transport.

3.5   TRANSPORT WITH FLOW

The equations above describe relative displacement which, as noted before, is displacement through (not with) the medium. Recall that bulk displacement (or transport with the medium), mainly in the form of flow, can occur also. The latter generates an additional flux density of

$$J = vc \qquad (3.22)$$

where v is the flow velocity through the plane in question. This flux density adds to that generated by relative displacement, described by Eq. 3.18. The total is

$$J = (U + v)c - D\frac{dc}{dx} \qquad (3.23)$$

To be more complete all terms above should be written as vector quantities. However, most applications in separations are one-dimensional. For present purposes it will be sufficient to keep in mind that each term contributing to J has a directional characteristic.

3.6   TRANSFORMATION TO CONCENTRATION UNITS

The ultimate goal of a basic study of separations is to obtain a description of how component concentration pulses (zones or peaks) move around in relationship to one another. The flux density J tells how solute moves across boundaries into and out of regions, but it does not detail the ebb and flow of concentration. To do the latter we must transform J into a form that directly yields concentration changes. The procedure followed below for this is standard in many fields. It is followed, for example, in treatments of heat conduction and diffusion [14, 15]. We shall continue to simplify our treatment to one dimension.

Let us imagine two parallel planes of unit area, each normal to the flux of component, one at position x and the other at $x + dx$. The two planes are illustrated in Figure 3.3. The flux entering at plane x minus that leaving at plane $x + dx$ represents the accumulation, in moles/second, of component between the planes

$$J_x - J_{x+dx} = \text{mol gained}/\text{s} \qquad (3.24)$$

This gain (or loss if it comes out with a negative sign) occurs between the planes, in a volume equal exactly to dx (inasmuch as the planes have unit area). We recognize the moles gained divided by the volume dx as simply the increment in concentration Δc. All of this acquires a simple form if we multiply the right-hand side of Eq. 3.24 by $dx/dx = 1$

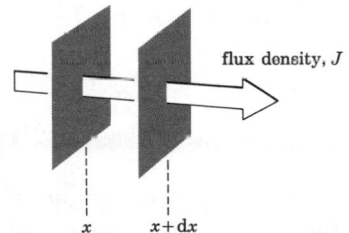

flux density, J

Figure 3.3. Movement of solute across two planes of unit area.

x $x+dx$

$$J_x - J_{x+dx} = \overbrace{\frac{\text{mol gained}}{dx}}^{\Delta c} \frac{dx}{s} \tag{3.25}$$

We can write the increment in concentration per second, $\Delta c/s$, as dc/dt, giving

$$J_x - J_{x+dx} = \frac{dc}{dt} \, dx \tag{3.26}$$

We now simplify the left-hand side by first making a Taylor's expansion of J_{x+dx} around point x

$$J_{x+dx} = J_x + \frac{dJ}{dx} \, dx \tag{3.27}$$

in which higher-order terms become negligible as dx approaches zero. The substitution of this into Eq. 3.26 gives

$$\frac{dc}{dt} = -\frac{dJ}{dx} \tag{3.28}$$

which is known as the *equation of continuity*. This equation is very general, applicable to any form that J may take (mass flux, heat flux, etc.), because it is based on the very general "inventory" considerations just developed.

The substitution of Eq. 3.23 into Eq. 3.28 yields

$$\frac{dc}{dt} = -\frac{d}{dx}(U + v)c + \frac{d}{dx}\left(D \frac{dc}{dx}\right) \tag{3.29}$$

When D, U, and v are constant, this becomes

$$\frac{dc}{dt} = -(U + v)\frac{dc}{dx} + D \frac{d^2c}{dx^2} \tag{3.30}$$

We note that U, v, and D may vary with x in some systems, but these variations are negligibly slow compared to the rapid variation of c with x associated with sharp concentration pulses. More complicated are variations of U, v, and D with concentration. However, since sample amounts are generally small in analytical separations, the concentration dependence can usually be ignored.

When flow and mean displacement velocities are zero ($U = 0$ and $v = 0$), the above reduces to Fick's second law of diffusion

$$\frac{dc}{dt} = D \frac{d^2c}{dx^2} \qquad (3.31)$$

Equations 3.30 and 3.31 are the basic differential equations for separation systems; when solved, they show how concentration varies with time and distance, that is, the way in which concentration pulses move and evolve with the passage of time.

3.7   SUMMARY OF PRINCIPAL TRANSPORT EQUATIONS

The basic expression for molar flux density, Eq. 3.23, can be written as

$$J = Wc - D \frac{dc}{dx} \qquad (3.32)$$

where W is given by

$$W = U + v \qquad (3.33)$$

Quantity W is simply the sum of all direct displacement velocities—those caused by bulk displacement at velocity v plus those caused by chemical potential gradients which impel solute at velocity U.

The basic equation for concentration changes, Eq. 3.30, thus assumes the form

$$\frac{dc}{dt} = -W \frac{dc}{dx} + D \frac{d^2c}{dx^2} \qquad (3.34)$$

where once again W symbolizes the total of all nondiffusional displacement velocities.

3.8   PHYSICAL INTERPRETATION OF TRANSPORT EQUATIONS

Recall that Eq. 3.34 is derived from Eq. 3.32 by means of the equation of continuity. The respective terms on the right of the two equations therefore represent the same physical processes. Insofar as Eq. 3.32 is concerned, the term Wc represents, as we noted in Section 3.4, a mole flux across a fixed plane of unit area due to the steady displacement of component through that plane at velocity W. The term $-D(dc/dx)$ represents a simple diffusion flux proportional to the concentration gradient. These two processes give rise to the two terms on the right of Eq. 3.34 as well. To visualize the latter, consider a bell-shaped component concentration profile moving at velocity W and undergoing simultaneous diffusion. Suppose we wish to monitor the rise and fall of component concentration in a thin cross-sectional slice of the zone lying between x and $x + dx$, as indicated in Figure 3.4. We would like to understand the

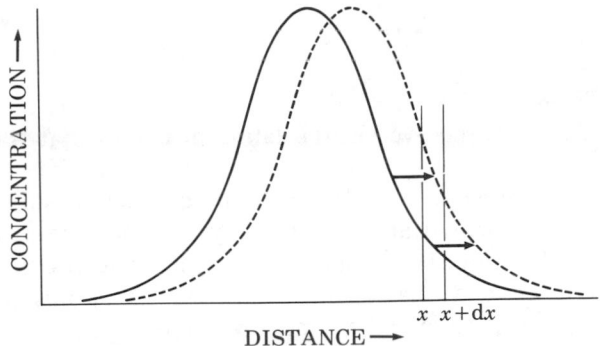

Figure 3.4. Bell-shaped concentration curve and its displacement right due to flow and/or field-induced drift at velocity W. The displacement of solute in and out of region dx is shown by arrows.

factors influencing dc/dt in that slice, or in any other volume element. First, it is clear that the overall displacement at velocity W will, after an interval, move component material from slice dx to a position somewhat right of dx, as shown by the lower horizontal arrow in the figure. Fresh component material will be moved in from the left as indicated by the upper arrow. However, inasmuch as dc/dx is negative—which means that the component is more concentrated on the left—the replacement material will be more concentrated and c will rise. In other words, dc/dt will be positive when dc/dx is negative (providing W is positive), and vice versa. In sum, the rate of increase of c will be proportional to the velocity W at which enriched component arrives and the magnitude of the negative concentration gradient $-dc/dx$, which expresses how quickly the concentration rises off to the left. Thus, we expect dc/dt to be proportional to $-W(dc/dx)$. In fact, it is exactly equal to $-W(dc/dx)$, as can be seen in Eq. 3.34. We can thus visualize the origin of the term $-W(dc/dx)$ using the above simple displacement arguments.

The term $D(d^2c/dx^2) = D(d/dx)(dc/dx)$ of Eq. 3.34 is also easy to rationalize. Since the concentration curve is concave up in the vicinity of position x, more component will diffuse in along the high concentration gradient at x than will diffuse out at $x + dx$, where the gradient is slightly less. Thus there will be an accumulation of component in element dx; this will be proportional to the abruptness with which gradient dc/dx is changing, that is, proportional to $(d/dx)(dc/dx)$. It is also proportional to D. Thus we arrive at the term $D(d^2c/dx^2)$ of Eq. 3.34. Extension of this argument shows that the diffusion mechanism leads to component accumulation in all parts of the concentration curve that are concave up and dissipation in all regions concave down.

3.9 TRANSFORMATION OF COORDINATES FOR ZONES IN UNIFORM TRANSLATION

The basic transport expression of Eq. 3.34 can be written in the partial derivative form

$$\frac{\partial c}{\partial t} = -W\,\frac{\partial c}{\partial x} + D\,\frac{\partial^2 c}{\partial x^2} \tag{3.35}$$

where, as indicated in Eq. 3.33, $W = U + v$. As noted above, the first term on the right describes the overall translation of the zone at velocity W and the second term accounts for the simultaneous diffusion of the zone outward from its center of gravity. Both W and D tend to be constant (or at most slowly varying) in analytical separation systems so the two processes, translation and diffusion, can usually be considered to occur steadily from the beginning to the end of a separative process. When this is not the case the treatment given here must be generalized [16, 17].

The translation and diffusion processes can be separated and the mathematics simplified by a change in coordinates. We define a coordinate system moving with the center of gravity of the zone, located at $x = X$, and thus in the case under consideration, moving forward at a velocity of W with respect to the old coordinate system. The distance along the axis of translation in the new coordinate system y is related to x by

$$y = x - X = x - Wt \tag{3.36}$$

Our object is to describe concentration c (and its derivatives) in this new coordinate system.

We begin by recognizing that c is a function of y and t

$$c \equiv c(y, t) \tag{3.37}$$

We make the transformation of coordinates by first obtaining the exact differential

$$dc = \left(\frac{\partial c}{\partial t}\right)_y dt + \left(\frac{\partial c}{\partial y}\right)_t dy \tag{3.38}$$

From this we obtain $(\partial c/\partial t)_x$, needed as the term on the left of Eq. 3.35

$$\left(\frac{\partial c}{\partial t}\right)_x = \left(\frac{\partial c}{\partial t}\right)_y - W\left(\frac{\partial c}{\partial t}\right)_t \tag{3.39}$$

The first term on the right of Eq. 3.35 reduces to

$$-W\left(\frac{\partial c}{\partial x}\right)_t = -W\left(\frac{\partial c}{\partial y}\right)_t \tag{3.40}$$

and the second term takes the form

$$D\left(\frac{\partial^2 c}{\partial x^2}\right) = D\left(\frac{\partial^2 c}{\partial y^2}\right) \tag{3.41}$$

Substituting the three expressions above into Eq. 3.35, we obtain the simple form

$$\frac{\partial c}{\partial t} = D\left(\frac{\partial^2 c}{\partial y^2}\right)$$ (3.42)

which represents diffusion alone as seen by comparison with Eq. 3.31. The translational motion represented by the first term on the right of Eq. 3.35 has been subtracted out by the transformation from x to y.

The advantage of Eq. 3.42 over Eq. 3.35, other than simplicity, is that the standard textbooks on diffusion [14, 15] and on heat conduction [18] (subject to an identical mathematical law) are replete with solutions to this fundamental equation. One of the most important solutions is the Gaussian function, which we will describe in some detail in Chapter 5.

Before looking at how Gaussian peaks arise from these transport equations we will devote our attention, by way of the next chapter, to the details of another form of transport, flow transport, and some related viscous phenomena.

REFERENCES

[1] H. J. V. Tyrrell, *Diffusion and Heat Flow in Liquids*, Butterworth, London, 1961.

[2] R. Haase, *Thermodynamics of Irreversible Processes*, Addison-Wesley, Reading, MA, 1969.

[3] S. Wisniewski, B. Stainszewski, and R. Szymanik, *Thermodynamics of Nonequilibrium Processes*, Reidel, Boston, 1976.

[4] Y. L. Yao, *Irreversible Thermodynamics*, Van Nostrand, New York, 1981.

[5] B. H. Lavenda, *Nonequilibrium Statistical Thermodynamics*, Wiley, New York, 1985.

[6] J. Keizer, *Statistical Thermodynamics of Nonequilibrium Processes*, Springer-Verlag, New York, 1987.

[7] E. L. Cussler, *Multicomponent Diffusion*, Elsevier, Amsterdam, 1976, p. 37.

[8] L. J. Gosting, in M. L. Anson, K. Bailey, and J. T. Esall, Eds., *Advances in Protein Chemistry*, Vol. 11, Academic, New York, 1956, pp. 430–548.

[9] C. Tanford, *Physical Chemistry of Macromolecules*, Wiley, New York, 1961, Chapter 6.

[10] E. Nelson, *Dynamical Theories of Brownian Motion*, Princeton University Press, Princeton, NJ, 1967.

[11] I. Prigogine, *Non-Equilibrium Statistical Mechanics*, Interscience, New York, 1962, Chapter 3.

[12] N. Wax, Ed., *Noise and Stochastic Processes*, Dover, New York, 1954.

[13] A. Fick, *Ann. Phys.* (Leipzig), **170**, 59 (1855).

[14] R. C. L. Bosworth, *Transport Processes in Applied Chemistry*, Wiley, New York, 1956, Chapter 1.

[15] J. Crank, *The Mathematics of Diffusion*, Oxford University Press, London, 1956, Chapter 1.

[16] P. R. Rony, *Sep. Sci.*, **3**, 425 (1968).

[17] G. H. Weiss and M. Dishon, in M. Ezrin, Ed., *Polymer Molecular Weight Methods* (Advances in Chemistry Series), American Chemical Society, Washington, DC, 1973, Chapter 19.

[18] H. S. Carslaw and J. C. Jaeger, *Conduction of Heat in Solids*, Oxford University Press, London, 1947.

EXERCISES

3.1(***)   If the differential equation of motion of a particle of mass m is written as

$$m \frac{d^2x}{dt^2} = -f' \frac{dx}{dt} + F' \tag{1}$$

then when the driving force F' and friction coefficient f' are assumed constant, a solution exists in the form

$$x = Ae^{-kt} + Bt + C \tag{2}$$

Assuming that velocity dx/dt and position x are both zero at time zero, obtain expressions for A, B, C, and k. Show that velocity dx/dt is

$$\frac{dx}{dt} = \frac{F'}{f'} \left[1 - \exp\left(-\frac{f't}{m} \right) \right] \tag{3}$$

3.2(*)   Equation 3 of Exercise 3.1 can be written as

$$\frac{dx}{dt} = \frac{F'}{f'} \left[1 - \exp\left(-\frac{t}{\tau} \right) \right] \tag{4}$$

where $\tau = m/f'$ is the relaxation time for achieving a steady-state velocity.

a. What is the value of τ for acetone in water at 20°C for which $f' = 3.49 \times 10^{-9}$ g s^{-1} molecule^{-1} ($f = Nf' = 2.10 \times 10^{15}$ g s^{-1} mol^{-1})?

b. What is τ for tobacco mosaic virus for which $M = 31,300,000$ and $f = 4.60 \times 10^{17}$ g s^{-1} mol^{-1} in water at 20°C?

3.3(**)   Transient time $\tau = m/f'$ (or M/f for a mole of molecules, Eq. 3.12) can be considered as measuring the duration (and thus the importance) of inertial effects in the motion of a body after applying a force to or changing an existing force on the body. Thus when a body is in motion and the force impelling it is suddenly removed, the body will drift to a halt due to frictional drag, retaining only the fraction $e^{-1} = 0.368$ of its momentum after time τ. More generally, show that the body's velocity decreases according to

$$\frac{\text{velocity at time } t}{\text{velocity at } t = 0} = e^{-t/\tau}$$

that is, show that this equation satisfies the equation of motion of the body.

3.4(*) A certain protein at a concentration of 1.0×10^{-4} mol/L sediments at a velocity of 5.0×10^{-4} cm/s through its buffer medium. What is the flux density of protein—the amount crossing a unit area perpendicular to the sedimentation path in unit time?

3.5(**) What is the value of the "diffusion velocity" (the mean molecular velocity caused by a concentration gradient) of a component found at a concentration of 0.010 mol/L and with a concentration gradient of 1.0 mol/L cm? The diffusion coefficient is 5.0×10^{-6} cm^2/s.

3.6(*) At 15°C the viscosity of water is 0.0114 g cm^{-1} s^{-1} (poises), and the measured diffusion coefficient for glucose in water is 0.52×10^{-5} cm^2/s. What is the friction coefficient f? What is the effective diameter of the glucose molecule, assuming that it behaves as a spherical particle in a continuous medium?

3.7(*) The viscosity of water is 0.0114 poises at 15°C and is 0.0072 poises at 35°C. Calculate f and D for glucose in water at 35°C using the results of the foregoing problem.

3.8(**) Estimate the molecular weight (MW) of bushy stunt virus, density 1.35 g/cm^3. Its diffusion coefficient in water at 20°C is 1.15×10^{-7} cm^2/s. At this temperature the viscosity of water is 0.0100 poises. Assume that the virus is spherical in shape.

3.9(***) Hemoglobin, MW 68,000, diffuses into water at a rate governed by $D = 6.90 \times 10^{-7}$ cm^2/s at 20°C, at which temperature water's viscosity is 1.00×10^{-2} poises.
a. What is the Stokes' law radius of the molecule?
b. What is the radius calculated on the basis that the molecule is a sphere whose density in solution is 1.33 g/cm^3?
c. If the discrepancy in (a) and (b) were due to a hydration shell moving along with the hemoglobin molecule, what would the shell thickness be? What would the MW of the total cluster

(molecule plus shell) be, assuming the hydration shell to have unit density?

3.10(***)   Repeat the preceding calculation (under the same conditions) for collagen, MW 345,000, $D = 6.90 \times 10^{-8}$ cm^2/s, and density 1.44 g/cm^3. How plausible is the hydration hypothesis? What else might cause the discrepancy?

3.11(**)   The diffusion coefficient of glucose-H$_2$O at 15°C is 0.52×10^{-5} cm^2/s. If we could attach a fine, invisible (and nonfrictional) string to each molecule of a mole of glucose in a dilute aqueous solution and tie them all to a team of horses that could pull a total of 4400 lb (which you can assume to be 2.00×10^9 dyn), how fast would the team drag the glucose through the water, assuming that the latter was firmly anchored? How many teams of horses would be needed to get this mere 192 g of glucose up to a speed of 0.2 mm/s?

3.12(**)   Imagine a concentration profile given by

$$c = c_0 + ax^2$$

in a region where $W = 175$ μm/s, $D = 910$ μm^2/s, and constant $a = 1.30 \times 10^{-5}$ (moles/L)/μm^2. At what coordinate position, x_s, is the concentration static (i.e., constant in time)? What is the value of dc/dt at $2x_s$?

3.13(*)   In an experiment done to separate proteins by electrophoresis at 20°C, the voltage is switched on at time $t = 0$. At what time does serum albumin, for which $D = 5.94 \times 10^{-7}$ cm^2/s and MW = 66,000, reach 99% of its terminal (steady-state) average velocity? (See Exercises 3.1 and 3.2.)

4

FLOW TRANSPORT AND VISCOUS PHENOMENA

Flow processes constitute a major mechanism of separative transport. Furthermore, the particular nature of flow—its magnitude, cross sectional distribution, origin, and limits—is largely responsible for the level of success in many attempts at separation.

As noted earlier, flow is a form of bulk displacement in which components entrained in a flowing medium are carried along nonselectively with the medium. Flow displacement thus stands in contrast to the other major transport mechanism—relative displacement—which is selective.

Flow transport is governed by Eq. 3.22, which expresses the flux density contributed by flow as

$$J = vc \tag{4.1}$$

To fully characterize transport, one clearly needs to determine flow velocity v at every point in the system. For simple flow systems (such as open capillary tubes), the v values can be calculated at each point whereas for complex systems (e.g., packed columns), the distribution in vs can only be estimated by statistical methods.

The strength of flow is that it provides the most powerful and versatile mechanism of transport available for separative displacement. The weakness of flow—other than its nonselectivity—is its nonuniformity. For most flow systems, v varies widely from point to point in the flow space. The different vs carry component molecules downstream at different rates, thus leading to the broadening of component zones. We must understand the fundamentals of flow in order to control this broadening while still enjoying the significant advantages of flow transport.

4.1   MECHANICS OF FLOW

The factors determining local v values are worth putting in perspective. Flow can be broken down conceptually into the motion of small elements of fluid volume. Each microscopic volume element is subject to a set of forces; that volume element responds to the forces in accordance with Newton's laws, just as do mechanical bodies (see Eq. 3.5). If there is net force, the fluid element will accelerate whereas if the various applied forces balance one another to give zero net force, the fluid element will maintain steady motion.

Several kinds of forces can act on fluid elements. These include:

1. External forces, such as gravitation or centrifugation.
2. Pressure forces, which are forces deriving from a pressure gradient. (A finite pressure gradient means that a slightly higher pressure exists on one side of the fluid element than on the other; this pressure inequality leads to a net force on the element.)
3. Viscous forces, resulting mainly from the viscous resistance encountered in sliding one fluid element over another.

In principle, one can write down all of these forces and formulate the Newtonian equations of motion for the fluid; this yields a complicated differential equation known as the Navier–Stokes equation [1–3]. A complete solution of the Navier–Stokes equation gives the exact trajectory and velocity of each fluid element. In practice, the calculations are often difficult because one must simultaneously account for all fluid elements and the interactions between these elements caused by the viscous drag forces. (The simultaneous motion of many interacting fluid elements is analogous to the simultaneous motion of many interacting mechanical objects, the latter being so complicated that it is described as the "many body problem.") However, in certain cases, the Navier–Stokes equation is reduced to a tractable form by the existence of steady low-velocity flow and high symmetry in the flow conduit (e.g., capillary tubes of circular cross section). We will examine such simple cases shortly.

4.2   SOURCES OF FLOW

There must be a driving force to make flow occur; otherwise all flow would be brought to a halt by viscous forces. The driving force is different in different kinds of systems. For example, in gas chromatography an outside source of pressure, such as a tank of pressurized gas, forces the gas phase through the column. Pumps are most often used in liquid chromatography, field-flow fractionation, ultrafiltration, and so on, because of the precise controllability of flow. Gravity sometimes supplements or replaces outside driving forces; centrifugal forces have also been used. An unusual but promising source known as electroosmotic flow has been proposed for liquid chromatography [4].

In methods where flow is not confined by a column (such as paper and thin-layer chromatography), flow is usually driven by capillarity—the tendency of a liquid to flow into vacant capillaries having a wettable surface. Flow by capillarity will be discussed in Section 4.7.

4.3 VISCOSITY

Viscous forces are of great importance in separative flow. Viscous forces are the drag forces that resist shear, which is the sliding of one fluid element (or layer) past another [5]. These forces are proportional to the area A of contact between fluid layers and to the *shear rate, dv/dy*. (The shear rate can be interpreted as the difference Δv in the velocity of adjacent layers divided by the distance Δy between these layers.) Thus the viscous force F_η can be written as

$$F_\eta = \eta A \, dv/dy \qquad (4.2)$$

where η is a constant termed the *viscosity*. The greater η, the greater the force F_η resisting the shear motion of adjacent fluid elements. The viscosities of common liquids are roughly 100 times larger than the viscosities of gases. Some typical viscosity values are shown in Tables 4.1 and 4.2. These viscosities correspond to pressures of about one atmosphere; they are not very pressure sensitive until much higher pressures are reached [6].

TABLE 4.1 Viscosity η of Some Common Liquids at 20°C Expressed as g/cm s, or Poises

Liquid	η (Poise) $\times 10^2$
Water	1.00
Hexane	0.33
Ethyl acetate	0.46
n-Octane	0.54
Chloroform	0.58
Toluene	0.59
Carbon tetrachloride	0.97
Ethyl alcohol	1.20
Butyric acid	1.54
n-Butyl alcohol	2.95
Aniline	4.40

TABLE 4.2   **Viscosity η of Various Gases at 20°C in Poises**

Gas	η (Poise) $\times 10^4$
Air	1.83
Argon	2.22
Helium	1.94
Nitrogen	1.75
Xenon	2.26
Propane	0.80
Ammonia	1.09
Carbon dioxide	1.48
Hydrogen	0.87
Methane	1.09

4.4 FLOW IN CAPILLARIES

It is instructive to consider steady fluid flow (sometimes called *Poiseuille flow*) in a thin capillary tube. This example has many purposes; it provides (1) a model flow calculation, (2) an illustration of how velocity profiles arise, (3) an explanation of the nature of flow in capillary chromatography, and (4) a foundation for capillary flow models of packed beds.

Flow in a capillary can be maintained by a steady pressure difference Δp applied between inlet and outlet ends. We assume gravitational (and other external) forces to be negligible (true for a horizontal tube or for any tube with a large Δp). With the application of Δp, the fluid in the tube accelerates to a flowrate at which the viscous drag forces balance the applied pressure forces. For thin tubes the Newtonian acceleration forces are significant for only a brief moment before steady flow is achieved.

By ruling out gravitational and acceleration forces, we are left with a simple balance: pressure acting against viscous forces. By considering a symmetric cylindrical tube, we have a geometry so simple that this balance can be easily formulated and the flow equations readily solved.

Consider the tube shown in Figure 4.1. The pressure-derived force pushing fluid through the tube, equal to cross sectional area πr_c^2 times pressure drop Δp, is balanced by the drag force exerted by the tube wall. The fluid immediately adjacent to the tube wall is generally held stationary

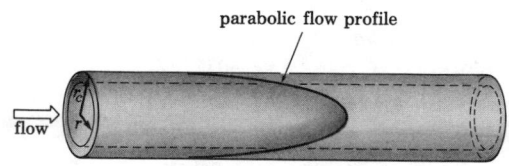

parabolic flow profile

flow

Figure 4.1. Flow in a capillary tube of radius r_c and length L.

by adhesion to the wall; inner fluid layers slide over this outer stationary layer and thus create a shear rate $-(dv/dr)_0$ at the wall. (Since v decreases as radial distance r increases, the sign is negative.) The viscous forces resisting that shear (and thus resisting the fluid motion) can be calculated from Eq. 4.2; area A is the area of contact between the tube wall and the fluid, $2\pi r_c L$. When we equate the above pressure forces and viscous forces, we get

$$\pi r_c^2 \Delta p = -2\pi r_c L\eta \left(\frac{dv}{dr}\right)_0 \tag{4.3}$$

This equation yields a shear rate at the wall of magnitude

$$-\left(\frac{dv}{dr}\right)_0 = \frac{r_c \Delta p}{2L\eta} \tag{4.4}$$

To unravel the details of flow away from the wall, we must apply a similar procedure to an imaginary inner cylinder (arbitrary radius r) like that shown in Figure 4.1. The pressure-derived force exerted on the face of the inner cylinder (the shaded area of Figure 4.1), $\pi r^2 \Delta p$, is balanced by the viscous forces, $-2\pi r L\eta (dv/dx)$, resisting the sliding of that inner cylinder past the adjacent layer of fluid outside the cylinder. Equating these two forms in analogy with Eq. 4.3 yields a balanced-force equation applicable at any radius r, not just at the wall

$$\pi r^2 \Delta p = -2\pi r L\eta \frac{dv}{dr} \tag{4.5}$$

This is a differential equation whose variables v and r can be separated to give

$$dv = \frac{-\Delta p}{2L\eta} r \, dr \tag{4.6}$$

an equation that integrates to

$$v = \frac{-\Delta p}{4L\eta} r^2 + \text{const} \tag{4.7}$$

The condition that $v = 0$ at the wall (where $r = r_c$) fixes the constant; $\text{const} = \Delta p r_c^2 / 4L\eta$. We then have

$$v = \frac{\Delta p}{4L\eta} (r_c^2 - r^2) \tag{4.8}$$

which expresses velocity $v = v(r)$ at any radius r.

Equation 4.8 shows that v varies with r; the resulting bullet-shape (parabolic) velocity profile is shown in Figure 4.1. The velocity reaches its peak value at the tube center, $r = 0$, where

$$v = v_{max} = \frac{\Delta p r_c^2}{4L\eta} \tag{4.9}$$

The average velocity $\langle v \rangle$ is obtained by weighting each v value by the small increment $2\pi r \, dr$ of cross-sectional area over which that particular v is found, then averaging

$$\langle v \rangle = \frac{\int_0^{r_c} v(r)2\pi r \, dr}{\int_0^{r_c} 2\pi r \, dr} \tag{4.10}$$

With v from Eq. 4.8 substituted in the upper integral, the integration yields

$$\langle v \rangle = \frac{\Delta p r_c^2}{8L\eta} = \frac{v_{max}}{2} \tag{4.11}$$

which shows that the average velocity increases with tube radius squared for a constant pressure gradient $\Delta p / L$ and viscosity η.

The ratio of the velocity $v(r)$ at radial position r (obtained from Eq. 4.8) to mean velocity $\langle v \rangle$ (Eq. 4.11) can be expressed as

$$\frac{v(r)}{\langle v \rangle} = 2\left(1 - \frac{r^2}{r_c^2}\right) \tag{4.11a}$$

This equation shows that $v(r)$ at any r is scaled to $\langle v \rangle$; this means that any factor leading to an increased $\langle v \rangle$ (such as a change in η, L, Δp, or r_c) will cause a corresponding increase in velocity v at any given radial position r/r_c.

Finally, the total fluid flux Q (volume per unit time) through the single capillary is $\langle v \rangle$ times the cross-sectional area πr_c^2 of the tube

$$Q = \frac{\pi \Delta p r_c^4}{8L\eta} \tag{4.12}$$

which is known as the *Hagen-Poiseuille equation* [5, 7]. This equation shows that the single capillary flux increases with the fourth power of radius.

The last few equations show that all the fluid flux and velocity terms—Q, $\langle v \rangle$, v_{max} and v—increase significantly with increasing capillary radius r_c for a given pressure gradient. It appears, therefore, that large capillaries are generally conducive to flow while small capillaries substantially impede fluid movement. The physical meaning of this is clear: in smaller capillaries there is a larger surface area s (per unit of fluid volume) tending to drag the fluid to a standstill. This conclusion can be generalized: in any pore space, irrespective of pore geometry, pressure-driven flow is inhibited by small pores because of the companion condition of a large surface area over which the fluid must drag.

Flow in a thin rectangular channel (Figure 4.2), such as that used in field-flow fractionation, can be treated in a manner similar to that used for cylindrical capillary tubes. If the drag at the edges of the channel is neglected (infinite parallel plate model), then the force balance expression (corresponding to Eq. 4.5 for capillary tubes) becomes

$$2by'\Delta p = -2Lb\eta \frac{dv}{dy'}$$ (4.13)

which, with the boundary condition that $v = 0$ at the wall ($y' = \pm w/2$), integrates to a parabolic velocity profile of the form

$$v = \frac{\Delta p}{L\eta} \left(\frac{w^2}{8} - \frac{y'^2}{2} \right)$$ (4.14)

From this we find that the maximum velocity (at $y' = 0$) is

$$v_{max} = \frac{\Delta p w^2}{8L\eta}$$ (4.15)

while the cross sectional average reduces to

$$\langle v \rangle = \frac{2}{3} v_{max} = \frac{\Delta p w^2}{12L\eta}$$ (4.16)

If we change the coordinate system to $y = y' + (w/2)$, so that the new coordinate y is the distance from one wall instead of from the center (see Figure 4.2), Eq. 4.14 becomes

$$v = \frac{\Delta p w^2}{2L\eta} \left(\frac{y}{w} - \frac{y^2}{w^2} \right) = 6\langle v \rangle \left(\frac{y}{w} - \frac{y^2}{w^2} \right)$$ (4.17)

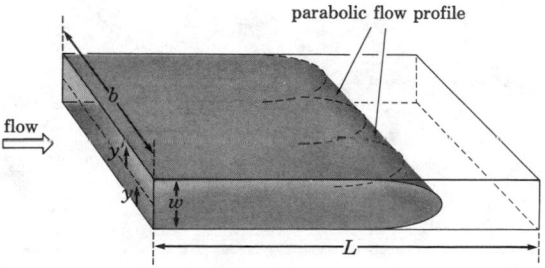

Figure 4.2. Flow in a thin rectangular channel of thickness (width) w, breadth b, and length L.

4.5 FLOW IN PACKED BEDS

Separative flow often occurs in a packed bed—typically a tube filled with a granular material. Chromatography in packed columns is the most important example of packed-bed flow. Similar flow is found in porous membranes used for membrane separation. The fluid flowing through such media can be a gas, a liquid, or a supercritical fluid.

Small volume elements of fluid take a tortuous path on flowing through the interstitial pore space of porous materials. The fluid elements must dodge from side to side in order to avoid the solid matter in their path. Such tortuous flow is illustrated in Figure 4.3 and brings to mind a boulder-filled river in which the current veers erratically to find channels between the randomly placed rocks.

Along with erratic sideways motion, the forward progress of any small fluid element is also very irregular. Upon encountering a major constriction in the flow path, the fluid element will slow down considerably because of increased viscous drag, only to speed up again as the channel widens. (The opposite trend—an increased velocity through constrictions—characterizes flow in single (as opposed to multiple interconnected) channels because in single conduits the flow cannot escape around constrictions but must funnel through them.) Thus by any measure, the microscopic flow through a granular material is unpredictable and capricious. The flow displacement is best thought of as a random process.

The irregularity of flow on a microscopic scale can be traced to the complex geometry of granular materials. The solid elements (particles) of such materials acquire a random arrangement as they are packed together. This explains the dodging and skirting that must be done by fluid in flowing through such materials; there are no straight paths.

A fluid flowing through any finely divided material meets with a good deal of viscous resistance. Small particles pack together more tightly than large and present a larger surface area to resist fluid motion. In order to maintain flow, a pressure gradient (greater for small particles than large) or

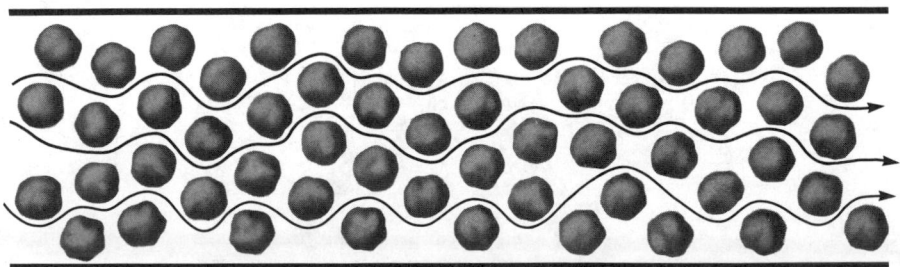

Figure 4.3. Typical streampaths through a section of packed column.

other means for forcing the fluid through the column or bed must be continuously applied.

The relationship between flow and pressure drop is expressed by Darcy's law [7, 8], an empirical equation applied (starting in the last century) to numerous forms of porous media. This law may be thought of as the Ohm's law of fluid flow, another linear law of transport expressing, in this case, the flux density q of fluid (analogous to current density) in the form

$$q = K' \frac{\Delta p}{L} \tag{4.18}$$

where $\Delta p / L$ is the pressure increment (like voltage) per unit length and K' is a constant of proportionality (the conductivity). Quantity q is simply the volume of fluid crossing a unit area normal to the flow axis in unit time. Since resistance to flow is due to viscous forces, the fluid flux is inversely proportional to viscosity η, a fact expressed by the modified equation

$$q = \frac{K_0}{\eta} \frac{\Delta p}{L} \tag{4.19}$$

where K_0 is known as the *specific permeability* and has the dimensions of length squared. If flow is caused by gravity acting on the fluid in a vertical column, the pressure gradient $\Delta p / L$ may be replaced by ρg, the product of fluid density and gravity ($g = 980$ cm/s^2).

Equation 4.19 can be related to the flow velocity in the pores as follows. In a column of unit cross-sectional area, the volume of fluid passing a given point in one second is q cm^3, following the definition of q. The q cm^3 would fill the column to a depth of q cm if the column were empty. Since it is not empty because of occupancy by the packing, the depth becomes greater. If a fraction ϵ of the volume is available to the fluid (with the fraction $1 - \epsilon$ occupied by solid matter), q cm^3 would fill the column to the depth q/ϵ cm. Now since q cm^3 pass a point in a second, fluid must be displaced in that second over the distance q/ϵ cm, the length of segment containing that fluid. The distance traveled per second, q/ϵ, is simply the cross-sectional average pore flow velocity $\langle v \rangle$. Combining this with Eq. 4.19, we find the average flow velocity to equal

$$\langle v \rangle = \frac{K_0}{\eta \varepsilon} \frac{\Delta p}{L} \tag{4.20}$$

The fractional void space ϵ is known as the *porosity*.

Unfortunately, Darcy's law says nothing about the microscopic flow profile in the packing. The most serious weakness of Darcy's law, however, is that K_0 varies strongly with the kind of packing material with no hint on the nature of the dependence. This shortcoming is addressed below.

In the so-called *capillaric models*, it is assumed that flow through the interstices (void spaces) between particles is analogous to flow through a bundle of fine capillary tubes [7, 8] as shown in Figure 4.4. The flow in each capillary obeys the Hagen-Poiseuille law, in which the volume flowing through a tube in a second is given by

$$Q = \frac{\pi d_c^4}{128\eta} \frac{\Delta p}{L} \tag{4.21}$$

This is the same expression as Eq. 4.12 except that capillary radius r_c has been replaced by $d_c/2$, where d_c is the capillary diameter.

If we consider a capillary bundle of unit area, directly comparable to a unit area of the packing material, then the number of tubes will be proportional to $1/d_c^2$ (the larger the tube area, proportional to d^2, the fewer the tubes in a unit area). Thus the total volume flowing per second through the bundle's unit area, denoted by q as before, is proportional to the flow through a single capillary (αd_c^4) times the number of capillaries in the bundle (αd_c^{-2}). Thus total flow q is proportional to d_c^2. (In agreement with Darcy's equation and its modifications, q is also proportional to pressure gradient and inversely proportional to viscosity.) Assuming that the "effective" diameter of the interstitial channels, and thus of the analogous capillaries, is proportional to the mean particle diameter d_p of the packing, it may be concluded that flowrate q and flow velocity v are proportional to d_p^2. This is generally observed. Thus a more comprehensive flow equation, incorporating the effect of particle size, is [8]

$$\langle v \rangle = \frac{1}{\phi} \frac{d_p^2}{\eta} \frac{\Delta p}{L} \tag{4.22}$$

where ϕ (which replaces 2ϕ in the original work) is the dimensionless *flow resistance parameter*, found empirically to be 500–1000. Comparing this equation with Eq. 4.20, we find for the specific permeability

$$K_0 = \frac{\epsilon d_p^2}{\phi} \tag{4.23}$$

One other factor influences flow. Even with the same particle size, a loose

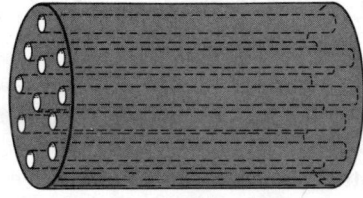

Figure 4.4. Capillaric model of a packed column. The flow space is assumed to consist of a bundle of parallel capillaries.

packing of high porosity will have larger channels and be more permeable than a tightly packed material of low porosity. Most packing materials used in chromatography are quite compact so that porosity ϵ remains at its minimum, about 0.4, for random packings. When ϵ varies significantly it is best to invoke the *Kozeny–Carman equation*, which can be written as

$$\langle v \rangle = \frac{\epsilon^2}{180(1 - \epsilon)^2} \frac{d_p^2}{\eta} \frac{\Delta p}{L} \tag{4.24}$$

(This equation appears to work better for chromatographic materials if 180 is replaced by 270.)

The above discussion not only shows how various factors affect mean flow properties in packed beds but it provides background needed to understand the microscopic flow pattern. The observation that velocity increases with channel diameter squared is applicable to single channels as well as to the packed medium as a whole. While the fact that single channels are very short and discontinuous (linking into a network of other channels on all sides) prevents an exact formulation, as a rough rule the local flow velocity increases with the diameter squared of the particular channel under consideration. Thus poorly packed columns, in which the range of pore sizes is large, tend to have highly variable flowrates (that is, an erratic velocity profile) within the medium [8]. This leads to poor performance in techniques like chromatography.

For many packed and porous media (chromatographic columns, ultrafiltration membranes) the pressure drop Δp associated with flow is important. From Eq. 4.22 we get

$$\Delta p = \frac{\phi \eta \langle v \rangle L}{d_p^2} \tag{4.24a}$$

Thus for a given mean flow velocity $\langle v \rangle$, Δp increases rapidly with decreasing particle (or pore) size d_p.

4.6 COMPRESSIBLE FLOW: BASIS OF GAS CHROMATOGRAPHY

Both gases and liquids flow according to the universal laws of flow mechanics discussed above. However, gases and liquids vary substantially in some physical constants and characteristics, which leads to different pressure and velocity distributions along the length of a tube or column. For example, gases are about 100 times less viscous than liquids (Section 4.3), which means that the same flow velocity can be established in the same flow medium with a pressure gradient that is 100 times smaller than for liquids. Gases are also highly compressible, which leads to rather substantial gradients in pressure and velocity along the flow path.

As a gas stream is driven down a column from a position of high pressure to one of low pressure, it expands considerably. Since in steady flow the same mass of gas must pass through each cross section in unit time, a greater volume flux of gas will be observed at points where the gas is more expanded. This means that the flow velocity, which is proportional to volume flux, is largest where the pressure is least—at the column outlet. In view of the proportionality of the mean velocity $\langle v \rangle$ and the molar volume V, Boyle's law can be rewritten as $p\langle v \rangle = \text{const}$, where $p\langle v \rangle$ replaces pV.

To determine the pressure and velocity distributions, we must relate mean cross-sectional flow velocity $\langle v \rangle$ to the pressure gradient. This is done by means of Eq. 4.11 for capillary tubes and Eq. 4.22 for packed columns. To work with these equations, we must replace the overall pressure gradient $\Delta p / L$ by the local pressure gradient $-dp/dx$, where the negative sign arises because pressure decreases as one moves along the positive flow coordinate x. With this substitution, Eq. 4.11 can be rearranged to

$$-\frac{dp}{dx} = \frac{8\eta}{r_c^2} \langle v \rangle \tag{4.25}$$

Since according to our modified Boyle's law the product $p\langle v \rangle$ is constant throughout the column or tube, we may write

$$p\langle v \rangle = p_i \langle v \rangle_i \tag{4.26}$$

where subscript i stands for inlet values. If $\langle v \rangle$ from Eq. 4.26 is substituted into Eq. 4.25 and the latter rearranged, we get

$$-p \, dp = \left(\frac{8\eta p_i \langle v \rangle_i}{r_c^2} \right) dx = I \, dx \tag{4.27}$$

where I is a constant given by the expression in parentheses. (We note that I is constant because viscosity η for an ideal gas is independent of pressure.) We integrate this expression from the inlet where $p = p_i$ and $x = 0$ to some arbitrary point x having pressure p

$$-\int_{p_i}^{p} p \, dp = I \int_{0}^{x} dx \tag{4.28}$$

The integration yields

$$p_i^2 - p^2 = 2Ix \tag{4.29}$$

which gives the pressure distribution as a function of the distance x from the inlet. The velocity distribution can be obtained from this by Eq. 4.26. We get

$$\frac{\langle v \rangle}{\langle v \rangle_i} = \left(\frac{p_i^2}{p_i^2 - 2Ix} \right)^{1/2} \tag{4.30}$$

Equation 4.29 can be used to obtain a more convenient expression for constant I. Recognizing that at the outlet p becomes p_0 and x becomes L, Eq. 4.29 yields, upon the substitution of these values, the expression

$$I = \frac{p_i^2 - p_0^2}{2L} \tag{4.30a}$$

When this is used to replace the I of Eq. 4.30, the latter yields upon rearrangement

$$\frac{\langle v \rangle}{\langle v_i \rangle} = \left\{ \frac{P^2}{P^2 - [(P^2 - 1)x/L]} \right\}^{1/2} \tag{4.30b}$$

where P is the *compression ratio*, p_i/p_0. A plot of this velocity profile is shown in Figure 4.5 for various P values. This shows explicitly how the flow velocity increases as the gas stream makes its way along the column. We will find later that chromatographic elution time and efficiency depend upon flow velocity, making it necessary to account for these velocity variations in gas chromatography.

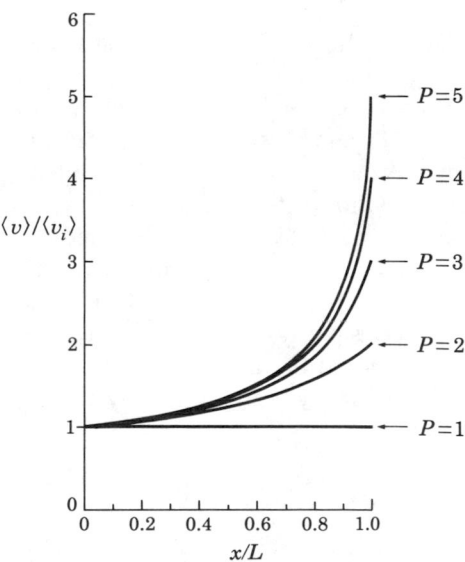

Figure 4.5. Profiles of gas velocity from inlet $(x/L = 0)$ to outlet $(x/L = 1)$ of column.

In the case of packed columns, we start with Eq. 4.22 instead of the capillary expression given by Eq. 4.11. The treatment is identical, but for packed columns I assumes the value

$$I = \frac{\phi \eta p_i \langle v \rangle_i}{d_p^2} \tag{4.31}$$

The pressure and velocity profiles derived above are applicable to steady (time-independent) gas flow. Equations describing nonsteady gas flow (pressure pulses and transients, etc.) can be found elsewhere [9].

Compared to gases, liquids are relatively incompressible. However, liquid chromatography is normally carried out in columns having much finer support particles, and liquids are more viscous than gases. Therefore, for a given flow velocity, the pressure gradient is much greater for liquids. This conclusion is supported by Eq. 4.22, applicable to packed columns. Compressions of the order of 10% will be found for pressures exceeding 1000 atm. Therefore, for careful quantitative work at high pressure, compression effects must be allowed for in liquid chromatography as well [10]. The procedure for obtaining velocity and pressure profiles is much like that used above except that the appropriate equation of state for the liquid must replace our modified Boyle's law, Eq. 4.26.

4.7 CAPILLARITY: FLOW ORIGIN IN PAPER AND THIN-LAYER CHROMATOGRAPHY

Normally, separative flow is driven along a confined channel by a pressure drop Δp produced by a pump or other outside source. By contrast, flow in paper and thin-layer chromatography occurs in thin porous beds which are not usually confined by walls and therefore cannot be driven by an external source. The driving force is instead capillarity, the tendency of wetting fluids to migrate by a wicking action into empty capillary or pore space [11]. Through capillarity, liquids such as water spontaneously flow from a reservoir into and along a dry paper strip or a thin layer of dry adsorbent (or other dry and wettable porous materials). The advance of the liquid into the vacant pore volume is responsible for carrying solutes along the chromatographic bed leading to their separation.

The driving force for capillarity is measured by the capillary pressure

$$\Delta p = \frac{2\gamma}{r} \tag{4.32}$$

where γ is the surface tension and r is the radius of curvature of the liquid meniscus advancing into the capillary space. If we assume that the chromatographic bed is made up of uniform capillary tubes of radius r_c, as shown in Figure 4.6, then $r = r_c$ for zero contact angle. If the liquid has a contact angle of θ, then $r = r_c/\cos\theta$ and

$$\Delta p = \frac{2\gamma\cos\theta}{r_c} \qquad (4.33)$$

The mean flow velocity in the capillary is given by Eq. 4.11; if we replace Δp in this equation by Eq. 4.33 and if we replace bed length L by the distance X_f that the liquid front has advanced, we get

$$\langle v \rangle = \frac{r_c\gamma\cos\theta}{4x_f\eta} \qquad (4.34)$$

This equation shows that the capillary flow velocity increases with capillary radius r_c and surface tension γ; it decreases, as do all flow velocities, with viscosity η. The equation also shows that $\langle v \rangle$ decreases with the length X_f of penetration of liquid into the capillary space; that is, flow diminishes as the liquid progresses further and further into the pore space. This diminution can be calculated as a function of time as follows.

Average velocity $\langle v \rangle$ is a direct measure of how rapidly the liquid front is pushed into the empty capillary. Thus if we replace $\langle v \rangle$ by dX_f/dt in Eq. 4.34 and rearrange, we get a differential equation for X_f

$$X_f\frac{dX_f}{dt} = \frac{r_c\gamma\cos\theta}{4\eta} \qquad (4.35)$$

Upon separation of variables and integration, and use of the initial value condition that $X_f = 0$ at $t = 0$, we get

$$X_f^2 = \kappa t \qquad (4.36)$$

where

$$\kappa = \frac{r_c\gamma\cos\theta}{2\eta} \qquad (4.37)$$

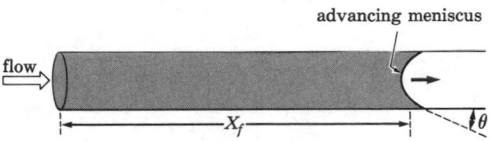

Figure 4.6. Fluid flow in capillary tube driven by capillarity. This is a simple model of flow in a paper or thin-layer chromatography bed.

The general features of the *parabolic flow law* expressed by Eq. 4.36 have been widely confirmed [12]. However, the flow process itself is more complicated than that implied above. The irregular pore space of most granular materials may better be thought of as a bundle of nonuniform capillaries, all interconnected to one another. Near the advancing solvent front the liquid is pulled into the smaller pores because of the lower r_c and thus r. The large pores are left dry until more liquid arrives from the rear, at which time the cavities fill one by one from smaller to larger, but by then the semidry front has moved on. Thus the level of liquid saturation of the porous bed decreases continuously as one advances toward the front.

The interconnected capillary model, when developed theoretically, shows that the entire saturation profile (not just the front) expands parabolically along the porous bed without changing shape [10]. This is confirmed in Figure 4.7 by noting that a group of saturation (concentration) profiles

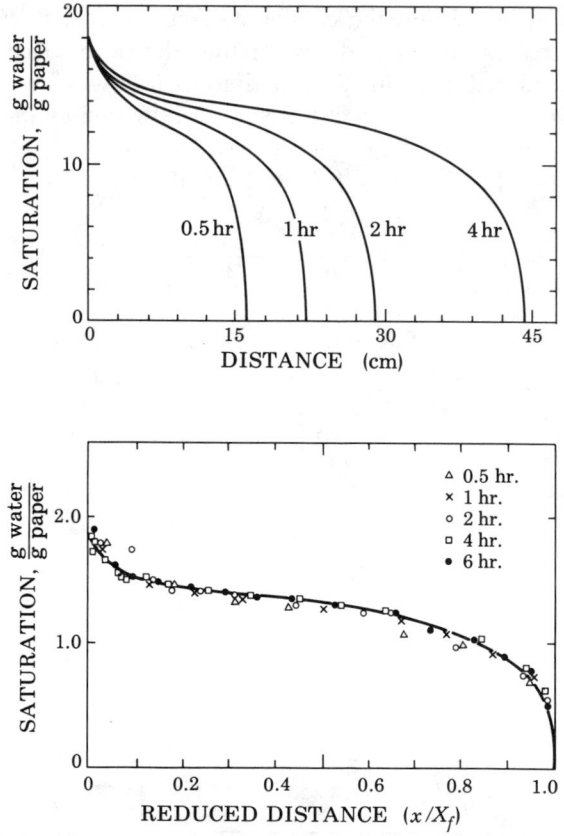

Figure 4.7. Saturations as a function of distance along a paper strip after different elapsed times (top). Superposition of saturation data when plotted as a function of reduced distance (bottom).

observed at different times (top) in a paper strip all superimpose on one another (bottom) when plotted against the fraction of the distance x/X_f to the front [13].

The above situation is complicated even more if the chromatographic bed is oriented vertically so that gravitational forces enter or if centrifugal forces are used. Complexities also arise if geometries other than rectangular are used.

Since chromatographic migration rates depend on the amount of liquid phase and on the liquid velocity, the above noted gradients, established by the actions of capillarity, have important chromatograhic effects [14]. More details on the chromatographic consequences of capillary flow can be found in the literature (12) and in subsequent chapters.

4.8 CONVECTIVE FLOW

Natural convection is the flow induced by the unequal pull of gravity on fluid elements of different densities. For example, if we inject a globule (or layer) of dense aqueous solution marked with a dye into a beaker of water, the dense globule will be observed to sink under the influence of gravity, as illustrated in Figure 4.8. That sinking motion is actually a form of bulk displacement or flow, specifically natural convective flow.

The density differences leading to natural convection most often have a thermal origin. The lofting of warm air masses at the earth's surface and their replacement by cooler and denser masses from above is an example of convective flow (thermal convection) that profoundly affects our atmosphere. Likewise, heated fluid elements in a separation chamber will rise convectively while cooler elements descend. The resulting transport of solute can profoundly affect separations.

In some cases, natural convective flow plays an integral role in separations. For example, thermogravitational (TG) columns rely on a combination of thermal convective flow and relative (selective) displacement by thermal diffusion.

A section of a simple TC cell is illustrated in Figure 4.9. This consists of two parallel metal plates, one heated and one cooled, held rigidly a distance w apart to form a channel between. The fluid in the channel (gas or liquid) rapidly forms a temperature gradient due to heat conduction from the walls.

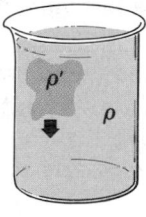

Figure 4.8. Convective flow (sinking) of globule of density ρ' through a carrier of density ρ, where $\rho' > \rho$.

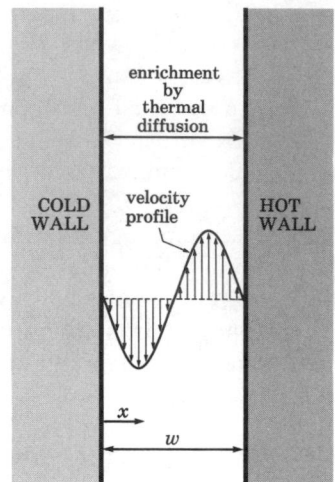

Figure 4.9. Example of a themogravitational column or cell in which differentially heated walls lead to convective flow. The coupling of convective flow with component enrichment by thermal diffusion leads to separation along the flow axis.

Convective motion then begins: fluid near the hot wall rises and the cooled fluid descends. Meanwhile, thermal diffusion enriches different components in either the hot or cooled regions, and the different flow direction in the two regions leads to component separation.

We note here that natural convective flow is countercurrent; the vacation of space by any descending volume elements of fluid is immediately compensated by an ascending volume.

The velocity profile in thermal diffusion cells (see Figure 4.9) is calculated from the same balance-of-force mechanics used earlier in this chapter to analyze flow in capillaries. However, gravitational forces replace pressure forces as the driving influence balanced against viscous forces. The resulting velocity profile is found to be [15–17]

$$v(y) = \frac{\rho\gamma g w^2 \Delta T}{12\eta} \frac{y}{w}\left(1 - \frac{y}{w}\right)\left(1 - \frac{2y}{w}\right) \tag{4.38}$$

where ρ is the mean density, γ is the coefficient of thermal expansion, g is gravitational acceleration, and ΔT is the temperature difference between the two walls. The general form of this velocity profile is illustrated in Figure 4.9.

Equation 4.38 is, of course, essential to design effective TG columns. However, it has a much wider use in aiding our understanding of the role of natural convection in other separation methods.

Natural convection generally has a negative effect on separation; its countercurrent motion leads to the disruption and smearing of component zones. Electrophoresis is particularly subject to such convective flow because the passage of electrical current generates heat that in turn sets up temperature gradients in the electrophoretic medium.

Equation 4.38 provides some guidelines for reducing the strength of convective flow, represented by the magnitude of $v(y)$. Convective flow can, for example, be discouraged by increasing the viscosity η (which inhibits all flow), but this action also slows separative transport. One can also attempt to reduce the driving factor of thermal convection, the temperature increment ΔT, although this may also interfere (through electrical current reduction, etc.) with separation. However, using a medium of high thermal conductivity will reduce ΔT by encouraging thermal equilibration.

A highly effective means of reducing natural convection is to reduce channel thickness w. Not only does $v(y)$ at any relative position y/w depend on the square of w, but the reduction of w in some situations will lead to the more rapid dissipation of heat, thus reducing ΔT as well.

The reduction of the thickness of the flow channel, as discussed earlier, is equivalent to introducing more surface area per unit volume of medium. High surface areas inhibit all flow, including natural convective flow. One can increase relative surface areas by going to thinner tubes or channels, or by using a fine granular or porous support medium. Both approaches are used in electrophoresis as discussed in a subsequent chapter.

In chromatography, also, there are some heat sources and sinks, such as the enthalpy of sorption and desorption of solute in the stationary phase and the dissipation through viscous drag of the energy used to drive flow. However, the high surface-to-volume ratio of the fine tubes and packings used in chromatography generally suppresses convective flow.

More subtle tricks can sometimes be used to fight convection. For example, water reaches its maximum density at 4°C, at which temperature the thermal expansion coefficient γ goes to zero. Thus, an aqueous separation medium at 4°C is particularly stable against thermal convection. A "far out" approach is separation in space where gravity is zero; electrophoretic separations have already been carried out in orbiting satellites.

4.9 ELECTROOSMOTIC FLOW

Electroosmotic flow is a flow induced by an electrical potential. When a voltage is applied across a packed bed or a capillary tube filled with an aqueous electrolyte solution, the solution begins to flow along the field axis. Such flow is known as electroosmotic flow.

Electroosmotic flow is a consequence of the way ions are distributed near surfaces. Nearly all surfaces are charged; the surface charge attracts a cloud of oppositely charged ions (counterions) into adjacent layers of liquid, forming a double layer. If there is an electrical field component parallel with the surface, that field will pull the counterions along the surface, dragging the solution with it. Thus flow is induced.

Whether its role is positive or negative, electroosmotic flow is common in electrophoresis because surface charges cannot be totally eliminated. Elec-

troosmotic flow often has a nuisance role in electrophoretic methods because it tends to smear the bands of components being separated. For this
reason, electroosmotic flow requires careful consideration in nearly all
electrophoretic separations [18, 19]. However, electroosmotic flow is beginning to play a constructive role in electrophoresis [20], with some promise in
chromatography as well [21, 22].

4.10 LAMINAR AND TURBULENT FLOW

At low flow velocities the carrier fluid travels straight down its confining
tube or weaves its way smoothly through support particles. Each microscopic fluid element follows a fixed path known as a *streamline* or
streampath (see Figure 4.3). Any subsequent fluid element, started at the
same point, will follow the same streampath along its entire course. Such a
constant pattern of flow is characteristic of *laminar flow*. As flow velocity
increases to a certain level, some streampaths begin to fluctuate erratically
and fluid elements no longer follow one another along the same precise
course. This is the beginning of *turbulent flow*. Turbulence is due to the
dominance of inertial or acceleration forces over viscous forces at high
velocities. The transition from laminar (or streamline) flow to turbulent flow
is most often discussed relative to flow through open tubes. For tube flow it
is found that smooth laminar flow changes rather abruptly to frenzied
turbulent flow. The transition velocity is found by applying Reynold's
criterion: if the dimensionless Reynold's number

$$Re = \frac{d_c \rho \langle v \rangle}{\eta} \tag{4.39}$$

exceeds about 2000 then the flow velocity $\langle v \rangle$ is sufficiently large for
turbulence. The other terms influencing the velocity at which turbulence
begins are tube diameter d_c, fluid density ρ, and viscosity η.

   In granular materials like those found in column packings in chromatography, turbulence starts more readily as a result of the irregularity of the
flow pattern. However, turbulence does not suddenly occupy all regions of
flow space. Instead, turbulent conditions may be envisioned as beginning in
a few of the larger void spaces between particles and gradually spreading
throughout the remaining void regions as the velocity increases. This growth
occurs mainly in the range $Re = 1$–100, where Re is now modified to read

$$Re = \frac{d_p \rho \langle v \rangle}{\eta} \tag{4.40}$$

in which mean particle diameter d_p replaces tube diameter d_c.

Most experimental work in separations, particularly in chromatography, involves laminar flow. However, some high flow systems in gas chromatography may well encounter turbulence. (Turbulence has been intentionally introduced in some experiments [23].) For example, a nitrogen mobile phase at 30°C has a viscosity of approximately 1.8×10^{-4} poises (where the poise is the cgs unit of viscosity, $g\,cm^{-1}\,s^{-1}$) and a density at atmospheric pressure of 1.1×10^{-3} g/cm^3. If its velocity through a column packed with particles of mean diameter 0.020 cm is 50 cm/s (higher than the typical velocity), the *Re* value calculated from the above equation is 6.1, high enough for some degree of turbulence.

We note that our previous descriptions of flow processes have tacitly assumed laminar flow. For example, flow in capillaries was described by balancing pressure-derived forces against viscous forces, ignoring acceleration (inertial) effects. Darcy's law, Eq. 4.18, is also based on laminar flow. With turbulence, flow resistance increases; the pressure gradient is no longer linearly related to flow (see Eqs. 4.18 through 4.20) but increases more rapidly as expressed by

$$\frac{\Delta p}{L} = a\langle v \rangle + b\langle v \rangle^2 \tag{4.41}$$

4.11 VISCOSITY: PIVOTAL ROLE

Early in this chapter we defined viscosity and discussed its role in opposing flow. By resisting flow, viscosity often limits the flowrate to less than optimal values. Higher pressure drops can be used to offset viscous drag but the pressure drop necessary to force fluid at the desired rate through narrow pores is often unattainable. Thus the rate of flow is often viscosity limited. For example, the rate of fluid transfer through both membranes and chromatographic columns is often limited by viscous resistance; this in turn limits the speed of separation.

Along with limiting flow, our analysis has shown that viscous effects control the detailed form of flow. This form, which exhibits itself in the various flow velocity profiles already noted, is important to many methods of separation.

However, the role of viscosity does not end with flow. Viscosity also controls and limits the rate of relative (selective) transport. It does this through the friction coefficient f. The reason viscosity affects f is that a solute species moving through a solvent drags part of the solvent along and in so doing sets up velocity gradients (shear) in the solvent which are resisted by viscous forces. The magnitude of the resistance, and thus the magnitude of f, is controlled by viscosity.

In the previous sections of this chapter we have discussed flow in relationship to viscous effects. In the next few sections we will discuss the effect of viscosity on f. Finally, we shall conclude this chapter by looking more closely at viscosity and the means to control it.

4.12   SEPARATION SPEED, FRICTION COEFFICIENTS, AND VISCOSITY

In Chapter 3 we found that all relative transport processes, whether induced by external fields or diffusion, proceed at a rate inversely proportional to the friction coefficient f. Since virtually all separation methods require a certain level of completion of transport, or a certain number of transport steps, the time scale of the separation is linked to the time scale of the required transport: both ultimately hinge on the magnitude of f. This conclusion is valid whether one is using methods such as chromatography where the transport processes must maintain the distribution of components between phases at a point near equilibrium, or electrophoresis where transport proceeds only fractionally to equilibrium.

By learning more about the nature of f, we can discover methods for changing solvent properties and other parameters of the system so that f can be minimized. This is an important consideration in the design of optimal high-speed separation systems.

Fundamental to the description of the friction coefficient in liquids is Stokes law, which shows that f is proportional to the radius a of spherical particles undergoing transport and proportional to the viscosity η of the medium [24, 25]

$$f = 6\pi\eta\mathcal{N}a \qquad (4.42)$$

This equation provides the important link between f and η noted in the last section.

Avogadro's number $\mathcal{N}$ appears in Eq. 4.42 because f is for one mole, as noted in Chapter 3. This equation leads to $f \sim 2 \times 10^{15}$ g/s mol for small organic molecules in water. Because of a larger radius a, f may be one or two orders of magnitude greater than this for macromolecules.

Stokes law is based on the hypothesis that the spherical particle is moving through a stationary viscous continuum. To obtain Stokes law, friction coefficient f is defined (see Eq. 3.7) as the force/velocity ratio

$$f = \mathcal{N}F_\eta/U \qquad (4.43)$$

where F_η is the force needed to maintain the particle in steady motion at velocity U against the drag of the viscous medium. The multiplication of F_η by Avogadro's number $\mathcal{N}$ gives the force per mole.

Unfortunately, the derivation of an expression for F_η (and thus for f) is made difficult by the complex pattern of viscous displacement around the moving sphere. A simplified derivation has been presented by Lauffer [26]; below a still simpler treatment is given.

An approximate derivation of Stokes law is useful in showing the origin of the dependence of f on η and a. First we recall, Eq. 4.2, that viscosity η is defined in relationship to the force F_η resisting the sliding of liquid layers of area A past one another at a shear rate of dv/dy. Since the entire sphere is moving, area A may be taken as the surface area of the sphere. The motion of this surface drags the adjacent fluid along with it at velocity U. This fluid pulls neighboring fluid elements along. This disturbance in the otherwise stationary fluid extends out about one particle radius a. (Clearly, the bigger the particle, the more extensive the disturbance created by its motion.) Therefore the fluid velocity, which is U at the particle surface, drops almost to zero at distance a. Thus the shear rate can be estimated by

$$\frac{dv}{dy} = \frac{\text{change in velocity}}{\text{distance}} \sim \frac{U}{a} \tag{4.44}$$

If we use this and if we replace A in Eq. 4.2 by the surface area of the particle, $A = 4\pi a^2$, we get

$$F_\eta = 4\pi\beta\eta aU \tag{4.45}$$

where β is a constant of order unity introduced to correct for the approximations made. According to Eq. 4.43, f becomes

$$f = 4\pi\beta\eta Na \tag{4.46}$$

an equation identical in form to Stokes law, Eq. 4.42. The equations are numerically identical if correction factor β is taken as 1.5.

4.13   FRICTION COEFFICIENTS AND MOLECULAR PARAMETERS

Upon substituting Eq. 4.42 for f into $D = \mathcal{R}T/f$ (Eq. 3.21), we obtain an equation for diffusion coefficient D

$$D = \frac{\mathcal{R}T}{6\pi\eta Na} \tag{4.47}$$

This expression is known as the *Stokes–Einstein equation*. This formula correctly relates diffusivity to molecular dimensions and viscosity for cases in which Stokes law is applicable.

Stokes law—although strictly valid only for spheres—yields a good approximation for the dependence of f on the size of small particles and molecules. However, a few modifications and limitations must be noted. First, as stated above, Stokes law was derived for macroscopic spheres in motion in a continuous medium, not for molecular sized bodies moving among discrete solvent molecules of similar size. Despite this, the law works reasonably well for the transport of small molecules; calculated values of D are of the correct order of magnitude but numerically too small by about 1.5–2.5. Better estimates of D for small molecules can be obtained from various semiempirical equations [27].

Stokes law works exceedingly well for spheres of diameter greater than that of the surrounding solvent molecules. Unfortunately, few sample molecules are perfect spheres. An effort to evaluate the role of asphericity was made by Perrin [28], who evaluated f for prolate (cigar-shaped) and oblate (disk-shaped) ellipsoids of revolution. His results can be expressed in terms of f/f_0, the friction coefficient of an ellipsoidal body compared to that of a sphere of the same volume. This ratio is a function of the ratio of the semimajor axis a to the semiminor axis b of the ellipsoid. These results are illustrated in Figure 4.10. They show the rather surprising result that f changes only slowly as the departures from sphericity $[(a/b) = 1]$ intensify. To an approximation (29), f/f_0 for prolate ellipsoids is linear in a/b, specifically $f/f_0 = 0.94 + 0.06(a/b)$.

Molecules, of course, are generally no closer to ellipsoids than to spheres. Nonetheless, the results of Perrin clearly show the order of magnitude of the effects of molecular elongation and flattening—and by implication the effects of other geometric distortions as well.

In that molecules of similar geometries have f values proportional to some linear dimensions (such as maximum length or radius), we find that f is related to molecular weight M by the approximate expression

$$f = \text{const} \times M^{1/3} \tag{4.48}$$

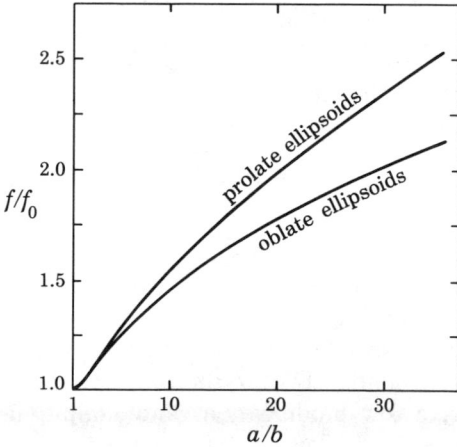

Figure 4.10. Increase in friction coefficient for spherical bodies when reshaped into ellipsoids of different axial ratios a/b.

However, for linear polymers existing as random coils in solution, f increases in rough proportion to $M^{1/2}$.

We note here that any appreciable solvation will increase the effective size of the molecule and thus increase f.

Because the radius of a nonspherical molecule cannot be defined precisely, molecular friction coefficients and diffusion coefficients are often related to the Stokes radius (or Stokes diameter). This is defined as the radius (or diameter) of a sphere having f and D values identical to those of the molecule under consideration.

In the case of transport in gases, Stokes law applies only if particle radius a is much larger than the mean free path (0.065 μm in air at 20°C and 1 atm). For smaller particles and molecules, a correction factor must be applied to Stokes law [30]. Alternatively, the friction coefficient may be approximated using Eq. 3.21 and the Stephan–Maxwell equation for diffusion in an ideal gas mixture composed of solute of molecular weight M_1 in a carrier of molecular weight M_2

$$f = 380 \mathcal{R} p \sigma_{12}^2 \left[\frac{2}{T} \left(\frac{1}{M_1} + \frac{1}{M_2} \right) \right]^{1/2} \tag{4.49}$$

when $M_1 \gg M_2$, f approaches

$$f = 380 \mathcal{R} p \sigma_{12}^2 \left(\frac{2M_2}{T} \right)^{1/2} \tag{4.50}$$

where p is the pressure in atmospheres and σ_{12} is the collision diameter of a solute-carrier pair in angstroms. These equations show that f can be reduced by decreasing pressure, by increasing temperature, by decreasing the molecular diameter of the carrier (which contributes to σ_{12}), and by decreasing carrier molecular weight M_2. Thus H_2 or He at low pressure appear ideal for gas-based processes such as gas chromatography.

The foregoing conclusion is subject to another constraint for gas chromatography: a substantial pressure drop is sometimes needed to force the gas phase through the column. In this case the mean pressure, unable to approach zero, should at least be reduced to a minimum, which for true optimization will require vacuum outlet conditions [31].

Often, as above, the simplest optimization criteria (e.g., low pressure) for separations must be tempered by practical requirements (e.g., high pressure drop). A wider knowledge of the many variables affecting separations must then be brought to bear in selecting true optimizing conditions. We will discuss optimization more fully in later chapters when we deal with specific methods.

More precise ways for estimating f in terms of solute and solvent characteristics for both gases and liquids can be obtained from various diffusion equations in the book by Reid, Prausnitz, and Sherwood [27]. Knowing D, one can calculate f by the rearranged form of Eq. 3.21: $f = \mathcal{R}T/D$.

4.14 VISCOSITY REDUCTION IN LIQUIDS

Insofar as liquid solvents are concerned, the most important factor governing f is viscosity η, as the Stokes equation clearly demonstrates. Therefore any systematic effort to increase separation speed requires close attention to viscosity, with an emphasis on finding solvents and conditions for which viscosities are minimal. The reduction of viscosity can be pursued systematically in place of a hit or miss search for low viscosity solvents. The approach below was developed by the author and his colleagues [32] for use in optimizing size-exclusion chromatography, but the conclusions are generally applicable to separations.

It has long been known that liquid viscosity decreases exponentially with absolute temperature [27]

$$\eta = A \exp(E_\eta / \mathcal{R} T) \tag{4.51}$$

In this equation A is a constant and E_η is an effective activation energy for molecular displacement. This equation suggests that viscosity can be decreased indefinitely (normally by 1–2% per °C) by increasing temperature, although eventually the liquid will boil and the equation will no longer hold. One can combat boiling and thus go well above the normal boiling point by increasing pressure a few atmospheres. For example, one can typically extend the boiling point by 60°C at 5 atm [33]. With each increment in temperature so gained, viscosity can be incrementally reduced. We have suggested, in fact, that the temperature might profitably be increased into the supercritical region (which lies above the critical temperature of the solvent) if a few hundred atmospheres were applied [32].

We observe that high temperature is not the main objective. If increasing the temperature were the only goal, one could choose a high boiling solvent and avoid altogether the use of elevated pressures. However, high boiling solvents are disadvantageous because they are more viscous to begin with. It is clearly necessary to consider both temperature and intrinsic solvent viscosity to reduce η and thus f to the lowest possible values. Fortunately, there is a simple rule to go by.

It has been observed that the viscosities of many liquids converge to a common value, $\eta_b = 0.29$ centipoises, at their normal boiling point T_b [32–34]. Thus one is starting at roughly the same point on the viscosity scale for different liquids at the boiling point of each, whether the boiling point is high or low. The value of viscosity depends, therefore, not on the absolute temperature but on the temperature relative to the boiling point, the reference point at which viscosities are found to have a common value. Thus if solute decomposition is a problem at elevated temperatures, one can often choose a solvent with a low boiling point and, by using low temperature, gain the same advantage in reducing viscosity as would be possible working at a higher temperature with a high boiling solvent.

The reduction in viscosity as one approaches and then exceeds the normal boiling point can be easily approximated. To do this we write Eq. 4.51 in the following form

$$\eta = A \exp\left[\left(\frac{E_\eta}{\mathcal{R} T_b}\right) \frac{T_b}{T}\right] \tag{4.52}$$

It is known that E_η is approximately $0.3 \, \Delta H_v$ where ΔH_v is the enthalpy of vaporization. Trouten's rule gives $\Delta H_v / T_b = 21 \, \text{cal/mol deg}$, which means that $E_\eta / T_b \simeq 6 \, \text{cal/mol deg}$. Writing the gas constant as $\mathcal{R} = 1.987 \, \text{cal/mol deg}$, we have

$$\frac{E_\eta}{\mathcal{R} T_b} \simeq 3 \tag{4.53}$$

When Eq. 4.53 is substituted into Eq. 4.52, we get

$$\eta \simeq A \exp\left(\frac{3 T_b}{T}\right) \tag{4.54}$$

The value of the viscosity relative to its value at the boiling point is therefore roughly approximated by

$$\frac{\eta}{\eta_b} = \exp 3\left(\frac{T_b}{T} - 1\right) \tag{4.55}$$

where typically $\eta_b \simeq 0.29$ centipoises. This equation shows, for example, that when T is only $0.75 T_b$, $\eta \simeq e \eta_b \simeq 2.7 \eta_b$. When T is extended from below to above T_b, viscosity drops below η_b. Extended to the critical temperature, normally $1.5 T_b$, $\eta \simeq e^{-1} \eta_b \simeq \eta_b / 2.7$. These changes in η are substantial and can have a major impact on separation speed and effectiveness.

The relative viscosity curve predicted by Eq. 4.55 is plotted in Figure 4.11. Experimental viscosities for several common solvents are also shown in the figure; these data points verify the trends predicted above.

Figure 4.11 is a useful guide for the reduction of viscosity and thus of friction coefficient f. It is clear from the figure that small increases in the temperature relative to the boiling point can lead to substantial decreases in η.

The stratagem of reducing viscosity without undue temperature gains by substituting a lower boiling solvent should be widely applicable to separation systems. However, a few solvents are so unique that they cannot be replaced by substitutes. The special solvent properties of H_2O, for example, cannot be found in any other solvent, particularly one of lower boiling point (NH_3 is perhaps the closest low-boiling substitute). In this case the only recourse is to work at maximum permissible temperatures.

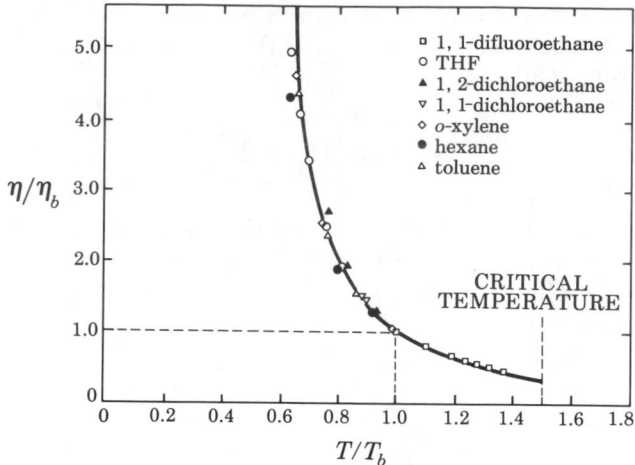

Figure 4.11. Plot of Eq. 4.55 showing reduction in liquid viscosity η with increasing temperature T. Both are expressed relative to the boiling point values η_b and T_b; the former is ~0.29 centipoises for most liquids. Curve is applicable up to the critical temperature providing sufficient pressure is applied to suppress boiling.

REFERENCES

[1] A. Sommerfeld, *Mechanics of Deformable Bodies, Lectures on Theoretical Physics*, Vol. II (trans. G. Kuerti), Academic, New York, 1950, Chapter III.

[2] J. Happel and H. Brenner, *Low Reynolds Number Hydrodynamics*, Prentice-Hall, Englewood Cliffs, NJ, 1965, Chapter 2.

[3] R. E. Meyer, *Introduction to Mathematical Fluid Dynamics*, Wiley-Interscience, New York, 1971, pp. 171–174.

[4] V. Pretorius, B. J. Hopkins, and J. D. Schieke, *J. Chromatogr.*, **99**, 23 (1974).

[5] F. M. White, *Viscous Fluid Flow*, McGraw-Hill, New York, 1974.

[6] R. C. Reid, J. M. Prausnitz, and T. K. Sherwood, *The Properties of Gases and Liquids*, 3rd ed., McGraw-Hill, 1977, Chapter 9.

[7] A. E. Scheidegger, *The Physics of Flow Through Porous Media*, University of Toronto Press, Toronto, 1960.

[8] J. C. Giddings, *Dynamics of Chromatography, Part I, Principles and Theory*, Marcel Dekker, New York, 1965.

[9] P. D. Schettler, Jr. and J. C. Giddings, *Anal. Chem.*, **37**, 835 (1965).

[10] M. Martin and G. Guiochon, *Anal. Chem.*, **55**, 2302 (1983).

[11] A. L. Ruoff, G. H. Stewart, H.K. Shin, and J.C. Giddings, *Kolloid- Zeitschrift*, **173**, 14 (1960).

[12] F. Geiss, *Fundamentals of Thin Layer Chromatography (Planar Chromatography)* (Chromatographic Methods Series XVIII), Huethig Publishing, New York, 1987.

[13] A. L. Ruoff, D. L. Prince, J. C. Giddings, and G. H. Stewart, *Kolloid-Zeitschrift*, **166**, 144 (1959).

[14] J. C. Giddings, G. H. Stewart, and A. L. Ruoff, *J. Chromatogr.*, **3**, 239 (1960).

[15] R. B. Bird, W. E. Stewart, and E. N. Lightfoot, *Transport Phenomena*, Wiley, New York, 1960, p. 297.

[16] W. Jost, *Diffusion in Solids, Liquids, and Gases*, Academic, New York, 1960, p. 510.

[17] J. C. Giddings, M. Martin, M. N. Myers, *Sep. Sci. Technol.*, **14**, 611 (1979).

[18] C. J. O. R. Morris and P. Morris, *Separation Methods in Biochemistry*, 2nd ed., Pitman, London, 1976.

[19] Z. Deyl, Ed., *Electrophoresis. Part A: Techniques*, Elsevier, Amsterdam, 1979.

[20] J. W. Jorgenson, K. D. Lukacs, *Anal. Chem.*, **53**, 1298 (1981).

[21] T. S. Stevens, H. J. Cortes, *Anal. Chem.*, **55**, 1365 (1983).

[22] J. H. Knox and I. H. Grant, *Chromatographia*, **24**, 135 (1987).

[23] J. C. Giddings, W. A. Manwaring and M. N. Myers, *Science*, **154**, 146 (1966).

[24] A. G. Marshall, *Biophysical Chemistry, Principles, Techniques, and Applications*, Wiley, New York, 1978.

[25] C. Tanford, *Physical Chemistry of Macromolecules*, Wiley, New York, 1961.

[26] M. A. Lauffer, *J. Chem. Ed.*, **58**, 250 (1981).

[27] R. C. Reid, J. M. Prausnitz, and T. K. Sherwood, *The Properties of Gases and Liquids*, 3d ed., McGraw Hill, New York, 1977, Chapter 11.

[28] F. Perrin, *J. Phys. Radium*, **7**, 1 (1936).

[29] J. C. Giddings, M. N. Myers, F. J. F. Yang, and L. K. Smith, in M. Kerker, Ed., *Colloid and Interface Science*, Vol. IV, Academic, New York, 1976, p. 381.

[30] C. H. Murphy, *Handbook of Particle Sampling and Analysis Methods*, Verlag, Deerfield Beach, 1984.

[31] J. C. Giddings, *Anal. Chem.*, **36**, 741 (1964).

[32] J. C. Giddings, L. M. Bowman, Jr., and M. N. Myers, *Anal. Chem.*, **49**, 243 (1977).

[33] J. C. Giddings, L. K. Smith, and M. N. Myers, *Anal. Chem.*, **47**, 2389 (1975).

[34] J. H. Hildebrand, *Viscosity and Diffusion*, Wiley, New York, 1977.

EXERCISES

4.1(*)   A velocity of 4 cm/s is maintained by a pressure drop of 2 psi in a gas chromatographic column packed with spherical particles of diameter $d_p = 0.01$ cm. What pressure drop would you expect to use to reach the same velocity if the glass beads were of diameter 0.02 cm, everything else being constant?

4.2(**)   A packed and a capillary column are observed to have flowrates of Q_1 and Q_2, respectively. If the diameter of both tubes are tripled while all other parameters (e.g., pressure drop, particle diameter, viscosity) are held constant, what are the new flowrates?

4.3(**)   An ethyl ether mobile phase flows through a vertical column of 150 cm length filled with particles of diameter 0.0050 cm. The flow is caused by gravity acting over the column length only. At the operating temperature, 20°C, the viscosity and density of ether are 2.3×10^{-3} poises and 0.87 g/cm^3, respectively. The flow resistance parameter, ϕ, is found empirically to be 500. What is the flow velocity, v? What is v for a similar column 75 cm in length?

4.4(*)   A benzene mobile phase flows at 2.0 cm/min through an alumina column at 30°C with particles of mean diameter 10 μm. The viscosity and density of benzene at 30°C are 5.6×10^{-3} poises and 0.87 g/cm^3, respectively. What is the Reynolds number? Is the flow turbulent?

4.5(*)   Water wicked into a strip of Whatman 3 mm paper by capillarity at 30°C exhibited a visible front at 16.2 cm at $t = 30$ min. How far would you expect the liquid front to advance in $t = 4$ h? (The measured distance was 44.0 cm; see ref. 13.)

4.6(*)   A stream of helium carrier gas at 20°C and exactly 3 atm inlet pressure flows through an 8.4 m capillary GC column and emerges at 1 atm. What is the pressure at the column midpoint, 4.2 m from the inlet?

4.7(**)   For an ideal gas flowing through a column or channel we have modified Boyle's law, $pV = $ constant, to the form $p\langle v \rangle = $ constant. Prove that this modification is valid at constant temperature under steady flow conditions using the general flux expression (Eq. 3.14) in the form

$$J = c\langle v \rangle = \text{constant}$$

where the "constant" expresses the steady flow condition which requires that the amount of gas (in moles or mass units) passing through a unit cross section of flow space per second is constant up and down the column.

4.8(***)   For an ideal gas flowing through a column along which a linear temperature gradient is imposed, $T = T_i(1 + \beta x)$, prove that the pressure distribution is given by

$$p^2 = p_i^2 - \theta[(1 + \beta x)^{5/2} - 1]$$

Express ρ in terms of T_i. (Hint: see previous exercise. Also remember that the viscosity of an ideal gas increases with the square root of absolute temperature.)

4.9(*)   A pressure of 50.0 psi is used to drive hexane through a 4.00 m length of capillary of 7.50 μm diameter. Calculate the average flow velocity $\langle v \rangle$.

4.10(**)   Water at 20°C is pumped through a thin rectangular (ribbon-shaped) field-flow fractionation channel at a flowrate of 1.00 mL/min. The channel dimensions (see Figure 4.2) are length $L = 40.0$ cm, breadth $b = 2.00$ cm, and thickness $w = 254$ μm. Calculate the average and maximum flow velocities, $\langle v \rangle$ and v_{max}, and the pressure difference Δp in atmospheres needed to drive the flow.

4.11(**)   A liquid chromatographic column packed with 5.00 μm diameter solid support particles and having a porosity $\epsilon = 0.4$ is found to have a flow resistance parameter $\phi = 750$. Calculate the specific permeability K_0. Assume that the flowrate/pressure drop relationship is identical to that of a bundle of identical parallel capillaries whose axes are spaced (in a square cross-sectional array) a distance of 5.00 μm from one another. What is the single capillary diameter of the hypothetical bundle?

4.12(*)   Calculate the diffusion coefficient for a latex sphere of diameter 0.210 μm suspended in water at 20°C.

4.13(*)   Equation 4.55 (with η_b a constant) can be used to very roughly approximate the viscosities of nonpolar liquids based only on boiling point data. Calculate the viscosities of (a) hexane (b.p. 68°C), (b) chloroform (b.p. 61°C), (c) ethyl acetate (b.p. 77°C), (d) toluene (b.p. 111°C) and (e) carbon tetrachloride (b.p. 77°C) at 20°C and compare these values to those reported in Table 4.1.

5

ZONE FORMATION AND RESOLUTION

In most analytical separation methods the sample components become distributed over the separation path as discrete zones. This is illustrated by most forms of chromatography and electrophoresis. In such cases the zones spread continuously outward as the separation process advances. Yet the success of the separation hinges on keeping the zones reasonably narrow to avoid overlap and cross contamination with neighboring zones. Thus the containment of zone spreading is a major goal of separation science. Its implementation, however, requires an understanding of the processes underlying the formation and dispersion of concentration pulses. This chapter is intended to establish the basic laws governing the structure and evolution of zones and to relate zone broadening to indices measuring the effectiveness of separation such as resolution, plate height, number of plates, and peak capacity.

5.1 GAUSSIAN ZONES

Equation 3.42 is a partial differential equation that will, upon solution, yield concentration as a function of time and distance for any sample pulse undergoing uniform translation and diffusion. In theory, we need only specify the initial conditions (i.e., the mathematical shape of the starting peak) along with any applicable boundary conditions and apply standard methods for solving partial differential equations to obtain our solutions. These solutions tend to be unwieldy if the initial peak shape is complicated. Fortunately, a majority of practical cases are described by a relatively simple special case, which we now describe.

Components to be separated are in most cases confined to narrow initial zones in order to keep the final band width at a minimum. (Since zones tend to broaden with time, the narrower the initial zone, the narrower the final zone.) The narrow starting bands may be approximated mathematically by a δ-function. This function is, effectively, an infinitely narrow zone of unit area. These starting zones evolve by diffusion or diffusion-like processes into Gaussian zones; thus the Gaussian function becomes the desired mathematical solution to the differential equation. Rather than prove the Gaussian outcome by direct integration, we shall demonstrate its validity by assuming that the zones acquire a Gaussian form—an assumption generally in accord with observation—and by showing the consistency of the Gaussian assumption with the basic differential equation (Eq. 3.42) and the assumed δ-function starting profile.

We begin, then, by writing concentration c as a Gaussian (or normal distribution) profile along coordinate y

$$c(y, t) = F(t) \exp[-g(t)y^2] \tag{5.1}$$

where $F(t) \equiv F$ and $g(t) \equiv g$ are unspecified functions of time only. This expression for c has the characteristic bell shape of all functions of the mathematical form $\exp(-ay^2)$ (see Figure 5.1a). If the mathematical form of Eq. 5.1 is valid, c will satisfy Eq. 3.42 and will reduce to a δ-function as $t \to 0$. This we shall check.

We obtain the following partial derivatives from Eq. 5.1

$$\frac{\partial c}{\partial t} = c\left(-y^2 \frac{dg}{dt} + \frac{1}{F} \frac{dF}{dt}\right) \tag{5.2}$$

$$\frac{\partial c}{\partial y} = c(-2gy) \tag{5.3}$$

$$\frac{\partial^2 c}{\partial y^2} = c(-2g + 4g^2 y^2) \tag{5.4}$$

When these are substituted into Eq. 3.42, we get

$$-y^2 \frac{dg}{dt} + \frac{1}{F} \frac{dF}{dt} = -2Dg + 4Dg^2 y^2 \tag{5.5}$$

This equation—which is in fact the basic differential equation $\partial c/\partial t = D(\partial^2 c/\partial y^2)$ applied to our assumed solution—must be valid at all points of space; that is, the two sides must be equal for all values of space coordinate y. Therefore terms containing various powers of y can be grouped to form individual equations. First, the coefficients of y^2 are equated

$$-\frac{dg}{dt} = 4Dg^2 \tag{5.6}$$

When the variables are separated we get the form

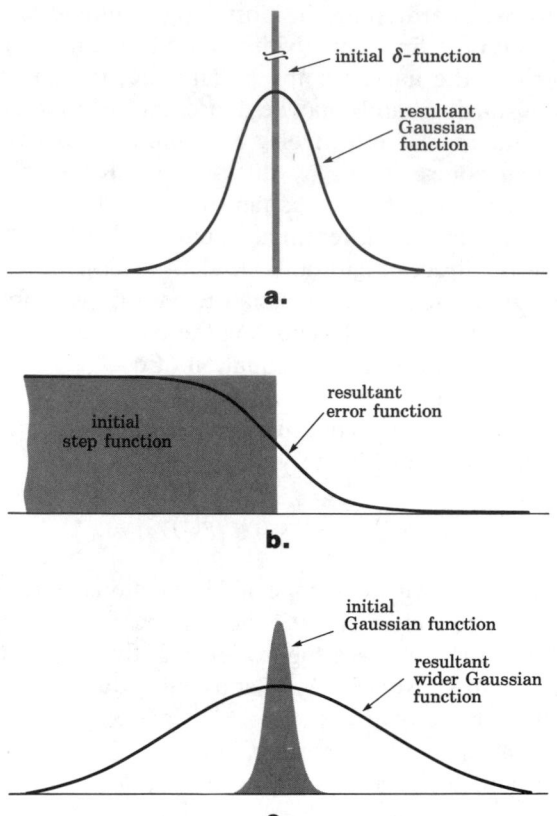

Figure 5.1. These diagrams show how the form of several different concentration profiles resulting from diffusion depend on the shape of the initial profile.

$$-\frac{dg}{g^2} = 4D\, dt \tag{5.7}$$

which integrates directly to

$$\frac{1}{g} = 4Dt + \text{const} \tag{5.8}$$

However, since $g(t)$ must approach infinity at $t = 0$ to meet the postulated δ-function starting condition, the constant of integration must be zero. Therefore, $g(bt)$ is simply

$$g = \frac{1}{4Dt} \tag{5.9}$$

The second group of terms is free of dependence on y. The collection of these terms yields

$$\frac{1}{F}\frac{dF}{dt} = -2Dg \tag{5.10}$$

With $g(t)$ substituted into this from Eq. 5.9, we get

$$\frac{1}{F}\frac{dF}{dt} = -\frac{1}{2t} \tag{5.11}$$

Separation of variables yields

$$\frac{dF}{F} = -\frac{1}{2}\frac{dt}{t} \tag{5.12}$$

which integrates directly to

$$\ln F = \ln(\text{const} \cdot t^{-1/2}) \tag{5.13}$$

and thus gives

$$F = \frac{\text{const}}{t^{1/2}} \tag{5.14}$$

The substitution of $F(t)$ and $g(t)$, Eqs. 5.14 and 5.9, back into Eq. 5.1 yields

$$c = \frac{\text{const}}{t^{1/2}} \exp\left(\frac{-y^2}{4Dt}\right) \tag{5.15}$$

Equation 5.15, as the foregoing proves, is a solution to Eq. 3.42. Furthermore, it becomes infinitely narrow at $t = 0$ and thus acquires the necessary δ-function form at the beginning. It can be shown that this expression normalizes to unit area (which means it applies to one mole of component, or one molecule if one chooses this unit of concentration, per unit area of cross section) when the constant in Eq. 5.15 equals $(4\pi D)^{-1/2}$. Thus the normalized Gaussian profile is

$$c = \frac{1}{(4\pi Dt)^{1/2}} \exp\left(\frac{-y^2}{4Dt}\right) \tag{5.16}$$

When n moles (or molecules) are applied to a unit cross-sectional area the concentration becomes

$$c = \frac{n}{(4\pi Dt)^{1/2}} \exp\left(-\frac{y^2}{4Dt}\right) \tag{5.17}$$

Finally, it is a simple matter to transform back to the original coordinate system by substituting $x - Wt$ for y, Eq. 3.36, a step that yields

$$c = \frac{n}{(4\pi Dt)^{1/2}} \exp\left[\frac{-(x - Wt)^2}{4Dt}\right] \tag{5.18}$$

The above treatment shows that the Gaussian curve is a valid solution to the basic diffusion equation. More importantly, it shows in detail how the Gaussian evolves in time and space. In general terms, the last two equations show that the Gaussian pulse becomes broader and lower (more dilute) with the passage of time.

For zones that cannot be adequately approximated by a δ-function at $t = 0$, the final zone can be described as a sum of Gaussians, one evolving from each thin slice of the initial zone. This superposition of solutions is possible because of the linearity of the basic differential equation, Eq. 3.42. In this way an initial step function (e.g., like that in frontal chromatography) becomes an error function (see Figure 5.1b). An initial Gaussian profile emerges as a Gaussian profile, but with increased width (Figure 5.1c). In practice the final profile is obtained as a convolution integral. The mathematical details are beyond the scope of this work but can be found in the literature [1, 2].

5.2   STATISTICAL MOMENTS

Every distribution function (Gaussian or otherwise) has a set of *statistical moments* that convey important information about the location and shape of the function. The nth *moment about the mean* is defined as the following average quantity

$$\langle y^n \rangle = \langle (x - \bar{X})^n \rangle = \langle (x - Wt)^n \rangle \tag{5.19}$$

where the angled brackets, $\langle \ \rangle$, denote the average of the quantity contained inside weighted by the distribution function. Thus

$$\langle y^n \rangle = \frac{\int_{-\infty}^{\infty} y^n c(y, t) \, dy}{\int_{-\infty}^{\infty} c(y, t) \, dy} \tag{5.20}$$

The first moment, which is simply the position of the center of gravity of the concentration pulse, is zero because the zero of coordinate axis y is arbitrarily fixed at the center of gravity (see Eq. 3.36). In fact all odd moments are zero if $c(y, t)$ is a Gaussian or other symmetrical function about $y = 0$. For asymmetric zones the value of the third moment becomes a measure of the asymmetry. More specifically, asymmetry is often measured by a normalized third moment called the *skew*, defined as $\langle y^3 \rangle / \langle y^2 \rangle^{3/2}$. The "flattening" of a zone is measured by *excess*, $(\langle y^4 \rangle / \langle y^2 \rangle^2) - 3$. Both skew and excess are zero for Gaussians as shown in Figure 5.2.

The most studied moment is the second, called the *variance*, σ^2. The square root of this, σ, is called the *standard deviation* and is a measure of the

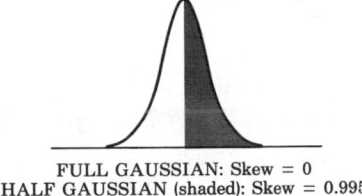

FULL GAUSSIAN: Skew = 0
HALF GAUSSIAN (shaded): Skew = 0.995

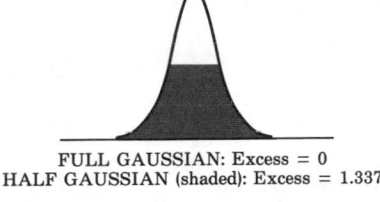

FULL GAUSSIAN: Excess = 0
HALF GAUSSIAN (shaded): Excess = 1.337

Figure 5.2. Skew and excess for normal Gaussian peaks and bisected Gaussian peaks. (From ref. 3. Reprinted with permission from E. Grushka, M. N. Myers, P. D. Schettler, and J. C. Giddings, *Anal. Chem.*, **41**, 889 (1969). Copyright 1969 American Chemical Society.)

overall width of the zone. For Gaussian zones, σ is the distance from the zone center to the point of inflection, as shown in Figure 5.3. The values of σ^2 and σ for the Gaussian distribution are found by substituting Eq. 5.16 or 5.17 into Eq. 5.20 and integrating with $n = 2$

$$\sigma^2 = 2Dt \tag{5.21}$$

$$\sigma = (2Dt)^{1/2} \tag{5.22}$$

The latter equation reflects the way in which peak width, proportional to σ, increases with time: the dependence is square root in form.

When the Gaussian function of Eq. 5.17 is written in terms of σ^2 by using Eq. 5.21, we have

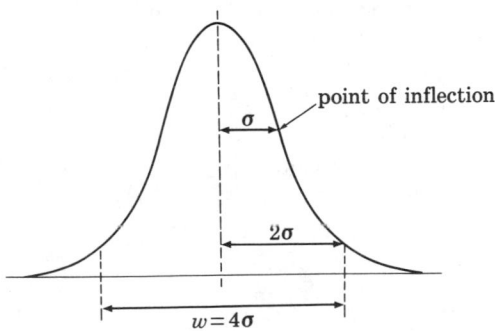

point of inflection

σ

2σ

$w = 4\sigma$

Figure 5.3. Gaussian zone showing standard deviation σ (from the center line to the point of inflection), 2σ, and effective zone width w.

$$c = \frac{n}{(2\pi\sigma^2)^{1/2}} \exp\left(\frac{-y^2}{2\sigma^2}\right) \tag{5.23}$$

It can be shown that 68.3% of a Gaussian area lies between $-\sigma$ and σ, 95.6% lies between -2σ and 2σ, and 99.7% lies between -3σ and 3σ. The distance from -2σ to 2σ, a total of 4σ, can be arbitrarily taken as the width w of the zone; we can term this the *effective zone width* (see Figure 5.3).

More information on statistical moments can be found in the literature [3] or in a plethora of books on statistics and probability theory, only a few examples of which are cited here [4–6].

5.3   RANDOM PROCESSES, VARIANCE, AND DIFFUSION

The fundamental events which give rise to zone broadening are generally of a random nature. Even molecular diffusion is based on the random jumping of molecules back and forth in their medium. Each molecule undergoes a succession of random excursions or steps with the passage of time. Because of statistical fluctuations, individual molecules will end up taking more excursions in one direction than another, and sometimes excursions of different lengths in the two directions, so that they spread out erratically from their starting point. In this way a narrow initial zone widens into a Gaussian concentration profile.

The degree of zone spreading for random processes depends on the length and frequency of the random steps and on the length of time over which they occur. A simplified model which describes the nature of this dependence is the *random walk model* [4, 7, 8].

The one-dimensional random walk consists of a sequence of steps of fixed length l along a given axis (in our case, the separation axis). The direction (positive or negative) of each step is determined randomly, as by the flip of a coin. After executing a large number of such random steps, some of the molecules involved in this hypothetical process will have taken more positive than negative steps (the same as coming up with more "heads" than "tails"), thus moving toward the front end of the molecular distribution. Those taking more negative than positive steps will arrive near the rear of the band of molecules. The probability of a molecule ending at a given position is given by the binomial distribution. The large number of molecules normally involved will ensure the presence of a near-continuous concentration profile of binomial form. It can be shown that the variance of a binomial follows the relatively simple mathematics expressed by [4, 8]

$$\sigma^2 = l^2 n \tag{5.24a}$$

or

$$\sigma = l n^{1/2} \tag{5.24b}$$

A physical rationale for this particular dependence [8] is summarized below.

First we note that σ, which is proportional to the width of the peak (see Figure 5.3), is directly proportional to l. We can explain this by observing that if the length of every step doubled, the final distance of a randomly walking molecule from the origin would also double. Quantity σ, which is a root-mean-square average of such displacement distances, would also double. Clearly, σ, peak width, or any other length parameter of interest, will be scaled in direct proportion to elementary step length l.

Somewhat more puzzling is the observation that σ increases with the square root of n rather than linearly with n. At first glance one might expect that twice as many steps imply twice as much net displacement or spreading from the origin, which would give a linear dependence of σ upon n. However, we easily find fault with this linear model. Linearity implies that if a molecule ends up at some distance x from the origin after n steps, it will end up at distance $2x$ after $2n$ steps. However, the ongoing random walk is as likely to send the molecule that has arrived at x scurrying back toward the origin as it is to send it on to $2x$. Thus the action of independent random steps often cancel previous displacements; the net result is a much slower growth of σ with n than a linear dependence would suggest. The actual dependence turns out to be a square root form as shown in Eq. 5.24b.

There is a basic law of statistics known as the *central limit theorem*, which guarantees that a sequence of random events of the most general kind will lead to a Gaussian distribution function [4]. It is only necessary that single random displacements be small compared to the final mean (or root-mean-square) displacement. A random walk fulfills this requirement if the number of steps is large; mathematically the binomial distribution resulting from a random walk converges to a Gaussian at large n. In general, then, nearly all random processes lead to Gaussian zones. The experimental observation of a Gaussian shape is no assurance that molecular diffusion, alone, is the cause.

Inasmuch as we have seen that the basic diffusion equations lead to a Gaussian profile, we may in most cases describe Gaussian spreading as an *apparent diffusion process*, with an effective diffusion coefficient related to σ through Eq. 5.21

$$D = \frac{\sigma^2}{2t} \tag{5.25}$$

If a random walk process underlies the apparent diffusion, D can be obtained in a precise form by substituting Eq. 5.24 into Eq. 5.25

$$D = \frac{l^2 \dot{n}}{2} \tag{5.26}$$

where $\dot{n}$ is the number of steps taken per unit of time. Inasmuch as $\dot{n}$ and l are generally constant with the passage of time, D does not vary with time, but σ^2 increases in proportion to the time because the total number of steps of Eq. 5.24 accumulate in proportion to time.

It is a fundamental statistical law that when several independent processes contribute to zone spreading, the variances are additive (4). Thus the total variance σ_T^2 is of the form

$$\sigma_T^2 = \sigma_1^2 + \sigma_2^2 = \sigma_3^2 + \cdots \tag{5.27}$$

It follows that the apparent diffusion coefficients, which Eq. 5.25 shows to be proportional to the respective variances, are additive in the same manner

$$D_T = D_1 + D_2 + D_3 + \cdots \tag{5.28}$$

To summarize, the observation of a Gaussian profile usually implies that transport is governed mathematically by the diffusion equations and mechanistically by one or more multistep random processes. Below we examine some of the random mechanisms operative in separations.

5.4   MECHANISMS OF ZONE BROADENING

Several kinds of random events other than molecular diffusion contribute to zone spreading in separation systems. With each increment in zone spreading there is a corresponding loss of resolution; it is thus important to understand these processes in order to minimize them. The random processes described below are responsible for effective diffusion in many engineering and chemical systems as well as in separations.

1. *Flow through packed beds* (*eddy* or *multipath diffusion*). In chromatography, component zones are carried through a bed of randomly packed particles. The streamlines in such flow veer back and forth to find passage between the particles (see Figure 5.4) and fluctuate in velocity

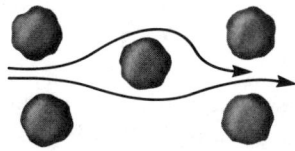

Figure 5.4. Random walk caused by streamlines erratically veering around particles in a packed bed. Molecules following different streamlines may move ahead of or fall behind one another just as in any other random walk. Above shows streampaths followed by two different molecules for an equal period of time, but suffering different displacements.

according to the narrowness of the passage. Hence a molecule transported in such a flow will undergo many random changes in velocity and thus in position. A detailed analysis of this phenomenon has been given [8]; more details will be found later in our treatment of chromatographic band broadening. Basically, the process can be treated as a random walk and once l and n are determined from a flow model, the variance can be estimated from the preceding equations.

2. *Displacement in porous beds and gels.* Electrophoresis often occurs in a bed of support material such as paper, powder, or gel. Here the conductance pathways veer randomly in response to random changes in conductance due to encounters with the support matrix. For a conducting matrix, the conductance pathways can lead through the solid obstacles, but macromolecules may be forced to stop and diffuse sideways before resuming migration. Such a stop-go process is highly random and therefore contributes to zone broadening (9).

3. *Nonequilibrium phenomena.* Here molecules exist in two or more interconverting physical or chemical states, each with a different rate of displacement. For example, in chromatography the sorbed molecules are one form and the desorbed molecules are another. In electrophoresis certain ions take on a variable number of charges, each representing a different form and having a different velocity. In each case the ion or molecule transforms back and forth from one form (velocity state) to another at random intervals. Each random transformation represents a change in velocity and thus leads to a change in position with respect to the zone center. These random events, like all others, lead to zone spreading. The theory, especially for chromatography, has been treated in some detail [8, 10].

It is important to note that the displacement of sample components in all the above processes is described by the basic mass transport equations developed earlier. However, some special considerations are needed to properly account for the random component of transport. Thus the basic equation of flow and transport, Eq. 3.35, is now expressed as

$$\frac{\partial c}{\partial t} = -W' \frac{\partial c}{\partial x} + D \frac{\partial^2 c}{\partial x^2} \tag{5.29}$$

where the displacement velocity W', which replaces the constant value W, incorporates the random velocity fluctuations. Velocity W' can therefore be broken into two parts

$$W' = W + W_{\text{err}} \tag{5.30}$$

where the erratically fluctuating part of W' is shown as W_{err}, leaving behind the constant average drift velocity W. Equation 5.29 is now in the form

$$\frac{\partial c}{\partial t} = -W \frac{\partial c}{\partial x} - W_{err} \frac{\partial c}{\partial x} + D \frac{\partial^2 c}{\partial x^2} \qquad (5.31)$$

The random nature of the second term on the right assures us, through the central limit theorem, that it contributes an effective diffusion term to zone spreading [11]. Thus, this term must have the equivalent form

$$-W_{err} \frac{\partial c}{\partial x} = D_{err} \frac{\partial^2 c}{\partial x^2} \qquad (5.32)$$

for which D_{err} (which, unlike W_{err}, is a constant) can be evaluated by a random walk model, Eq. 5.26, or by more elaborate methods that need not concern us here.

When Eq. 5.32 is used in Eq. 5.31, we have

$$\frac{\partial c}{\partial t} = -W \frac{\partial c}{\partial x} + (D_{err} + D) \frac{\partial^2 c}{\partial x^2} \qquad (5.33)$$

which gives us an equation in the same basic form as Eq. 5.29 and which yields mathematical solutions, $c(x, t)$ or $c(y, t)$, of the same Gaussian form discussed in Section 5.1. However, in these solutions W can be interpreted as an average displacement velocity. More importantly, the effective diffusion rate is enhanced because the coefficient of $\partial^2 c / \partial x^2$ is now the sum of the two terms [11]

$$D_T = D_{err} + D \qquad (5.34)$$

where D_{err} is the contribution of the random processes mentioned above. Hence, in the presence of these processes, the outward diffusion of zones is increased. Much effort has gone into the development of systems in which D_{err} is reduced by some means or another. For example, in chromatography the reduction of particle size reduces D_{err} and thus reduces zone spreading. Uniform gels (such as polyacrylamide) in electrophoresis reflect smaller D_{err} terms than do granular media and are thus favored for high-resolution work.

5.5 PLATE HEIGHT: A ZONE SPREADING INDEX

It has been emphasized that zones universally diffuse outward during their transport through a separation system. Corresponding to the total effective diffusion coefficient D_T, each zone acquires a variance which the form of Eq. 5.25 shows to equal

$$\sigma^2 = 2D_T t \qquad (5.35)$$

This equation shows that σ^2 is proportional to time t providing D_T is constant.

For a zone in uniform translation at a constant velocity W, the distance X traversed by the zone in time t is simply

$$X = Wt \tag{5.36}$$

Substituting $t = X/W$ from this expression into Eq. 5.35, we get

$$\sigma^2 = \left(\frac{2D_T}{W}\right)X \tag{5.37}$$

which shows that σ^2 is also proportional to zone migration distance X in uniform systems.

Equation 5.37 is important in showing how zones spread out (and consequently overlap) as they migrate through a separation system. When the coefficient $2D_T/W$ is large, spreading (measured by σ^2) is excessive relative to the distance X in which separation occurs. When $2D_T/W$ is small, the zones are comparatively narrow. Therefore we can use the coefficient $2D_T/W$ as an index expressing the rate of growth of σ^2 along the separation path. Employed in this way, it is an extremely important index defining separative power, and it is therefore given the special symbol H

$$H = \frac{2D_T}{W} \tag{5.38}$$

When this is substituted into Eq. 5.37, it yields the simple proportionality

$$\sigma^2 = HX \tag{5.39}$$

which means that the coefficient H is related as follows to the observable parameters σ^2 and X

$$H = \frac{\sigma^2}{X} \tag{5.40}$$

Parameter H, in the form described by Eq. 5.40, is identical to the *height equivalent to a theoretical plate*, or simply *plate height*, commonly used to assess column efficiency in chromatographic systems. However, H has been obtained here without reference to the theoretical plate model, in which the separation path is assumed to be divided into a series of linked stages or plates. The plate model is both cumbersome and inflexible in describing zone spreading in chromatographic columns or other separation systems. Despite the absence of the theoretical plate model in our derivation, H retains the name *plate height* for historical continuity.

The concept of theoretical plates evolved from studies of distillation and countercurrent distribution, where distinct stages frequently exist. The widespread adoption of plate concepts in chromatography can be traced largely to the seminal work of Martin and Synge [12]. Plate models, however, are neither very appropriate nor very useful in describing the continuous transport processes of chromatography, field-flow fractionation, electrophoresis, sedimentation, and related zonal methods [8, 13, 14]. They yield little of value but they do correctly predict σ^2 proportional to X, just like Eq. 5.37, which was derived on different grounds. Since continuous zonal systems such as chromatography share this key proportionality (σ^2 to X) with staged systems, there is nothing wrong with retaining the name "plate height" for the constant of proportionality—H in Eq. 5.38—whether the system is staged or continuous. In fact, the adoption of a common name for this constant makes H a nearly universal parameter whose evaluation permits the comparison of many diverse separation techniques (14). We shall therefore use H and related terms here for separations generally, recognizing that the plate model itself has little relevance to the discussion.

In order to use H broadly, we must define a derivative term N, called the *number of theoretical plates*. Quantity N is given by

$$N = \frac{X}{H} \tag{5.41}$$

which is the number of H units (plates) encountered in migration distance X. Since X and H both have the dimensions of length, N is dimensionless.

The substitution of Eq. 5.40 into 5.41 yields for N

$$N = \frac{X^2}{\sigma^2} = \left(\frac{X}{\sigma}\right)^2 \tag{5.42}$$

This equation shows that N increases with X/σ, the ratio of the migration distance in which separation can occur to the zone width parameter σ, which reflects zone overlap. The overall "goodness" of a separation obviously increases with this ratio, so N is clearly a valid index of separation power, as substantiated more fully elsewhere [8, 15]. In chromatography, N is often taken as a measure of *column efficiency*, a definition that can be extended to other zonal separation methods as well.

If we substitute Eq. 5.38 into 5.41, we get N in terms of D_T and W

$$N = \frac{WX}{2D_T} \tag{5.43}$$

This equation provides a useful way to calculate N in terms of the underlying diffusion process reflected in D_T.

We return now to note that the validity of the proportionality of σ^2 to X leading to the definition of H as σ^2/X, Eq. 5.40, hinges on the assumption of uniform conditions along the separation path or column, which implies that D_T/W is constant, Eq. 5.37. When conditions vary along the separation path, each small increment of migration distance dX leads to a specific increment in variance $d\sigma^2$ which may vary with time or position. The *local plate height* then becomes

$$H = \frac{d\sigma^2}{dX} \tag{5.44}$$

and this can clearly vary along the separation path. Equation 5.44 replaces the definition for H, $H = \sigma^2/X$, when columns are nonuniform, but approaches $H = \sigma^2/X$ as the nonuniformities vanish.

An extended discussion of zone spreading and plate height in nonuniform separation systems is beyond the scope of the present chapter. However, later we account for flow nonuniformities due to gas compressibility in gas chromatography. More generally, the treatment developed by the author [8, 16] for chromatographic columns can be expanded to describe most other zonal systems.

One object of separation theory is to predict H in terms of controllable experimental parameters such as flow velocity, field strength, and support particle diameter. This theory is important in optimizing separations. Details will be provided in later chapters.

While H and the related parameter N generally characterize the efficiency of a separation system, their values vary somewhat from one component to another and they therefore cannot be considered true constants of the system. The variation with component identity occurs because the ratio D_T/W of Eq. 5.38 depends to some degree on diffusion coefficients and other component-specific properties.

5.6 PLATE HEIGHT IN ELUTION SYSTEMS

When zones are washed (eluted) off the end of a column, as in most forms of chromatography or field-flow fractionation, they emerge one by one with the passage of time. In this case time rather than distance becomes the coordinate along which we observe the separation unfold. Zone displacements in space (along the separation axis) are no longer seen, although the spatial disengagement in the column remains the (hidden) basis of separation. However, we observe, and thus must describe, the time-based appearance of zones.

Because zones have a finite physical width, they take a finite time to emerge—the leading edge first, followed by the center and then the trailing edge. Two steps in the sequence are shown in Figure 5.5. This figure shows that the zone is displaced a distance of one σ in the time τ. A detector placed at the end of the column would sense the emergence of the zone with time. When displaced at uniform velocity, the zone shape on the time axis (traced out on the chart paper of a recorder or on a computer screen) resembles, of course, the zone shape in the space of the column. If one is

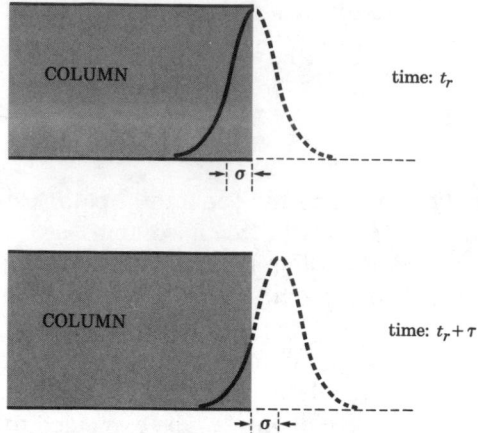

Figure 5.5. A Gaussian zone emerges from a column. In the time τ the zone is displaced one standard deviation σ through the column end. Time τ is thus the standard deviation in time units.

approximately Gaussian, so is the other. Thus a typical zone on a column will approach a Gaussian due to random processes, then emerge in time as a Gaussian or near-Gaussian profile as shown in Figure 5.6.

The time required for the emergence of a segment of a zone equal to one standard deviation is shown as τ in Figure 5.5. The time τ is equal to the time needed to transport the zone a distance σ through the column

$$\tau = \frac{\sigma}{W} \tag{5.45}$$

The elution time t_r, by analogy, is the time needed to carry the zone through the entire length L of the column at velocity W

$$t_r = \frac{L}{W} \tag{5.46}$$

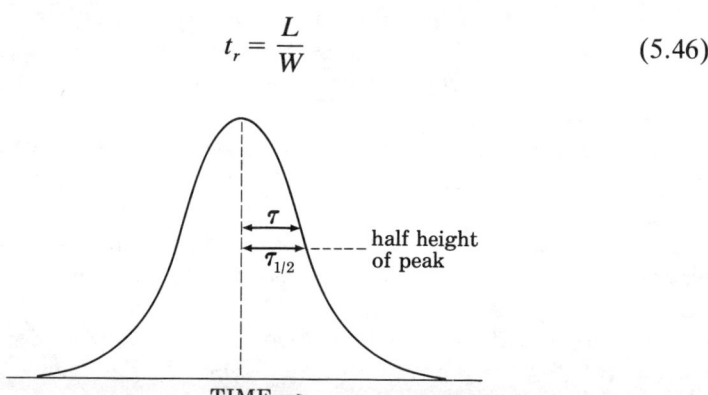

Figure 5.6. Concentration versus time for eluting peak, showing standard deviation τ and one-half the width at half-height $\tau_{1/2}$ in time units.

For elution, the plate height H, which by Eq. 5.40 is σ^2/X, becomes σ^2/L because the zone migration distance X ends up as column length L upon emergence. From Eq. 5.45, $\sigma^2 = \tau^2 W^2$, which gives

$$H = \frac{\sigma^2}{L} = \frac{\tau^2 W^2}{L} \tag{5.47}$$

From Eq. 5.46 we get $W = L/t_r$, which when substituted into Eq. 5.47 yields

$$H = \frac{L\tau^2}{t_r^2} \tag{5.48}$$

This is the basic equation for measuring plate height in elution systems. In practical use, H must be modified by a numerical constant if one chooses to use width at half-height ($2\tau_{1/2}$, see Figure 5.6) in place of τ as a measure of peak broadening. Thus

$$H = \frac{L\tau_{1/2}^2}{(2\ln 2)t_r^2} = \frac{0.72135 L\tau_{1/2}^2}{t_r^2} \tag{5.49}$$

The number of theoretical plates N generated in column length L is simply $N = L/H$ or, from Eqs. 5.48 and 5.49

$$N = \frac{t_r^2}{\tau^2} = 1.38629\,\frac{t_r^2}{\tau_{1/2}^2} \tag{5.50}$$

an equation analogous to Eq. 5.42.

Parameter N is dimensionless; it is shown above as the ratio of two times, each squared. Because of this, the ratio t_r/τ (or $t_r/\tau_{1/2}$) can be replaced by the ratio of two proportionate distances (the corresponding chart distances for the peak as it unfolds on the chart paper of a recorder) or two proportionate volumes (elution volume and the volume increment observed during the emergence of one standard deviation of the peak).

5.7 RESOLUTION

The most important index of success for the analytical separation of *two* specific components is the *resolution* R_s (see Section 1.4). This parameter categorizes the overlap (or lack of it) of two specified component zones (see Figure 5.7). If the centers of gravity of the two zones are found at locations X_1 and X_2, respectively, then the resolution can be defined [8, 17, 18] as

$$R_s = \frac{X_2 - X_1}{2(\sigma_1 + \sigma_2)} \tag{5.51}$$

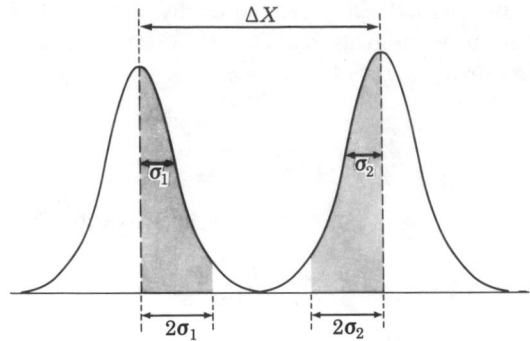

Figure 5.7. The resolution of two peaks, defined as $R_s = \Delta X/(2\sigma_1 + 2\sigma_2)$, measures peak separation ΔX relative to the zone dispersion ($2\sigma_1$ and $2\sigma_2$) leading to overlap and contamination.

where σ_1 and σ_2 are the respective standard deviations. Inasmuch as close-lying components generally have similar properties which are reflected in comparable σ values, we may replace σ_1 and σ_2 by an average value σ, giving

$$R_s = \frac{\Delta X}{4\sigma} \tag{5.52}$$

or by virtue of the fact that 4σ is the average effective zone width w

$$R_s = \frac{\Delta X}{w} \tag{5.53}$$

which shows that when two peak centers are separated by distance $w = 4\sigma$, R_s is unity.

Figure 5.8 shows pairs of Gaussian zones at different resolution levels. The top row shows how peaks of equal height disengage from one another as R_s increases. Two distinct maxima are found only when $R_s > 0.5$. "Baseline" resolution is found only for $R_s > 1.5$.

The center row shows the emergence of zones with a 2:1 peak height ratio. The bottom row shows how increasing resolution brings out zones with a 5:1 height ratio.

If we solve for σ from Eq. 5.39, we get the form

$$\sigma = (HX)^{1/2} \tag{5.54}$$

where X is the mean distance traveled by the two zones. Substitution of this into Eq. 5.52 yields

$R_s=0$ $R_s=0.25$ $R_s=0.50$ $R_s=0.75$ $R_s=1.00$ $R_s=1.50$

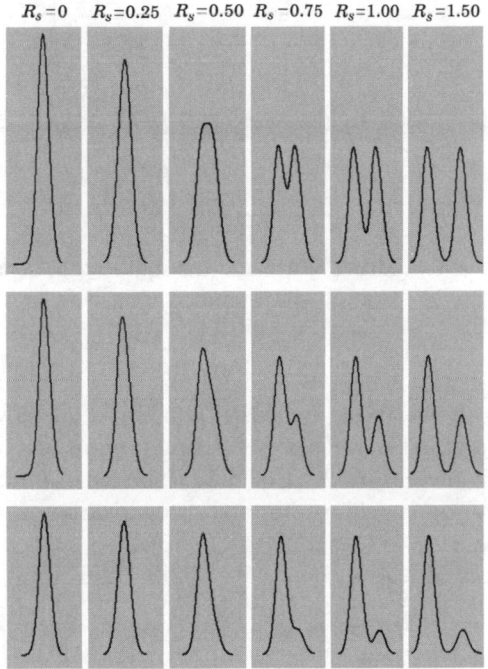

Figure 5.8. Left to right sequence gives the profile of a pair of Gaussian zones of equal σ at increasing levels of resolution, as shown. Top row shows this sequence for two zones of equal peak height; middle and bottom rows show the sequence for 2:1 and 5:1 peak height ratios (profiles courtesy of Joe M. Davis).

$$R_s = \frac{\Delta X}{4(HX)^{1/2}} \tag{5.55}$$

which shows that resolution improves with decreasing plate height H.

The separation of two zones occurs because they are traveling at the different velocities W_1 and W_2 for a time t, giving

$$\Delta x = (W_2 - W_1)t = \Delta W t \tag{5.56}$$

Because t is mean distance over mean velocity, X/W, we have

$$\Delta X = \frac{\Delta W X}{W} \tag{5.57}$$

A comparison of Eqs. 5.57 and 5.54 shows that ΔX increases linearly with displacement distance X while 4σ increases only as $X^{1/2}$. These functions of X are shown in Figure 5.9. Since R_s is the ratio of ΔX to 4σ, unit resolution can be reached only when X gets large enough that ΔX catches up with 4σ

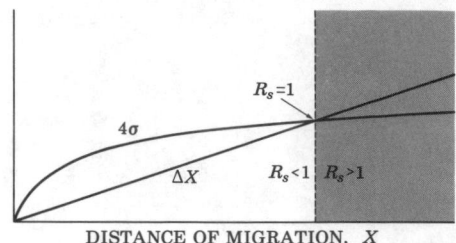

Figure 5.9. The linear dependence of ΔX **and the square root dependence of** 4σ **on** X lead to their inevitable crossing, at which point $R_s = 1$. Beyond the crossing (shaded area), $R_s > 1$.

and crosses it, as shown in the figure. When the relative velocity difference $\Delta W/W$ of the two zones increases, reflecting enhanced selectivity, ΔX rises more rapidly with increasing X (Eq. 5.57) and crosses the 4σ line earlier, that is, unit resolution is achieved in a shorter distance X. This is illustrated in Figure 5.10.

The substitution of Eq. 5.57 into Eq. 5.55 yields

$$R_s = \frac{\Delta W}{W}\left(\frac{X}{16H}\right)^{1/2} = \frac{\Delta W}{W}\left(\frac{N}{16}\right)^{1/2} \tag{5.58}$$

where we have used $N = X/H$, Eq. 5.41, to obtain the latter expression. This equation shows that high resolution requires a combination of good selectivity, reflected in incremental velocity ΔW, and/or good system efficiency (narrow peaks), reflected in N. These factors are involved in essentially all zonal separations, including chromatography and electrophoresis, and constitute major criteria for separation system design.

Unit resolution, $R_s = 1$, is adequate for most (and more than sufficient for some) analytical purposes. We need recognize, however, that Gaussian

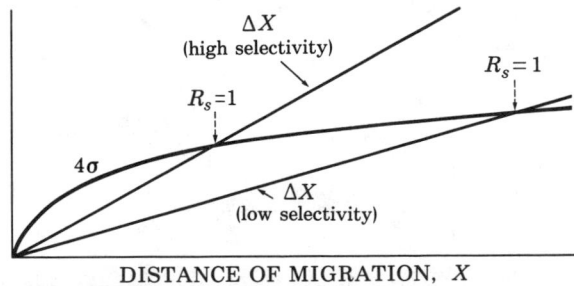

DISTANCE OF MIGRATION, X

Figure 5.10. With increasing selectivity, incremental distance ΔX increases more rapidly with X, crossing the 4σ line sooner and thus generating unit resolution in a shorter migration distance.

zones do not disengage from one another completely as they move further apart because the concentration at the edges, while dropping off dramatically, does not reach zero. For example, two equal Gaussian zones having unit resolution still penetrate one another to the extent that 2.2% of each zone's content extends past the midpoint between them into the other zone.

5.8 PEAK CAPACITY

In the separation of complex mixtures having many components, attention often focuses on the total number of zones or peaks separable rather than on the resolution of specific pairs. In this case we use a different index: the *peak capacity* n_c [18–20]. The peak capacity is the maximum number of separated peaks that can be fit (with adjacent peaks at some specified R_s value) into the path length or space provided by the separation method (see Figure 5.11). If a path of length L is available over which to distribute zones, the peak capacity for adjacent peaks separated at $R_s = 1$ is [21]

$$n_c = \frac{L}{w} = \frac{L}{4\sigma} \tag{5.59}$$

where w and σ are again the mean values of zone width and standard deviation, respectively, for all the zones separated. Length L may be a column length in chromatography or the total path length for electrophoretic or sedimentation migration, and so on.

If peaks can be adequately distinguished (depending on analytical goals and methods) at some R_s value less than one, then more peaks can be crowded into the separation space and n_c increases in accordance with the equation

$$n_c = \frac{L}{4\sigma R_s} \tag{5.60}$$

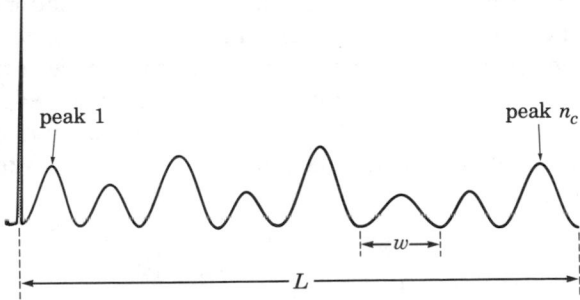

Figure 5.11. Peak capacity is defined as the number n_c of peaks or zones that can be separated (at a specified R_s) over path length L (or elution volume range $V_{max} - V_{min}$) provided by a separation system.

If we use σ from Eq. 5.54, where we replace migration distance X by L (although $L/2$ might be better as a mean value), Eq. 5.59 reduces to

$$n_c \sim \frac{N^{1/2}}{4} \qquad (5.61)$$

which is in approximate agreement with several equations for n_c derived on the basis of different variations of σ with X [14].

In elution systems like chromatography, n_c can be significantly larger than the value indicated in Eq. 5.61 because each successive volume sweeping the column can bring forth and resolve a new group of peaks. Analysis shows that the peak capacity (for adjacent peaks at $R_s = 1$) is approximately [19]

$$n_c = 1 + \frac{N^{1/2}}{4} \ln \frac{V_{\max}}{V_{\min}} \qquad (5.62)$$

where $V_{\max}$ and $V_{\min}$ are the largest and smallest volumes in which zones can be eluted and detected. Thus n_c increases not only with N, but also with the workable elution volume range $V_{\max}/V_{\min}$. The latter factor is responsible for the limited peak capacity of size exclusion chromatography (SEC) where $V_{\max}/V_{\min}$ is restricted to ~ 2 or less by the fundamental nature of the SEC process [19]. By contrast, $V_{\max}/V_{\min}$ may reach a value of 50 or more in normal gas and liquid chromatography and in field-flow fractionation.

The equations of this section show that resolution and peak capacity are inversely proportional to σ and w (usually reflected in H and N). These equations illustrate how the capacity for separation is diminished, using any reasonable measure, by increases in zone width. This conclusion reemphasizes our deep concern with zone spreading phenomena and the fundamental transport processes that underlie them.

We should add that while resolution and peak capacity are excellent criteria of merit for the separation of multicomponent mixtures into discrete zones, other criteria exist, some very general, for judging the efficacy of separation and purification in any separative operation (see Section 1.4). Various terms such as *impurity ratio* and *purity index* abound. Rony has developed a criterion termed the extent of separation [22]. Stewart, as well as de Clerk and Cloete, have shown that entropy can be formulated as a very general measure of separation power, as we might expect from the discussion of Section 1.6 [23, 24]. An excellent discussion of separation indices, with an emphasis on non-Gaussian zones (below), is found in Dose and Guiochon [25].

5.9   NON-GAUSSIAN ZONES

A Gaussian zone is the model around which most discussions of zonal separation methods revolve. However, there are frequent departures from the Gaussian profile because of various perturbations or nonidealities. Sometimes the departures are barely observable and of negligible importance and sometimes they are overwhelming, leaving behind little of the original Gaussian shape.

The Gaussian zone is a limiting form, valid only when the contribution of individual random events to zone width is negligible compared to the final magnitude of that width, as noted in our discussion of the central limit theorem. Thus for a true Gaussian we need an infinite number of random steps, each with infinitely short duration and displacement. Any random step of finite displacement will therefore perturb the Gaussian by a finite degree. Such processes include adsorption-desorption steps, partitioning into a liquid phase of finite depth, chemical interconversions, trapping in void volumes, or simply winding around a support particle. It is a little-known fact that the rigorous theories of these processes—in the few cases where they can be developed—never produce a strictly Gaussian zone. Simple adsorption-desorption steps in a chromatographic column, for instance, lead to a profile expressed in terms of Bessel functions [8]. Only in the (unreachable) limit of infinite time (involving an infinite number of steps) do these models converge to a Gaussian zone, in accordance with the central limit theorem. Therefore, strict adherence to a Gaussian is not expected and is not, in general, very important. However, when processes are encountered with long time constants, such as slow adsorption-desorption cycles or the trapping of solute in large dead volumes (inside or outside of the separation column), the distortion can be serious [26]. Sometimes a large degree of tailing results, leading to excessive overlap with neighboring zones. This is particularly serious in trace analysis where a large tail can obscure the small peak of an important trace constituent.

Peaks departing from Gaussians are often described by exponentially modified Gaussian (EMG) functions, a combination (via a convolution operation) of a Gaussian and an exponential function [27, 28]. Other empirical functions have been used as well [29]. None of the above functions has a very sound theoretical footing, but they are nonetheless important for descriptive purposes.

The departures noted above arise, in a mathematical sense, because the treatment of finite random displacements as a diffusion process is only an approximation, subject to the limitations of the central limit theorem. However, non-Gaussian zones can also arise out of system nonlinearities caused by overloading, in which component molecules are so numerous that they interfere with one another's migration. This interference makes local zone velocity W (along with diffusion coefficient D) dependent on concentration, which voids the theory (starting with Eq. 3.29) leading to Gaussian zones. This problem has been treated by Guiochon et al. and by Wilson in reference to preparative scale chromatography (30, 31). Again, small departures are tolerable but large departures cause tailing and changes in zone migration rate that seriously hinder separation. These highly nonlinear effects can be reduced by using smaller samples, but sample size reduction is counterproductive for preparative scale operation.

Finally, we note that some component zones do not acquire Gaussian shapes because the controlling processes are quite unlike those described above. This situation applies to some of the steady-state zones described in the following chapter.

REFERENCES

[1] J. Crank, *The Mathematics of Diffusion*, 2nd ed., Clarendon Press, Oxford, 1975.

[2] J. C. Sternberg, in J. C. Giddings and R. A. Keller, Eds., *Advances in Chromatography*, Vol. 2, Dekker, New York, 1966, Chapter 6.

[3] E. Grushka, M. N. Myers, P. D. Schettler, and J. C. Giddings, *Anal. Chem.*, **41**, 889 (1969).

[4] W. Feller, *An Introduction to Probability Theory and Its Applications*, 3rd ed., Wiley, New York, 1967.

[5] E. R. Mullins and D. Rosen, *Concepts of Probability*, Bogden and Quigley, Tarrytown-on-Hudson, NY, 1972.

[6] J. F. Ratcliffe, *Elements of Mathematical Statistics*, 2nd ed., Oxford University Press, London, 1967.

[7] S. Chandrasekhar, in N. Wax, Ed., *Noise and Stochastic Processes*, Dover, New York, 1954.

[8] J. C. Giddings, *Dynamics of Chromatography, Part I: Principles and Theory*, Marcel Dekker, New York, 1965.

[9] J. C. Giddings, *Anal. Chem.*, **34**, 37 (1962).

[10] J. C. Giddings, *J. Chem. Educ.*, **44**, 704 (1967).

[11] J. C. Giddings, *Sep. Sci. Technol.*, **13**, 3 (1978).

[12] A. J. P. Martin and R. L. M. Synge, *Biochem. J.*, **25**, 1358 (1941).

[13] J. C. Giddings, *J. Chromatogr.*, **2**, 44 (1959).

[14] J. C. Giddings, *Sep. Sci.*, **4**, 181 (1969).

[15] J. C. Giddings, *J. Gas Chromatogr.*, **2**, 167 (1964).

[16] J. C. Giddings, *Anal. Chem.*, **35**, 353 (1963).

[17] B. L. Karger, *J. Gas Chromatogr.*, **5**, 161 (1967).

[18] A. S. Said, *Sep. Sci. Technol.*, **13**, 647 (1978).

[19] J. C. Giddings, *Anal. Chem.*, **39**, 1927 (1967).

[20] E. Grushka, *Anal. Chem.*, **42**, 1142 (1970).

[21] J. C. Giddings and K. Dahlgren, *Sep. Sci.*, **6**, 345 (1971).

[22] P. R. Rony, *Chem. Eng. Prog., Symp. Ser.*, **68**, 89 (1972).

[23] K. de Clerk and C. E. Cloete, *Sep. Sci.*, **6**, 627 (1971).

[24] G. H. Stewart, *J. Chromatogr. Sci.*, **14**, 69 (1976).

[25] E. V. Dose and G. Guiochon, *Anal. Chem.*, **62**, 174 (1990).

[26] J. C. Giddings, *Anal. Chem.*, **35**, 1999 (1963).

[27] V. Maynard and E. Grushka, *Anal. Chem.*, **44**, 1427 (1972).

[28] J. P. Foley, *Anal. Chem.*, **59**, 1984 (1987).

[29] A. N. Papas and T. P. Tougas, *Anal. Chem.*, **62**, 234 (1990).

[30] P. Rouchon, M. Schonauer, P. Valentin, and G. Guiochon, *Sep. Sci. Technol.*, **22**, 1791 (1987).

[31] D. J. Wilson, *Sep. Sci. Technol.*, **22**, 1835 (1987).

EXERCISES

5.1(**)   Find the *second moment around the mean* in the x coordinate system, $\langle (x - \bar{x})^2 \rangle$, for the Gaussian, $c = \text{const} \cdot \exp(-x^2/4Dt)$. From integral tables one finds that

$$\int_0^\infty x^2 \exp(-ax^2)\, dx = (\pi/16a^2)^{1/2}$$

5.2(**)   Solute is introduced into a chromatographic column as a square-shaped concentration pulse, which occupies an initial width w on the column. Calculate the second moment about the mean (the variance σ^2) in terms of w. Calculate σ for $w = 1.0$ cm.

5.3(***)   Derive an expression for the variance σ^2 for a symmetrical (around $y = 0$) triangular concentration pulse of base width w. Calculate σ for $w = 1.0$ cm.

5.4(**)   In the text it is stated that all odd statistical moments are zero for a Gaussian or any other symmetrical function about $y = 0$. Prove that this is so.

5.5(*)   The diffusion coefficient of glycerol in water at 15°C is 7.2×10^{-6} cm^2/s. If a narrow pulse of glycerol can be approximated by a δ-function at time $t = 0$, to what effective zone width 4σ will this pulse spread by diffusion in (a) one second? (b) one minute? (c) one day?

5.6(*)   The diffusion coefficient of isopropanol vapor in helium at 150°C is 0.677 cm^2/s. If a narrow pulse of isopropanol can be approximated by a δ-function at time $t = 0$, to what effective zone width 4σ will this pulse spread by diffusion in (a) one second? (b) one minute? (c) one day?

5.7(**)   The Gaussian exponential function, $\exp(-y^2/2\sigma^2)$, remains finite for all finite values of coordinate y. If interpreted literally, this means that absolute purification is not possible with Gaussian zones because, no matter what their separation, each is contaminated by the residual finite concentration of the other. To get perspective on this problem, assume that the concentration at the zone center ($y = 0$) is 1 mol/L. How far out along the y coordinate, expressed as a multiple of σ, must one go before the concentration falls to 1 molecule/L, an utterly negligible concentration by any measure?

5.8(*)   Repeat the above calculation for the exponential profile $\exp(-y/l)$, replacing multiples of σ with multiples of l.

5.9(**)   It is clear that as one moves outward from the center of a Gaussian zone, concentration falls off. It is less obvious that this concentration falls off at an accelerating pace. Imagine moving out from $y = 0$ in steps of length σ so that after n steps, $y = n\sigma$. Show that the concentration c_{n+1} observed after taking step $n + 1$ compared to the concentration c_n encountered in the previous (or nth) step is given by

$$\frac{c_{n+1}}{c_n} = \exp[-(n + 1/2)]$$

Construct and briefly interpret a table showing this ratio for all steps from $n = 0$ to $n = 10$.

5.10(*)   In a typical liquid, molecules jump randomly to one side or another about 10^{10} times per second in steps of about 3 Å length. Calculate the diffusion coefficient for such molecules. (For liquids, D commonly ranges from 10^{-6} to 2×10^{-5} cm^2/s.)

5.11(*)   A thin pulse of ethyl alcohol carried along at 0.040 cm/s by a uniformly flowing stream of water at 25°C spreads by diffusion, $D = 1.24 \times 10^{-5}$ cm^2/s. What is the plate height corresponding to this spreading?

5.12(**)   A band of molecules passing through a nonadsorptive packed bed of length 30 cm at an average velocity of 1.00 mm/s undergoes band broadening due to random velocity fluctuations at each particle encountered by a molecule in its flow path. (a) If the particle diameter is $d_p = 10$ μm, and each encounter is assumed to lead to a random step of length d_p, estimate the effective diffusion coefficient for the process. (b) Estimate the plate height contribution of this process. (c) Estimate the number of theoretical plates generated in the packed bed.

5.13(**)   If an injected δ-function at the head ($x = 0$) of a uniform 50-cm long separation column widens to a Gaussian with $\sigma = 2.0$ mm after migration at velocity 0.20 cm/s to $X = 4.0$ cm, what is the standard deviation τ in time units upon eluting from the column?

5.14(**)   Two components, each with plate height $H = 0.0025$ cm, are observed to migrate to positions $X = 10.1$ cm and $X = 9.9$ cm, respectively, along a uniform separation column. How long must the column be to achieve unit resolution?

5.15(**)   The plate height measured in terms of elution parameters is $H = L\tau^2/t_r^2$, Eq. 5.48. Prove that this is equivalent to $H = L\tau_{1/2}^2/(2 \ln 2)t_r^2$ as given by Eq. 5.49.

5.16(**)   Prove that the distance from the center of a Gaussian to the point of inflection is exactly σ.

5.17(**)   Prove that a tangent to the point of inflection of a Gaussian peak will intercept the baseline a distance 2σ from center. (Therefore the distance between points where tangents on each side intercept the baseline is $w = 4\sigma$, the effective zone width.)

5.18(**)   Toss a coin 80 times, recording the sequence of heads and tails. Break this sequence into 8 successive groups of 10 tosses, each group representing a random walk of 10 steps in which a step forward of unit length is taken with each head, and a unit step backward with each tail. Calculate σ for this somewhat limited data set and compare to the theoretical value of σ for an infinite number of random walks rather than the eight recorded here. (You may assume that the mean displacement is zero.)

5.19(*)   What is the final σ for four independent random walks having σs of 1, 2, 3, and 4 mm, respectively?

5.20(*)   Two independent random walks acting in combination lead to $\sigma = 2$ mm. One random walk has 100 steps of length 0.1 mm. What is σ_2, the σ of the second random walk?

5.21(*)   The plate height H is found to be 0.010 cm for a particular amino acid in an ion exchange chromatography column. What is the zone width, 4σ, after the amino acid has migrated 25 cm? What is 4σ after a migration of 100 cm?

5.22(*)   A pentane peak elutes from a 1-m GC column in 200 s. Just before elution its "width," 4σ, is 8 cm. What is the standard deviation, τ, of the eluted peak in time units? Calculate the plate height in two ways, first based on the σ value and second on the τ value.

5.23(**)   Given Eq. 5.48, prove the validity of Eq. 5.49.

5.24(**)   What is the signal height midway between two Gaussian peaks of identical width relative to the height of the taller peak, which is twice the height of the shorter peak? The resolution between peaks is 1.50.

5.25(***)   Prove the assertion in the text that two overlapping identical Gaussian peaks display two maxima only when $R_s > 0.5$.

6

STEADY-STATE, TWO-DIMENSIONAL, AND OVERLAPPING ZONES

Chapter 5 described the important details of dynamic (nonsteady-state) zone formation in one-dimensional systems. Omitted were other important aspects of zone formation and structure, including steady-state zones, two-dimensional zones, and the problem of statistical zone overlap. These topics will be examined in this chapter.

6.1 STEADY-STATE ZONES

The zone dynamics outlined in Chapter 5 describe concentration pulses in translation through rather uniform homogeneous separation systems in which diffusion and various effective diffusion processes act continuously to increase zone width. Such dynamics are characteristic of chromatography, zone electrophoresis, rate-zonal sedimentation, and field-flow fractionation. However, in other separation systems, restraints on transport in the form of intervening walls and focusing forces exist which (a) bring zone migration to a halt and (b) oppose zone broadening. After a period of time, the ubiquitous zone-broadening tendencies of diffusion reach a balance with the zone-narrowing tendencies of the focusing influences and a steady-state configuration is reached. We discuss the nature of such steady-state zones below.

In some steady-state methods of separation (isoelectric focusing, density-gradient centrifugation, and sometimes elutriation), component zones approach a stationary configuration centered about different points in space. Separation occurs by virtue of the different steady-state positions of the various solutes. In other systems (field-flow fractionation, zone refining, ultrafiltration, and reverse osmosis), steady-state zones or layers are formed next to a wall or interface. Separation then occurs by differential displacement along the wall or differential permeation through it. In each case the efficacy of the separation hinges directly on the distribution of solute in the steady-state zones and layers. It is therefore important to understand the factors underlying the formation and structure of these layers.

The unified treatment of this subject, presented below, is patterned after a paper previously published by the author [1].

A steady-state zone or layer is defined as one in which the concentration profile remains constant (or very close to constant) with the passage of time. Thus a necessary condition for the steady state is

$$\frac{dc}{dt} = 0 \tag{6.1}$$

which must apply everywhere in the zone.

The reader may recall that the equation of continuity, Eq. 3.28, relates the derivative of flux density J to the accumulation rate dc/dt: $dJ/dx = -dc/dt$. Since dc/dt is zero, dJ/dx must also be zero, a condition that is possible only for

$$J = \text{const} = J_0 \tag{6.2}$$

Thus in the steady state J must assume a value J_0 that is constant at all points along the transport path. In particular cases J_0 can be positive, negative, or zero.

A general equation that relates J to c and its derivative is

$$J = Wc - D_T \frac{dc}{dx} \tag{6.3}$$

which has been adapted from Eq. 3.32 by replacing molecular diffusion coefficient D with the sum total D_T of all effective diffusion coefficients, as in Eq. 5.28. Because J equals the constant J_0, Eq. 6.3 becomes

$$J_0 = Wc - D_T \frac{dc}{dx} \tag{6.4}$$

which is a linear first-order differential equation having a general solution expressing c as a function of coordinate x [1]. We derive here a special case of the general solution adequate for our purposes.

It is useful to rearrange Eq. 6.4 to the form

$$c - \frac{J_0}{W} = \frac{D_T}{W} \frac{dc}{dx} \tag{6.5}$$

and then to define a new concentration scale

$$c' = c - \frac{J_0}{W} \tag{6.6}$$

such that c' equals the left-hand side of Eq. 6.5. If we assume that J_0/W is a constant—true either when W is constant or J_0 is zero—we can equate the two derivatives

$$\frac{dc}{dx} = \frac{dc'}{dx} \tag{6.7}$$

Equation 6.5 then simplifies to

$$c' = \frac{D_T}{W} \frac{dc'}{dx} \tag{6.8}$$

If D_T/W is assumed independent of concentration (but not necessarily independent of position), the separation of variables and integration of Eq. 6.8 between arbitrary limits yields

$$\int_{c_0'}^{c'} \frac{dc'}{c'} = \int_{x_0}^{x} \left(\frac{W}{D_T}\right) dx \tag{6.9}$$

We now identify the arbitrary coordinate value x_0 with some reference position in or near the zone, usually at the center of the zone or at the boundary of the wall or physical barrier confining the zone. We then define a coordinate system specifically fixed to the zone analogous to that of Eq. 3.36

$$y = x - x_0 \tag{6.10}$$

such that $y = 0$ at the reference position just noted, $x = x_0$. When this relationship and $dy = dx$ are used on the right-hand side of Eq. 6.9 and the left-hand side is integrated, we get

$$\ln \frac{c'}{c_0'} = \int_0^y \frac{W}{D_T} dy \tag{6.11}$$

With the aid of Eq. 6.6 this can be made explicit in c

$$\ln \frac{c - (J_0/W)}{c_0 - (J_0/W)} = \int_0^y \frac{W}{D_T} dy \tag{6.12}$$

where c_0 is the concentration at $y = 0$. We can write this in the form

$$\frac{c - (J_0/W)}{c_0 - (J_0/W)} = \exp\left(\int_0^y \frac{W}{D_T} dy\right) \tag{6.13}$$

Equation 6.13 is the desired solution to Eq. 6.4, subject only to the two conditions that J_0/W is a constant and W/D_T is independent of c.

In most cases of interest, velocity W can be written in the form

$$W = -ay^n \tag{6.14}$$

where the negative sign reflects the focusing motion that brings any component with a positive y value back toward $y = 0$. We note that if J_0/W is to be constant as required, either J_0 must be zero or n must be zero so that W is a constant. The two cases that interest us most are $n = 0$ and $n = 1$, discussed below.

With Eq. 6.14 and the assumption of constant diffusivity D_T, the integral of Eqs. 6.12 and 6.13 acquires the form

$$\int_0^y \frac{W}{D_T}\, dy = -\frac{ay^{n+1}}{D_T(n+1)} \tag{6.15}$$

When this form is substituted back into Eq. 6.13, we get

$$\frac{c - (J_0/W)}{c_0 - (J_0/W)} = \exp\left(-\frac{ay^{n+1}}{D_T(n+1)}\right) \tag{6.16}$$

which shows that steady-state zones tend to acquire some kind of exponential distribution, the form depending on whether $n = 0$ (which gives a simple exponential) or $n = 1$ (Gaussian). The Gaussian of the latter case is largely unrelated to the Gaussians formed in uniform translation, discussed at length in Chapter 5.

6.2 EXAMPLES OF STEADY-STATE ZONES

We now concentrate on three classes of separations in which steady-state zones are described by Eq. 6.16. The specific separation techniques are discussed in more detail in subsequent chapters.

Steady-State Zones in Free Space (Isoelectric Focusing and Isopycnic Sedimentation)

Isoelectric focusing and some of its relatives are methods in which each distinct component is forced toward a unique position (which we take to be $y = 0$) along the coordinate axis of the system [2]. However, even as the component is driven toward the designated position, it is diffusing outward in opposition to the focusing force, establishing a steady-state concentration profile (see Figure 6.1). For this steady-state condition, we can write $J_0 = 0$ because no component enters or leaves the zone after its formation.

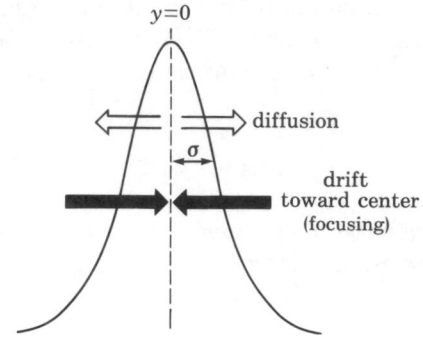

Figure 6.1. Steady-state Gaussian zone formed in methods such as isoelectric focusing and isopycnic sedimentation by the opposing interplay of a focusing force and diffusion. Different components focus at different locations to give separation.

For such zones we can assume $D_T =$ constant. Furthermore, for each zone we can (2) assume $W = -ay$, which is equivalent to Eq. 6.14 with $n = 1$. The linearity of the expression $W = -ay$ suggests that a linear restoring force or its equivalent is acting on the component molecules or particles to force them toward the origin at $y = 0$. The nature of this force for particular methods will be discussed in Chapter 8.

Mathematically, the above conditions reduce Eq. 6.16 to

$$\frac{c}{c_0} = \exp\left(-\frac{ay^2}{2D_T}\right) \tag{6.17}$$

which is, of course, a Gaussian profile with variance

$$\sigma^2 = \frac{D_T}{a} \tag{6.18}$$

The constants D_T and a can usually be estimated on physicochemical grounds, as shown in Chapter 8. Thus zone width can be calculated and resolution parameters estimated [2].

Steady Layers Formed at a Wall (Field-Flow Fractionation and Equilibrium Sedimentation)

In field-flow fractionation, a component undergoing flow transport through a thin channel is forced sideways against a wall by an applied field or gradient. The component is confined to a narrow region adjacent to the wall by a combination of the wall's surface, which it cannot pass, and the driving force, which prevents its escape toward the center of the channel. The component molecules or particles soon establish a thin steady-state distribution in which outward diffusion balances the steady inward drift due to the field. The structure and dimensions of this layer determine its behavior in the separation process.

Similar layers can be formed at the outside wall of a centrifuge tube when the centrifugation process is allowed to proceed to equilibrium.

For the above cases, the appropriate assumptions are $J_0 = 0$, $D_T =$ molecular diffusion coefficient D (usually a constant), and $W = -|W|$ (a constant). Quantity W is written as $-|W|$ to emphasize that it is negative: displacement occurs along the negative coordinate axis toward the wall. Since W is constant, the exponent n of Eqs. 6.14–6.16 is zero, and the resulting distribution is a simple exponential rather than a Gaussian. Equation 6.16 yields the form

$$\frac{c}{c_0} = \exp\left(\frac{-|W|y}{D}\right) = \exp\left(\frac{-y}{\ell}\right) \tag{6.19}$$

where the effective mean layer thickness ℓ is given by

$$\ell = \frac{D}{|W|} = \frac{D}{|U + v|} \tag{6.20}$$

Quantity ℓ has a profound effect on component migration and column efficiency in field-flow fractionation [3] and on system behavior in equilibrium sedimentation as well. We note that in field-flow fractionation the steady-state zones formed along axis y are transported dynamically along the flow axis at right angles to y. The connection between the steady and nonsteady processes is that the steady-state parameter ℓ determines the velocity of nonsteady migration at right angles to y. This matter is discussed in Chapter 11.

Steady Layers Formed at a Semipermeable Barrier with Influx (Ultrafiltration, Reverse Osmosis, Pressure Dialysis, and Zone Melting)

In the filtration-type methods (the first three techniques listed above), components accumulate as a steady-state (polarization) layer at a barrier or membrane [4]; this occurs in much the same way as in field-flow fractionation or equilibrium sedimentation. However, there are several complications. First, fresh solute is constantly brought into the layer by the flow of liquid toward and through the filter. This steady influx of solute components can be described by a finite flux density term J_0. Second, components can be removed from the outer reaches of the layer by stirring. Third, the membrane or barrier may be leaky and thus allow the transmission of a portion of the solute, profoundly affecting the attempted separation. In fact, one reason for our interest in layer structure is that leakiness depends on the magnitude of the solute buildup at the membrane surface. As solute concentration at the surface increases, more solute partitions into the membrane and is carried on through by flow.

A similar situation exists in zone melting where a molten zone passes through a solid bar gathering up impurities [5]. The trailing solid-liquid interface, which advances by refreezing, acts like a filter in differentially rejecting the impurities. The rejected components then accumulate in a thin layer next to the refreezing surface. The amount later recaptured and refrozen into the solid (equivalent to the leakage)—and thus the impurity content of the recrystallized solid—depends again on solute buildup. (It makes little difference in filtration or zone melting whether the solute-rejecting surface advances on stationary liquid solution or whether liquid advances on a stationary surface; only the relative motion is important. Usually, in zone melting, the zone and thus the interface is stationary, and the solid bar is in translation.)

With the initiation of the filtration or zone-melting process, solute begins to accumulate at the barrier and continues to build up in the solution until one of several things happens: precipitation or gelation occurs, the procedure is stopped, or stirring or leakage removes solute as rapidly as it arrives. It is not our object to describe all of these special cases here. Instead, we will discuss a simple steady-state model in which solute leakage reaches the same level as solute influx. The layer structure calculated on this basis is a reasonable approximation to the transient layer that exists when leakage is negligible. Every aspect of the separation, from plausible operating conditions (flowrate, duration, pressure, stirring methods) to the resulting level of concentration or purification, depends on the nature and form of this layer.

The limiting equations applicable at low concentrations are obtained by assuming a finite but negative J_0 which we write as $-|J_0|$, where $|J_0|$ is the influx, D_T is a constant (which equals molecular diffusion coefficient D if there is no stirring or convection), and $W = -|v|$, another constant equal to the relative velocity of the liquid and the surface. The absolute value forms are used to allow the positive coordinate axis to point into the flow from the barrier or interface to which we assign coordinate position $y = 0$.

With these assumptions, Eq. 6.16 becomes applicable (with $n = 0$) and yields

$$\frac{c - (|J_0|/|v|)}{c_0 - (|J_0|/|v|)} = \exp\left(-\frac{|v|\,y}{D_T}\right) \qquad (6.21)$$

Writing $|J_0|/|v|$ as J_0/v (both terms are negative and the ratio is therefore positive) and rearranging, we get

$$c = \frac{J_0}{v} + \left(c_0 - \frac{J_0}{v}\right)\exp\left(-\frac{|v|\,y}{D_T}\right) \qquad (6.22)$$

We note that the distribution is a simple exponential superimposed on the constant background concentration of the solute, J_0/v (see Figure 6.2). The effective thickness ℓ of the exponential component is seen to be identical in form (but with D_T replacing D) to that found for field-flow fractionation, Eq. 6.20: $\ell = D_T/|v|$.

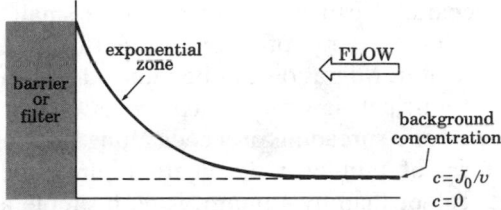

Figure 6.2. Formation of exponential zone superimposed on a background of constant concentration for solute piled up behind a partially rejecting barrier such as a filter.

The amount of excess solute E (in moles per unit area of filter) that can accumulate in the polarization layer is the excess concentration above that of the background solution, $c - (J_0/v)$, integrated over coordinate y

$$E = \int_0^\infty \left(c - \frac{J_0}{v} \right) dy \qquad (6.23)$$

With the substitution of Eq. 6.22 for c and integration, Eq. 6.23 yields

$$E = \frac{D_T}{|v|} \left(c_0 - \frac{J_0}{v} \right) = \ell \left(c_0 - \frac{J_0}{v} \right) \qquad (6.24)$$

Usually the separation is effective in proportion to E because E represents either the amount of contaminant removed if the rejected solute is undesirable or the amount of product concentrated if the solute is desirable. However, the degree to which the concentration at the interface c_0 can be increased to this end is limited because high values augment leakage, resistance to flow, and the risk of precipitation. Consequently, to increase E, efforts are generally made to increase effective diffusivity D_T, which is best done through stirring or convective processes. Thus these processes become important considerations to effective operation.

In summary, the mathematical similarity of different steady-state solute zones, critical to the success of seemingly unrelated separation methods, demonstrates the impressive unifying power of the basic transport approach to chemical separations. This unity is emphasized again in the next chapter, where we delve into the classification and comparison of separation methods.

6.3 ZONE FORMATION IN TWO DIMENSIONS

Chapter 5 dealt with the evolution of thin starting zones into concentration pulses distributed along a single axis. This treatment is appropriate for some important analytical separation methods, such as column chromatography carried out in thin one-dimensional tubes. However, as an alternate method the sample to be separated can be deposited as a small spot on a two-dimensional bed (such as a strip of paper or a thin layer of adsorbent material coated on a plate). Migration can be induced (by an applied field or by flow) along the principal separation axis x as shown in Figure 6.3. However, in this case zone spreading proceeds along both planar axes. This approach, in the form of thin-layer chromatography, where flow can be driven by capillarity rather than by a pump, is both simple and inexpensive. It has the advantage that a number of sample spots can be started simultaneously at different points along perpendicular axis y, giving multiple analyses in one run.

The spots of sample material are simply two-dimensional zones, with concentration highest at the center and falling off as one goes outward along either x or y axes. Various diffusion processes act on the gradients, both along x and y, leading to zone broadening in both dimensions. Axial (x-axis) broadening is usually largest, and is most detrimental because it leads to peak overlap, as in one-dimensional cases. Lateral broadening (along y) is also undesirable because it leads to the dilution of solute, thus affecting its detectability. Lateral broadening also interferes with true two-dimensional separations, which we shall discuss in the next section.

The two different diffusion processes act independently to form a Gaussian concentration profile along each axis, the same as along the single axis in one-dimensional systems. When a small spot containing n moles (or molecules) spreads simultaneously in two directions with diffusion coefficients D_x and D_y, the resulting concentration distribution $c(x, y)$ expressed in moles (or molecules) per unit area is given by n times the product of the two normalized Gaussian functions $c(x)$ and $c(y)$ [6]

$$c(x, y) = nc(x)c(y) \tag{6.25}$$

When $c(x)$ and $c(y)$ are both written in the Gaussian form of Eq. 5.18, we get

$$c(x, y) = \frac{n}{4\pi \bar{D}t} \exp\left[-\frac{(x - X)^2}{4D_x t} - \frac{(y - Y)^2}{4D_y t} \right] \tag{6.26}$$

where $\bar{D} = (D_x D_y)^{1/2}$. If we replace $2D_x t$ by

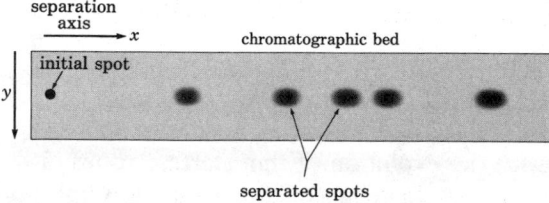

Figure 6.3. Separation of spot-like zones on a two-dimensional bed.

$$\sigma_x^2 = 2D_x t \tag{6.27}$$

and similarly

$$\sigma_y^2 = 2D_y t \tag{6.28}$$

both in accordance with Eq. 5.21, we get the bi-Gaussian distribution

$$c(x, y) = \frac{n}{2\pi\sigma_x\sigma_y} \exp\left[- \frac{(x - X)^2}{2\sigma_x^2} - \frac{(y - Y)^2}{2\sigma_y^2} \right] \tag{6.29}$$

A plot of Eq. 6.29 is shown in Figure 6.4. Note that the grid lines form Gaussian peaks. The ratio of σ_y/σ_x is 2.5. Thus the grid lines cutting through the peak along the y axis display Gaussian profiles that are 2.5 times broader than the Gaussian traced out along x-axis grid lines.

We have learned that Gaussian profiles have no sharp boundaries but instead trail off to infinity with steadily decreasing concentration. However, a visible spot on a two-dimensional chromatographic or electrophoretic bed often appears to have a more or less distinct boundary. The spot boundary is the line around the spot beyond which concentration drops below the level of visibility, or the level of detectability if instrumental detection is used (see Figure 6.5). The boundary is ordinarily rather sharp because Gaussian profiles drop rapidly in concentration at their outer extremities. If one assumes that the limit of visibility or detectability is concentration c_0, it can

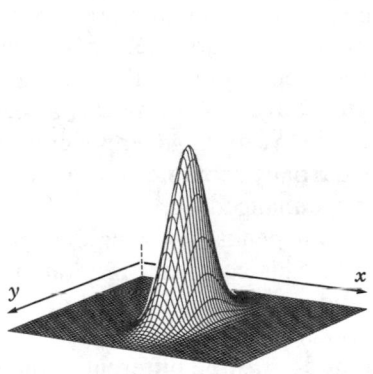 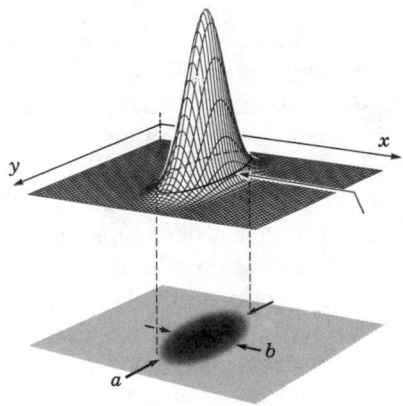

Figure 6.4. Bi-Gaussian peak, described by Eq. 6.29, with $\sigma_y/\sigma_x = 2.5$. (Computer plot courtesy of P. Stephen Williams, University of Utah.)

Figure 6.5. The visible boundary of this bi-Gaussian distribution is the line encircling its base at the threshold level of visibility. This boundary line forms an ellipse; thus spots appear generally to be elliptical in shape.

be shown that the asymmetric double Gaussian of Eq. 6.29 displays itself as an elliptical spot [6]. The length a of the spot along axis x and the breadth b along axis y are given by

$$a = 4[D_x t \ln(n/4\pi c_0 \bar{D} t)]^{1/2} = 2\sigma_x [2\ln(n/2\pi c_0 \sigma_x \sigma_y)]^{1/2} \qquad (6.30)$$

$$b = 4[D_y t \ln(n/4\pi c_0 \bar{D} t)]^{1/2} = 2\sigma_y [2\ln(n/2\pi c_0 \sigma_x \sigma_y)]^{1/2} \qquad (6.31)$$

The area A of the elliptical spot, given by $\pi(a/2)(b/2)$, therefore equals

$$A = 4\pi \bar{D} t \ln(n/4\pi c_0 \bar{D} t) = 2\pi \sigma_x \sigma_y \ln(n/2\pi c_0 \sigma_x \sigma_y) \qquad (6.32)$$

This equation shows that visible spot area A increases in proportion to the logarithm of the sample size n. This observation can be applied to quantitative analysis: a plot of A versus $\ln n$ (or $\log n$) yields a straight line from which values of n can be read for different experimental A values [7].

6.4   TWO-DIMENSIONAL SEPARATIONS

Various techniques exist and more are under development for separating components along two independent coordinate axes instead of one. The principal motivation for this development is that separation power is greatly multiplied by the addition of a second separation coordinate. The two-dimensional approach accordingly has great promise for future use [8].

Two-dimensional (2D) separations are typically carried out on or within a support structure having a rectangular configuration. The mixture is generally started as a small spot in one corner. It is then subjected to separative transport along one edge, leading to partial separation in the first dimension, which is taken to be axis x (see Figure 6.6). Following this a second driving force is applied along perpendicular axis y, forcing components away from the initial edge (along which they are partially separated) into the body of the rectangle where separation is further enhanced.

The distribution of a number of component peaks following 2D separation is shown more pictorially in Figure 6.7. This computer-generated plot shows seven peaks of different heights and σs. Two of the peaks show partial overlap, a common occurrence in practice (see Section 6.6).

The mechanism of separation in 2D methods *must* be different in the two directions. Otherwise, components that fail to resolve from one another along the first axis will be transported identically along the second axis and will fail to resolve along that axis also. If two specified components are capable of resolution by *either* of two independent separation mechanisms, then separation generally will be achieved in the 2D system. For complex mixtures having so many components that neither separative mechanism acting alone is adequate to resolve the mixture, the combination of mechan-

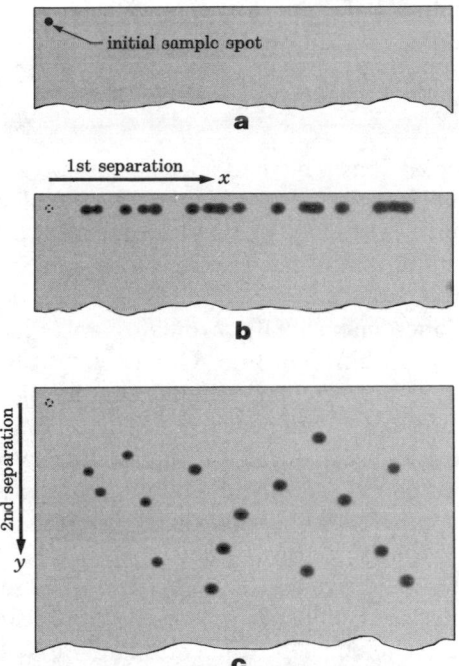

Figure 6.6. Sequential two-dimensional separation. Displacement and partial resolution of original sample spot along axis x is followed by a second displacement and much enhanced resolution along axis y.

isms provided by 2D separation is often effective. The mechanisms can lead to the development of either nonsteady-state zones or steady-state zones along their respective coordinates, or a combination of the two.

The method described above is called a *sequential* 2D separation because the two independent separation processes, one for each axis, are carried out in sequence (9). We have noted that the two separations must be independent (i.e., based on different factors); otherwise spots that failed to separate

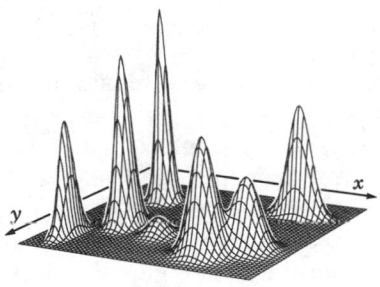

Figure 6.7. Separation of seven component peaks along the two axes of a 2D system. (Computer simulation courtesy of P. Stephen Williams, University of Utah).

on the first axis would also fail on the second, leading to many fused spots. If 2D chromatography is used, independence is provided by using two different mobile phases (solvents) along the two axes. If 2D electrophoresis is used, independence comes from the use of two different electrophoretic mechanisms.

Sequential 2D separations have had spectacular success in the separation of complex biological fluids using two highly effective electrophoretic mechanisms: isoelectric focusing along the first coordinate and gel electrophoresis (which separates by a different mechanism) along a second axis. Over 1000 proteins have been separated in a single run by this method (10–12), far more than existing (one-dimensional) chromatographic methods can provide (see Chapter 8). Along a different line, Guiochon and his co-workers have examined the 2D approach to column liquid chromatography [13, 14]. The column consists of a channel confined between two rectangular plates filled with an appropriate packing material for liquid chromatography. With one solvent, the sample is partially resolved along the first axis; with the second, the components are carried all the way through the second dimension and swept out (eluted) through a series of detectors.

Following a different approach, two *simultaneous* separative displacements are sometimes used in place of two sequential displacements. For example, by an old technique [15, 16], chromatography is used along one axis while electrophoresis is proceeding along the other. However, despite their experimental simplicity, simultaneous displacements will not work well in many circumstances, including those in which two chromatographic mechanisms are combined [8].

When the sample is applied as a single spot and separated into component spots along two axes, as described above, we are employing a *discrete* form of 2D separation. This form provides the enhanced resolving power so desirable for analytical-scale separations. Such discrete operations can be carried out with either sequential or simultaneous displacements, as noted above. However, there is another major class of 2D methods termed *continuous* separation. Here, a steady stream of sample is applied at a point along one edge. The stream is subjected to two right-angle displacement forces, which in this case must act simultaneously. The stream is broken by these forces into continuous filaments that strike off at different angles over the 2D surface. The separated filaments are then collected at different points along the opposite edge. The continuous throughput is advantageous for preparative-scale operation.

An old but still important example of continuous 2D separation is *deflection electrophoresis*. Here flow carries the sample stream in one direction while electrophoresis at right angles causes the differential deflection and separation of the stream into component filaments (see Chapter 8).

Chromatography can also be employed in a continuous 2D mode. An example is rotational chromatography, an old concept [17, 18] subject to ongoing developments (19). This technique is illustrated in Figure 6.8. Here

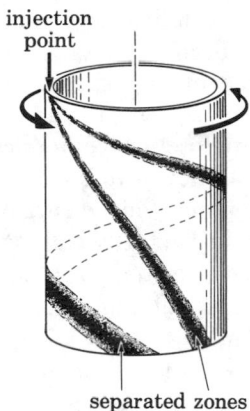

injection point

separated zones

Figure 6.8. Separation of two components by rotating continuous chromatography unit. (From ref. 18. Reprinted with permission from J. C. Giddings, *Anal. Chem.*, **34**, 37 (1962). Copyright 1962 American Chemical Society.)

the constant rotational displacement of an annular bed of packing combines with selective chromatographic displacement at right angles to split the sample stream into component filaments, each continuously collected upon emergence.

Other aspects of continuous separations have been covered in reviews by this author and by Wankat [8, 9].

The 2D approach to separation offers not only a great increase in separation power over one-dimensional (1D) methods, but also greater versatility. We have noted that 2D separation requires the use of pairs of 1D displacements. If N kinds of 1D displacements can be employed, then $\sim N^2$ different pairwise combinations can be found for 2D use. For example, dozens of 2D methods can be envisioned that use a field-flow fractionation (FFF) mechanism; these methods fall in four categories in which a given FFF mechanism can be combined with (1) another FFF subtechnique, (2) a form of chromatography, (3) an applied field (e.g., electrical), and (4) bulk flow displacement [20]. For separations generally, literally thousands of kinds of 2D separation systems are possible, although only a handful have been developed [8].

While 2D separations are technically more difficult to carry out than those in one dimension, and often subject to less convenient detection methods, their intrinsic resolving power and potential scope are sufficiently greater than those of one-dimensional systems that many important gains can be expected in the future.

From a broad perspective, 2D separation may be thought of as a subclass of *multidimensional separation methods*. In general, multidimensional techniques may be defined as those (1) employing two or more separative mechanisms and (2) achieving separation providing any one of the individual mechanism yields separation [21, 22]. A multidimensional system that is strikingly different from a 2D bed is based on the *coupled column* (or more generally the *coupled stage*) *method* in which fractions separated by one (a primary) column are fed into one or more secondary columns and thus subjected to a different separation mechanism. More specifically, samples are partially resolved in the primary column, then segments of sample stream with unresolved material are switched to another column for additional resolution. In this way, components traveling so closely together in one column that they cannot be resolved are subjected to new forces in the second column which often initiate differential migration and thus eventual resolution [21–26]. Components not resolved in two such stages can be subjected to a third or fourth separation stage, effectively providing a three- or four-dimensional separation. The means and sequences that can be used in coupling successive stages are enormously varied [22].

Coupled column systems are rapidly gaining adherents in chromatography [23–30]; initial use has been reported in field-flow fractionation as well [31]. The methodology is particularly promising for isolating active components from complex biological and environmental samples. The reader is referred to a recent book for more detail [32].

6.5   RESOLVING POWER IN TWO-DIMENSIONAL SEPARATIONS

To better understand the heightened resolving power of 2D systems, we need some measure by which 2D and 1D separations can be compared. Not all criteria of separation power lend themselves to ready comparison. The resolution of a specific pair is not a suitable criterion because this resolution varies widely for different separation mechanisms irrespective of 1D or 2D configurations. Plate height and plate number are not directly comparable because these are defined only for a single dimension. While H or N values can be found for each axis, it is not immediately obvious how to combine them for both axes in order to compare the overall separation effectiveness with that of a 1D system.

For steady-state zones, where H and N also lack definition, we turned to the peak capacity as a common denominator for different methods. We have learned how to estimate peak capacity for 1D separations; we now extend this concept to incorporate two axes. For this we must reconsider the matter of spot dimensions when migration occurs along both axes rather than just one.

Recall that a zone migrating along one axis in a 2D matrix ideally forms an elliptical zone or spot. The dimensions of the spot, as shown by Eqs. 6.30 and 6.31, are controlled by zone broadening along both axes x and y, measured by σ_x and σ_y, respectively. When two migrations occur in sequence along the two axes, both migration processes contribute independently to zone broadening. For spreading along axis x we have

$$\sigma_x^2 = \sigma_{x_1}^2 + \sigma_{x_2}^2 \tag{6.33}$$

where subscripts 1 and 2 denote the two successive migration processes. Likewise, for axis y

$$\sigma_y^2 = \sigma_{y_1}^2 + \sigma_{y_2}^2 \qquad (6.34)$$

As far as zone and spot dimensions for sequential two-dimensional separations are concerned, the equations of the last section are applicable if we use the two equations above to obtain σ_x and σ_y, respectively.

The peak capacity n_c, as usual, is the number of zones that can be crowded into the available separation space. However, in two dimensions "space" is measured by area, not length. We can thus estimate n_c as bed area $L_1 L_2$ over spot area A

$$n_c \sim \frac{L_1 L_2}{A} \qquad (6.35)$$

where L_1 and L_2 are the lengths of the two sides of the rectangular bed. Spot area A is given by Eq. 6.32. However, since elliptical spots cannot be packed without vacant space, it is more realistic to imagine each ellipse of length a and width b as occupying a small rectangle, as shown in Figure 6.9. The number of small rectangles of area $a \times b$ that fit in large rectangular area $L_1 \times L_2$ is

$$n_c \simeq \frac{L_1}{a} \times \frac{L_2}{b} \qquad (6.36)$$

This peak capacity, too, is only approximate because some spot overlap may be tolerable (giving an increase in n_c) depending upon analytical goals and instrumentation. Also, a slightly denser packing is possible with a staggered spacing of ellipses. However, the difference is of no great significance.

We now note that the peak capacity for a separation along the x coordinate alone is of the order of

$$n_x \sim \frac{L_1}{a} \qquad (6.37)$$

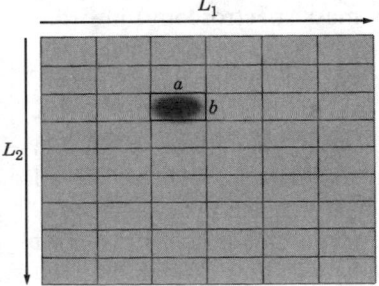

Figure 6.9. If a typical spot occupies area $a \times b$, then in area $L_1 \times L_2$ we can place $(L_2 \times L_2)/(a \times b)$ spots. This ratio is an approximation to the peak capacity.

where a, the spot length, approximates the minimum acceptable distance between spot centers. Likewise, the peak capacity along y is

$$n_y \sim \frac{L_2}{b} \tag{6.38}$$

Inspection of the last three equations shows that n_c is multiplicative

$$n_c = n_x n_y \tag{6.39}$$

Thus an approximate rule emerges for comparing 1D and 2D peak capacities: the single-dimensional peak capacities multiply together to yield the 2D peak capacity. This rule illustrates the enormous separation power of 2D (more generally, multidimensional) forms of separation.

While the multiplicative law for peak capacity is adequate as a guideline, we should be aware of its limitations. First, it requires a uniform flow pattern over the entire rectangular space. Second, the a and b in Eq. 6.36 refer to spot dimensions after migration along both axes, whereas a and b in Eqs. 6.37 and 6.38 refer to spot dimensions after migration along the separation axis only. The latter values will be smaller because diffusion encountered in the other displacement step of this two-step separation is not included in these terms. Finally, we note that some close-lying pairs of components are so chemically similar that they do not resolve well along either coordinate. In this case, the final resolution between the two spots is not as high as would be expected considering the large multiplicative peak capacity.

6.6 OVERLAPPING ZONES

Ideally, the component zones or peaks emerging from a separation process should be well isolated from one another. With isolation, each peak's center of gravity can be precisely located and used to establish or confirm the identity of the component forming the peak. The peak area or peak height can be accurately measured and used as a source of information on the amount of the component. Also, the peak contents can be collected in pure form and subjected to additional tests (such as mass spectroscopy) in order to confirm the identity of the component without risk of interference.

Unfortunately, the ideal separation, producing only well-isolated peaks, is rarely found in practice. This is particularly so with complex multicomponent mixtures. Very often there is not sufficient space along the separation coordinate (that is, there is not sufficient peak capacity) to isolate all components. However, even in circumstances in which the peak capacity is adequate to handle all components (i.e., when n_c exceeds the number of components), the components tend not to oblige; rather than filling the separation coordinate evenly, they frequently overlap strongly in some regions while leaving large tracts of vacant space scattered about elsewhere.

If peak overlap occurs, there is a loss in the quality of analytical information relative to that derived from isolated peaks. The extent of the loss depends on the extent of overlap. For slight overlap, one may simply lose information at the level of a few percent on the quantity of the components in the sample. With strong overlap, particularly between small peaks and large peaks, the small peak may be totally obscured; even its existence may be in doubt.

Reality, therefore, forces separation scientists to deal with various aspects of peak overlap. The subject is a complicated one. We present in this section a brief introduction to the nature, implications, and unexpectedly high frequency of overlap. Some theoretical methods for dealing with overlap are outlined in the following section. While this treatment is based on studies of peak overlap in chromatography [33, 34], the concepts are equally valid for electrophoresis and other zonal separation methods.

We return to the idea that a peak capacity of n_c will accommodate n_c resolved peaks if the peaks are evenly spaced (see Figure 5.11). For example, a family of homologs, depending on the technique used for separation, may have one member follow another at almost equal intervals and thus, depending on peak width, fill the separation space to capacity (see Figure 6.10 from ref. [35]). Such systematic differences in molecular structure within families can lead to simple ordered arrays of peaks or zones.

Complex samples, by contrast, consist of components from many chemical families. The members of different families tend to intermix randomly along the separation coordinate, frequently overlapping.

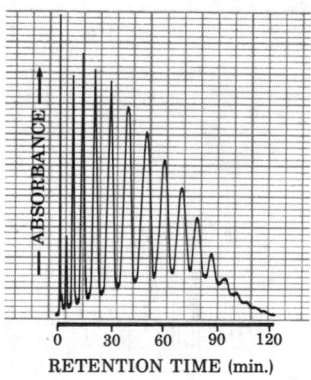

RETENTION TIME (min.)

Figure 6.10. Separation of a series of oligomers of polystyrene of average molecular weight 800 by liquid chromatography. The uniform spacing is a result of the constant difference (one monomer unit) between successive members of the series. (From ref. 35. Reprinted with permission from T.-B. Hsu, W. A. Howard and L. B. Rogers, *Sep. Sci. Technol.*, **17**, 1623 (1982–1983). Copyright 1982–1983 Marcel Dekker, Inc.)

If the centers of single-component peaks lie too close to one another, they appear as a single fused peak or zone. We call this fused peak an *apparent peak* because from all outward appearances there is only one peak, the several components being obscured by the overlap. This was illustrated by Figure 5.8 for pairs of single component peaks separated at different resolution levels. If the two component peaks are of equal height (top row of Figure 5.8), they are tightly fused under a single maximum up to a resolution of $R_s = 0.5$. Beyond $R_s = 0.5$, two maxima emerge and the existence of different component peaks becomes clear.

Following separation, the entire sequence of peaks or zones is usually revealed by a sensitive detector. The detector response plotted against the separation coordinate or elution time can be termed a *fractogram*; in the special case of chromatography the response versus time curve is a *chromatogram*. If we look at a peak on the fractogram or chromatogram of a complex mixture, we have no immediate way of knowing if the peak is a single-component peak or a closely fused peak consisting of two, three, or more single-component peaks. Fused peaks are thus often mistaken for single-component peaks, leading to erroneous analytical results.

Randomly overlapping component peaks in chromatography have been studied by Rosenthal [36], by Davis and this author [33, 34], and by Guiochon and co-workers [37, 38]. The results may be assumed to apply to other separation methods as well.

As part of this study, random peak overlap was simulated by computer-generated chromatograms in which single-component peaks of random heights were distributed at random positions along the separation coordinate [39]. An example is shown in Figure 6.11. The numbers by each apparent

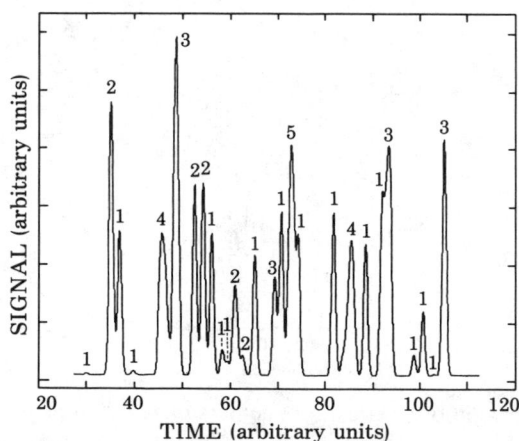

Figure 6.11. Computer-generated chromatogram of 50 randomly spaced component peaks of random peak height (150-fold range). Although this chromatogram is not unduly crowded ($\alpha = 0.5$ for $R_s^{\ddagger} = 0.5$), overlap is ubiquitous, as shown by the numbers indicating how many component peaks are associated with each observed maximum. (Simulation courtesy of Joe M. Davis, University of Southern Illinois.)

peak show how many single-component peaks have fused together under the single maximum. This synthetic chromatogram, in which the location of every component peak is exactly known, shows how fused peaks can be easily mistaken for single component peaks. This problem is serious for real-world chromatograms where component peaks are frequently obscured by overlap.

6.7 STATISTICS OF PEAK OVERLAP

With sufficiently complex samples, particularly biological and environmental samples, the frequency of overlap can be estimated by statistical means. In a statistical model developed by Davis and this author [33], far-reaching conclusions follow from a simple basic assumption: the probability that any small interval dx along the separation path x is occupied by a component peak center is $\lambda\,dx$, where λ is a constant. This assumption defines a Poisson process and leads to well-known statistical conclusions.

According to Poisson statistics the average spacing between adjacent component peaks is $1/\lambda$; the number $\bar{m}$ of components expected to occupy an interval of length X along coordinate x is thus X divided by $1/\lambda$, or

$$\bar{m} = \lambda X \tag{6.40}$$

We note that $\bar{m}$ is a statistical number related to basic constant λ; it may differ slightly from the true component number m. However, because of peak overlap, the true m cannot be obtained by counting the number of peaks appearing in the chromatogram. Thus $\bar{m}$, if obtainable, becomes our best approximation for m.

Whether two adjacent component peaks are fused or separated depends on their resolution. We assume that some critical resolution $R_s^{\ddagger}$, not necessarily unity, is needed to successfully resolve the two; the choice of $R_s^{\ddagger}$ will depend on the type and accuracy of analytical data required. To achieve the designated $R_s^{\ddagger}$, the distance ΔX between two component peaks must be some minimum value x_0. We get x_0 by using Eq. 5.52

$$x_0 = 4\sigma R_s^{\ddagger} \tag{6.41}$$

where 4σ is the average component peak width.

According to Poisson statistics, the probability that the gap ΔX between component peaks exceeds x_0 is

$$P(\Delta X > x_0) = \exp(-\lambda x_0) \tag{6.42}$$

Thus, with any given component peak, the probability that the following component peak is successfully resolved is $\exp(-\lambda x_0)$. If the two are not resolved they are considered to be fused, becoming part of a single apparent peak. Consequently, the probability that an emerging component marks the beginning of a new apparent peak is $\exp(-\lambda x_0)$. If there are approximately $\bar{m}$ components emerging in sequence, the number p of apparent peaks generated is $\bar{m}$ times this probability, or

$$p = \bar{m} \exp(-\lambda X_0) \tag{6.43}$$

Clearly p is always less than $\bar{m}$, which means we see fewer peaks (because of overlap) than we have components. The discrepancy between p and $\bar{m}$ increases with the product λx_0.

The probability that a given component peak is isolated from *all* other components requires two conditions. First, the component peak must be at least distance x_0 from the previous component peak. The probability of this is $\exp(-\lambda x_0)$ as just shown. However, the component peak must also be at least x_0 removed from the following peak, for which the probability is also $\exp(-\lambda x_0)$. The probability of isolating the component on both sides by distance x_0 or greater is the product of these two independent probabilities, or $\exp(-2\lambda x_0)$. Thus of $\bar{m}$ components, the number of isolated single component peaks or "singlets" equals

$$s = \bar{m} \exp(-2\lambda x_0) \tag{6.44}$$

By similar reasoning, the number of doublets (two fused peaks) and triplets (three fused peaks) are

$$d = \bar{m} \exp(-2\lambda x_0)[1 - \exp(-\lambda x_0)] \tag{6.45}$$

$$t = \bar{m} \exp(-2\lambda x_0)[1 - \exp(-\lambda x_0)]^2 \tag{6.46}$$

and so on [33].

The exponent λx_0 in the preceding expressions can be interpreted physically. From Eq. 6.40

$$\lambda = \frac{\bar{m}}{X} \tag{6.47}$$

which means that λ is a component peak density—the number of components per unit length of separation coordinate. Using this λ we get

$$\lambda x_0 = \bar{m} \frac{x_0}{X} \tag{6.48}$$

However, X/x_0 is simply the peak capacity n_c—the number of components spaced a distance x_0 apart able to fit in coordinate distance X. Thus

$$\lambda x_0 = \frac{\bar{m}}{n_c} = \alpha \qquad (6.49)$$

the ratio $\alpha = \bar{m}/n_c$—the number of components estimated to be present over the maximum number the system can possibly separate—is a measure of the saturation of the separation space by components. We term α the *saturation factor*.

Expressed in terms of α, the number p of apparent (observed) peaks in the chromatogram, Eq. 6.43, is simply

$$p = \bar{m} \exp(-\alpha) \qquad (6.50)$$

while the number of isolated component peaks (singlets) from Eq. 6.44 becomes

$$s = \bar{m} \exp(-2\alpha) \qquad (6.51)$$

Since the success of separation generally hinges on the isolation of singlets (which by definition are component peaks separated by minimum resolution $R_s^{\ddagger}$ on both sides), the "success ratio" is simply s over the number of components $\bar{m}$

$$\frac{s}{\bar{m}} = \exp(-2\alpha) \qquad (6.52)$$

This ratio, equal to unity when $\alpha = 0$, drops precipitously with increasing saturation, as shown by Figure 6.12. In a relatively uncrowded chromatog-

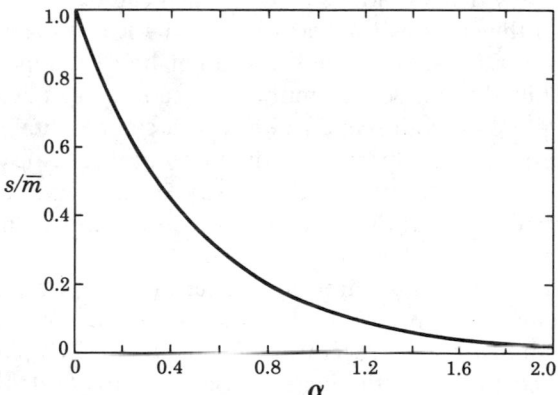

Figure 6.12. A plot showing the rapid decrease in $s/\bar{m}$ with saturation α. Since s is the number of isolated components or "singlets" and $\bar{m}$ the total number of components (some isolated, some not), the "success ratio" $s/\bar{m}$ indicates the fraction of components isolated.

ram of only 50% saturation ($\alpha = 0.5$), this ratio has dropped to $\exp(-1) = 0.368$. This means that only 36.8% of the components will be successfully isolated, or alternatively, that the probability of isolating a desired component is only 0.368. Such numbers reinforce the concept illustrated by Figure 6.11 that component peak overlap is far more pervasive than expected.

Figure 6.11 can be used as a concrete example of the foregoing concepts. This chromatogram consists of 50 components, a peak capacity of 100, and thus a saturation (all defined relative to $R_s^{\ddagger} = 0.5$) of $\alpha = 0.50$. Application of Eq. 6.50 would lead us to expect 30 apparent peaks, within a reasonable statistical range of the 27 maxima observed in the chromatogram. Equation 6.51 predicts 18 singlets; we observe 15. The success ratio is therefore only $15/50 = 0.30$, a disappointingly low number. For better success in isolating singlets we need an even higher peak capacity, although n_c is already 100, twice the value of $\bar{m}$.

The above cited numbers are all readily accessible for Figure 6.11 because it is generated by computer. Most real chromatograms do not, however, reveal m; they yield only p. Which of the p observed peaks are singlets is rarely clear, so s is also not known.

Some help in unraveling this complex situation can be found by substituting for α in Eq. 6.50 the form for $\alpha = \lambda x_0$ shown in Eq. 6.48, which gives

$$p = \exp(-\bar{m}x_0/X) \qquad (6.53)$$

This expression can be rewritten as

$$\ln p = \ln \bar{m} - \bar{m}\,\frac{x_0}{X} \qquad (6.54)$$

These equations show that variations in x_0—which can be caused by changes in peak width and thus in σ as shown in Eq. 6.41—lead to changes in p. A plot of $\ln p$ versus x_0/X should yield a straight line of slope $-\bar{m}$ and of intercept $\ln \bar{m}$. Thus the component number $\bar{m}$ can be estimated from both slope and intercept. With both p and $\bar{m}$ known, α can be obtained from Eq. 6.50 and s from Eq. 6.51. In this way all the important statistical parameters are obtained, although we still have no way of knowing which of the specifically observed peaks are singlets and which are higher multiplets.

The above procedure has been verified for computer-generated chromatograms (39). For real separations, series of chromatograms acquired at different column efficiencies (thus different values of σ) are rarely available, as required by the procedure. Fortunately, a simpler procedure has been developed to deal with single chromatograms [34]. For this, Eq. 6.54 is written in the form

$$\ln p' = \ln \bar{m} - \bar{m}\,\frac{x_0'}{X} \qquad (6.55)$$

where p', which replaces the peak count p, is the number of gaps between adjacent maxima that exceed the arbitrary distance x_0'. As x_0' is varied, p' varies and can be graphed in a $\ln p'$ versus x_0'/X plot. Again, $\bar{m}$ estimates are available from both slope and intercept, leading to the major statistical parameters noted earlier.

An example of the application of this procedure to a complex chromatogram is shown in Figure 6.13 [34]. The sample is a mixture of polynuclear aromatic hydrocarbons from river sediments fractionated by capillary gas chromatography in the laboratory of Dr. M. L. Lee [40]. A count of peak maxima yields $p = 145$. Component numbers estimated from the slope and intercept give $\bar{m} = 234$ and 267, respectively, yielding an average estimate of $\bar{m} = 250$ for the number of components, which exceeds the number of peaks

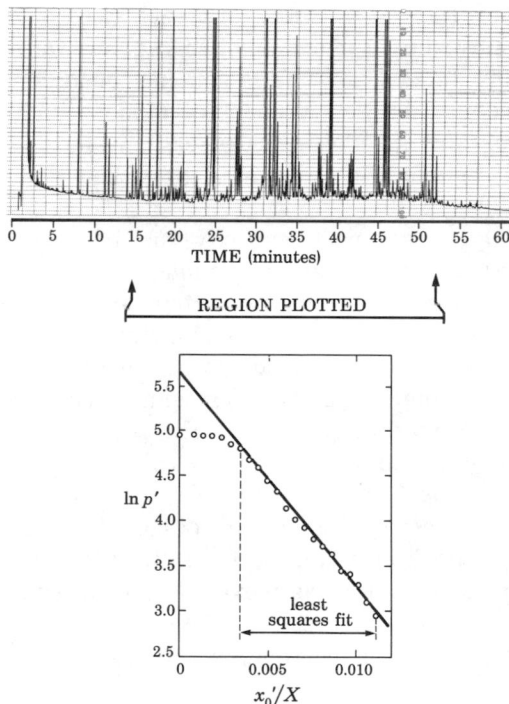

Figure 6.13. Capillary gas chromatogram of polynuclear aromatic hydrocarbons along with the associated plot from Eq. 6.55. The plotted region of the chromatogram contains 145 observed peaks or maxima. However, the plot suggests the presence of approximately 250 components (see text, also ref. 34). The large discrepancy is a consequence of peak overlap. (Chromatogram courtesy of M. L. Lee and R. West, Brigham Young University, published in ref. 34. Reprinted with permission from J. M. Davis and J. C. Giddings, *Anal. Chem.*, **57**, 2178 (1985). Copyright 1985 American Chemical Society.)

counted by 105. The saturation (relative to $R_s^{\ddagger} = 0.5$) is calculated from Eq. 6.50 as $\alpha = 0.545$. The number s of singlets (still referred to $R_s^{\ddagger} = 0.5$) from Eq. 6.52 is 84. If, however, we need a resolution of $R_s^{\ddagger} = 1.0$ to estimate component amounts from peak areas, then x_0 and α double and s falls to 28 and the success ratio $s/\bar{m}$ is only 0.11. This shows the difficulty of dealing with complex samples even with the most efficient of modern analytical separation systems.

A more recent evaluation of peak overlap statistics has been reported by Delinger and Davis [41].

6.8   APPROACHES TO SOLVING THE PEAK OVERLAP PROBLEM

Several lines of attack are possible to improve the gloomy prospect of frequent peak overlap. An obvious approach is to develop more efficient separation systems whose greater number of plates will resolve fused peaks more effectively. Indeed, this approach has been long used and is the driving force behind the development of modern high-efficiency systems. However, large gains in efficiency, increasingly hard to develop, provide only marginal benefits in reducing the frequency of peak overlap [21, 33, 39].

A substantial gain in peak capacity can be made by utilizing two-dimensional systems, as noted in Section 6.5. This approach has been successfully implemented in the form of two-dimensional electrophoresis systems, described in Section 6.4, but effective technology for two-dimensional column chromatography is still to be developed [13, 14].

Coupled column (multidimensional) systems in chromatography have also been developed to improve resolution. The coupled column procedure, as noted in Section 6.4, requires two or more columns of different kinds having different retention mechanisms.

The approaches described above are designed to increase peak capacity and thus reduce peak overlap. A radically different approach involves accepting component overlap as inevitable and directing attention at numerical rather than physical peak resolution. Numerical resolution allows the recovery of analytical information but not the recovery of purified components.

The shape of fused peaks normally differs from that of the underlying component peaks. Large shape differences are apparent to the eye; examples can be seen in the third and fourth peaks from the left-hand side in any of the rows of peaks of Figure 5.8. For these fused peaks the resolution is 0.5 and 0.75, respectively. Small shape differences, such as those found for the second peak ($R_s = 0.25$) from the left of Figure 5.8, may not be visually apparent, but their altered mathematical form can be detected as a difference in statistical moments [42].

While shape abnormalities suggest fused peaks, it is difficult to determine the number and size of the component peaks that have gone into the fused peak. This determination—needed to satisfy analytical goals—requires additional information about the fused peak. The best source of information is multichannel detection in which some kind of spectrum (mass spectrum, UV absorption, etc.) is acquired at many different points along the separation axis. With luck, the components of the mixture will differ substantially in their spectral fingerprints. In the most favorable case, each component will reveal itself at a unique wavelength; the fractogram or chromatogram displayed for that wavelength will thus reveal the corresponding component peak.

Inevitably, there is extensive spectral overlap along with separative overlap. The detector response plotted as a function wavelength (along one axis) and separation coordinate (another axis) will then fail to sharply resolve component peaks along either axis. Under these complicated circumstances, it is necessary to use special numerical techniques to break the fractogram (or that limited part of it occupied by a fused peak) into its component parts. Significant advances toward this goal have been made by Kowalski and co-workers [43] using factor analysis techniques and by Harris and co-workers [44, 45] using least-squares methods.

REFERENCES

[1] J. C. Giddings, *Sep. Sci. Technol.*, **14**, 871 (1979).

[2] J. C. Giddings and K. Dahlgren, *Sep. Sci.*, **6**, 345 (1971).

[3] J. C. Giddings, *J. Chem. Educ.*, **50**, 667 (1973).

[4] S. Sourirajan and T. Matsuura, *Reverse Osmosis/Ultrafiltration Process Principles*, National Research Council Canada, Ottawa, Canada, 1985.

[5] W. G. Pfann, *Zone Melting*, 2nd ed., Wiley, New York, 1976.

[6] J. C. Giddings and R. A. Keller, *J. Chromatogr.*, **2**, 626 (1959).

[7] J. M. Bobbitt, *Thin Layer Chromatography*, Reinhold, New York, 1963, Chapter 10.

[8] J. C. Giddings, *Anal. Chem.*, **56**, 1258A (1984).

[9] P. C. Wankat, *Sep. Sci. Technol.*, **19** (1984).

[10] P. H. O'Farrell, *J. Biol. Chem.*, **250**, 4007 (1975).

[11] N. G. Anderson and N. L. Anderson, *Anal. Biochem.*, **85**, 331 (1978).

[12] E. Jellum, in A. Zlatkis, Ed., *Advances in Chromatography*, Elsevier, Amsterdam, 1982, p. 139.

[13] G. Guiochon, M.-F. Gonnord, and M. Zakaria, *Chromatographia*, **17**, 121 (1983).

[14] M. Zakaria, M.-F. Gonnord, and G. Guiochon, *J. Chromatogr.*, **271**, 127 (1983).

[15] G. Haugaard and T. D. Kroner, *J. Am. Chem. Soc.*, **70**, 2135 (1948).

[16] E. L. Durrum, *J. Colloid Sci.*, **6**, 274 (1951).

[17] A. J. P. Martin, *Discussions Faraday Soc.*, **7**, 332 (1949).

[18] J. C. Giddings, *Anal. Chem.*, **34**, 37 (1962).

[19] J. M. Begovich, C. H. Byers, and W. G. Sisson, *Sep. Sci. Technol.*, **18**, 1167 (1983).

[20] J. C. Giddings, *J. Chromatogr.*, **504**, 247 (1990).

[21] J. C. Giddings, *J. High Res. Chromatogr. Commun.*, **10**, 319 (1987).

[22] J. C. Giddings, in H. J. Cortes, Ed., *Multidimensional Chromatography: Techniques and Applications*, Marcel Dekker, New York, 1990, pp. 1–27.

[23] D. H. Freeman, *Anal. Chem.*, **53**, 2 (1981).

[24] J. A. Apffel and H. McNair, *J. Chromatogr.*, **279**, 139 (1983).

[25] T. V. Raglione, N. Sagliano, Jr., T. R. Floyd, and R. A. Hartwick, *LC·GC*, **4**, 328 (1986).

[26] R. E. Majors, *J. Chromatogr. Sci.*, **18**, 571 (1980).

[27] J. F. K. Huber, E. Kenndler, and G. Reich, *J. Chromatogr.*, **172**, 15 (1979).

[28] H. J. Cortes, C. D. Pfeiffer, and B. E. Richter, *J. High Resolut. Chromatogr. Chromatogr. Commun.*, **8**, 469 (1985).

[29] G. Schomburg, *LC·GC*, **5**, 304 (1987).

[30] M. M. Bushey and J. W. Jorgenson, *Anal. Chem.*, **62**, 161 (1990).

[31] H. K. Jones and J. C. Giddings, *Anal. Chem.*, **61**, 741 (1989).

[32] H. J. Cortes, Ed., *Multidimensional Chromatography: Techniques and Applications*, Marcel Dekker, New York, 1990.

[33] J. M. Davis and J. C. Giddings, *Anal. Chem.*, **55**, 418 (1983).

[34] J. M. Davis and J. C. Giddings, *Anal. Chem.*, **57**, 2178 (1985).

[35] T.-B. Hsu, W. A. Howard, and L. B. Rogers, *Sep. Sci. Technol.*, **17**, 1623 (1982–1983).

[36] D. Rosenthal, *Anal. Chem.*, **54**, 63 (1982).

[37] D. P. Herman, M. F. Gonnard, and G. Guiochon, *Anal. Chem.*, **56**, 995 (1984).

[38] M. Martin and G. Guiochon, *Anal. Chem.*, **57**, 289 (1985).

[39] J. C. Giddings, J. M. Davis, and M. R. Schure, in S. Ahuja, Ed., *Ultrahigh Resolution Chromatography*, ACS Symp. Ser. No. 250, American Chemical Society, Washington, DC, 1984, pp. 9–26.

[40] M. L. Lee and R. West, personal communication, 1984.

[41] S. L. Delinger and J. M. Davis, *Anal. Chem.*, **62**, 436 (1990).

[42] E. Grushka, M. N. Myers, P. D. Schettler, and J. C. Giddings, *Anal. Chem.*, **41**, 889 (1969).

[43] D. W. Osten and B. R. Kowalski, *Anal. Chem.*, **56**, 991 (1984).

[44] F. J. Knorr, H. R. Thorsheim, and J. M. Harris, *Anal. Chem.*, **53**, 821 (1981).

[45] S. D. Frans, M. L. McConnell, and J. M. Harris, *Anal. Chem.*, **57**, 1552 (1985).

EXERCISES

6.1(*)   If a in the focusing velocity expression $W = -ay$ is given by 1.2×10^{-3} s^{-1} and $D_T = D = 5.0 \times 10^{-7}$ cm^2 s^{-1}, what is the effective width 4σ of the resulting steady-state Gaussian zone?

6.2(**)   Suppose that the focusing velocity $W = U = -ay$ originates as a relative displacement velocity U impelled by an external potential function of the following form having a minimum at $y = 0$

$$\mu^{ext} = \mu_0^{ext} + ky^2$$

Calculate velocity U at distance $y = 1.00$ mm for the protein hemoglobin in aqueous solution for which we can assume a Stokes' law radius of 62 Å, $\eta = 1.00 \times 10^{-2}$ poise, and $k = 1.20 \times 10^6$ cal mol^{-1} cm^{-2}.

6.3(**)   For albumin, assume that $k = 7.5 \times 10^5$ cal mol^{-1} cm^{-2} (refer to previous problem) and that molecular diffusion alone causes zone broadening $(D_T = D)$. Calculate the effective width 4σ of the resulting Gaussian zone at 20°C.

6.4(*)   What value is assumed by the effective mean layer thickness ℓ of a steady-state zone held next to a wall by virtue of displacement at velocity $|W| = 1.0 \times 10^{-3}$ cm/s? The zone consists of spherical particles of diameter 2.5 nm in an aqueous medium of viscosity $\eta = 1.0 \times 10^{-2}$ poise at 20°C.

6.5(**)   What value is assumed by the effective mean layer thickness ℓ of a steady-state zone held next to a wall at 20°C by virtue of a chemical potential gradient of 4.18×10^6 J mol^{-1} cm^{-1}? The zone consists of spherical particles of diameter 2.5×10^{-7} cm in an aqueous medium of viscosity $\eta = 1.0 \times 10^{-2}$ poise.

6.6(**)   Solute piles up against (but does not leak through) a membrane filter at an absolute flux rate of $|J_0| = 1.0 \times 10^{-6}$ mol cm^{-2} s^{-1} for a period of 3.6 s. If $D_T = D = 3.0 \times 10^{-6}$ cm^2/s and $|v| = 0.0020$ cm/s, calculate the value of c_0, the concentration at the membrane wall, after the stated period of accumulation. What would you expect to happen if one attempted to continue accumulation for 1 h?

6.7(***)   A plot of the visible area A of a spot on a thin layer plate against ln n ($n = $ no. moles) yields a straight line of the form

$$A = B \ln n + c$$

where the length/breadth ratio of the spot is observed to be 2.50 and $B = 0.0814$ cm^2. The solute in the spot becomes visible at a threshold concentration on the plate of 1.00×10^{-5} mol/cm^2. Find values for σ_x, σ_y, and the intercept C.

6.8(**)   Prove that the contour line representing concentration c_0 within a two-dimensional zone described by Eq. 6.29 is elliptical in shape.

Further, prove that the length and breadth of the ellipse are given by Eqs. 6.30 and 6.31, respectively.

6.9(**)   Assume that a Gaussian peak of maximum signal height S_m and standard deviation σ is superimposed on a sloping baseline that increases $0.1\ S_m$ units for each σ of distance. By what distance (relative to σ) does the observed peak maximum shift relative to its true value as a result of the sloping baseline?

6.10(**) Two equal Gaussian peaks of identical height and with equal σs partially overlap such that the resolution between them is 0.75. Calculate how far (relative to σ) the observed peak maxima are shifted inward (toward one another) from the true Gaussian peak centers.

6.11(**) One hundred components are eluted from a GC column of 200,000 plates having an elution volume ratio V_{max}/V_{min} of 8.0. How many of these components are expected to be isolated as pure component peaks ($R_s \geq 1$) based on statistical arguments?

6.12(**) Using the equations in the text in conjunction with Figure 6.9 and its caption, compare the observed number of doublet, triplet, and quadruplet peaks in the chromatogram and the number expected on statistical grounds.

6.13(*)   A small spot containing exactly 1 μg of leucine on a thin-layer plate has evolved into a two-dimensional Gaussian zone with $\sigma_x = 1.25$ mm and $\sigma_y = 0.720$ mm. The leucine is detectable at levels down to 1.00×10^{-7} g/cm^2. Calculate the apparent spot length a and width b. Compare these values to the "width," 4σ, of the underlying Gaussians.

6.14(*)   Figure 6.11 illustrates a chromatogram for which the peak capacity n_c is 100 for peaks crowded together with $R_s^{\ddagger} = 0.5$. What is n_c for $R_s^{\ddagger} = 0.8$?

7

CLASSIFICATION AND COMPARISON OF METHODS

In previous chapters of this book, we established a number of close basic relationships between groups of separation methods. We proceeded on the premise that the recognition and study of common features of scattered techniques are most worthwhile: it facilitates understanding, aids evaluation, allows comparison, provides a basis for prediction, and simplifies the theory to a set of common elements. However, the further codification of like and unlike properties requires a system of classification. Unfortunately, the development of a complete classification system for separations is made difficult by the great number of variables of interest and their imperfect correlation with one another. The problem has been discussed in a paper by the author; here we borrow liberally from that work [1].

This chapter begins with three sections on basic concepts of classification. In the last four sections of the chapter, we describe the role of flow in separations. This subject relates also to classification because flow is one of the major variables distinguishing one class of separation methods from another.

7.1 PROBLEMS AND OPPORTUNITIES IN CLASSIFICATION

Forging various relationships into orderly schemes of classification is an important part of science. The theoretical study of classification is itself a science and is termed taxonomy [2]. Its fruits pervade many fields. In chemistry, the classic example is the periodic table. While discovered empirically, it is based on the grouping of elements with a common number of outer-shell electrons. This electron count is so fundamental that many properties correlate with it: reactivity, electronegativity, ionization energy, bonding capacity, atomic size, toxicologic properties, metallic character, and so on. Still, there is the anomaly of hydrogen to remind us of the lack of well-defined categories for everything.

Separation methods are more difficult to classify and relate than are the chemical elements. In chemistry there are 100 or so invariant elements, each with discrete properties constant for all time. Separation methods have too many synthetic, open-ended, overlapping characteristics for simple classification. Among the characteristics to be established are the choice between external fields and nonmiscible phases, the types of fields or the types of phases, the presence or absence of flow, multiple stages, continuous throughput, high pressures, discrete zones, chemical reaction, elution, columns, porous support, large sample capacity, high energy efficiency, and so on. Each choice is important in its own right and may become critical for certain groups of scientists and engineers seeking particular goals [1].

Clearly, a complete classification of separations would require an accounting based on similarities and differences in many properties. In taxonomy, this would be termed a *polythetic* classification, as opposed to *monothetic* classifications based on only one property. Monothetic classifications are not necessarily erroneous—they are simply incomplete. In a complex field such as separations, many monothetic classifications exist, each one different, each with an element of validity, but each incomplete.

The focus on general taxonomy is made here to explain to the reader the reasons for the many different "logical" systems for classifying separations in the literature. The approaches range from the simple and traditional division into equilibrium and rate processes by Karger et al. [3] to the historically important classification of driving forces and resistive forces by Strain et al. [4]. We note also a contribution to classification by Rony [5] and an important study by Lightfoot and his co-workers [6]. Many other authors have discussed the matter [7–12].

It is not our object here to review the details of existing classificatory systems. Instead, we follow an alternate route to classification suggested by our emphasis on transport.

We found earlier that a theoretical consideration of displacement and transport exerts a unifying influence on separation science, bringing diverse methods under a common descriptive umbrella. The theory leads in a natural way to the formation of categories of separations which can be considered the beginning of a fundamental classificatory system. Here we generalize the results of transport theory to develop a fundamental basis for classification. While the resulting scheme will not be a complete polythetic classification, it will be based upon some of the most fundamental features of the separation process. These basic features, incorporated in the classification, should correlate well with other properties of separations in the same way that the number of outer-shell electrons is directly related to the diverse properties of the elements of the periodic table. This transport-oriented approach should not only help us assign individual separation techniques to reasonable categories, but should strengthen our understanding of the separative process. It should also help evaluate the relative strengths and weaknesses of separation when carried out by the various distinct methods.

7.2   CLASSIFICATION BASED ON TRANSPORT

The classification of separations should reflect the patterns of component transport and equilibrium that develop in the physical space of the system. The transport equations show that we have two broad manipulative controls that can be structured variously in space to affect separative transport. First is the chemical potential which controls both relative transport and the state of equilibrium. Chemical potential, of course, can be varied as desired in space by placing different phases, membrane barriers, and applied fields in appropriate locations. A second means of transport control is flow, which can be variously oriented with respect to the phase boundaries, membranes, and applied fields—that is, with respect to the structure of the chemical potential profile.

Viewed in this way, chemical potential profiles (along with flow) govern separation; different phases, membranes, and applied fields are simply convenient media for imposing the desired profiles. The media are selected on pragmatic grounds: chemical compatibility with the components and the system, selectivity between components, noninterference with detectability, ease of solvent removal (another separation process), facilitation of rapid transport, and so on.

While chemical potential can be varied in an enormous number of detailed ways along the coordinate(s) of a separation system by using different phases and fields, a few basic patterns or classes emerge. Variations within a class may alter separation speed and resolution but not the overall structure of the separation. For example, the substitution of one organic phase for another in a solvent extraction system will alter the increment in chemical potential at the interface for any given species, and it will therefore change the equilibrium distribution coefficient, but it will not fundamentally change the way in which the separation evolves. If, on the other hand, we establish our chemical potential increment by using a profile of an altogether different form, due, say, to an imposed electrical field, the separation (now electrophoretic) will lose all resemblance to the original solvent extraction separation. Going one more step, if we now replace the electrical field by a sufficiently strong sedimentation field, there will be many detailed changes, including a reordering of component transport velocities, but not a change in the essential pattern of multiple Gaussian solute zones emerging along the axis of the field. Thus the above examples illustrate the point that the general structure of the chemical potential profile, and not the particular phases and fields used to establish that profile, exerts the most decisive influence on the form of separation.

It could be argued, of course, that the differences and similarities cited above stem from the fact that solvent extraction is essentially a steady-state (equilibrium) process while electrophoresis and sedimentation are transient (rate) processes. However, such an argument would overlook the fact (to be explained later) that the different forms of the chemical potential profile determine which systems *can* be run successfully in the steady-state mode and which in the transient mode. Thus the chemical potential profile and associated flow structure emerge as dominant influences that should be classified at the very beginning of any attempt to organize separation phenomena into a cohesive discipline.

In the above discussion, we have ignored the role of diffusive transport as represented by D_T. However, diffusivity is a dissipative process that is always present, shows little variation in form, and is not subject to significant structuring—only to changes in magnitude. Diffusivity, rather than a structure-forming process, can be viewed as a structure-blurring process, a rather universal (and negative) response to other structuring elements.

7.3   CATEGORIES OF SEPARATIONS

The basic combinations of chemical potential profiles and flow structures are limited in number. The following major categories can be distinguished.

First, the overall chemical potential profile μ^* (representing the sum of external field effects and internal molecular interactions)

$$\mu^* = \mu^{\text{ext}} + \mu^0 \tag{7.1}$$

may be continuous, discontinuous, or a continuous-discontinuous combination

$$c = \text{continuous } \mu^* \text{ profile} \tag{7.2}$$

$$d = \text{discontinuous } \mu^* \text{ profile} \tag{7.3}$$

$$cd = \text{combined } \mu^* \text{ profile} \tag{7.4}$$

The c profile is usually generated by an external field or continuous gradient acting in one phase. The d profile assumes the form of a step function at the discontinuity between two immiscible phases or between a membrane and a solution. Typical c, d, and cd profiles are illustrated in Figure 7.1.

Second, the system may or may not employ flow to structure the separation

$$S = \text{static (nonflow) system} \tag{7.5}$$

$$F = \text{flow system} \tag{7.6}$$

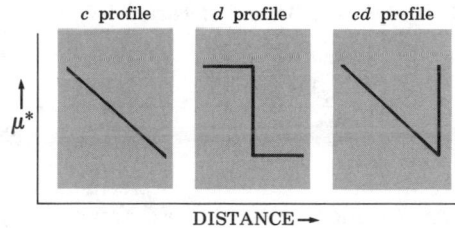

Figure 7.1. Three basic classes of chemical potential profiles in separation systems.

the word "flow" is used in a relative sense and denotes the movement of one phase with respect to another, with respect to an imposed field, or relative to a membrane. We do not include in the flow category those systems exhibiting only passive or parasitic flow (such as electroosmosis) not essential to the structure of separation (see Section 7.5). We also exclude mixing flow (like that used to intermix phases in solvent extraction), which may speed transport and rapid equilibration, but not otherwise affect the nature or outcome of separation. The basic functions of different types of flow are discussed in detail in Section 7.5.

Finally, in category F systems, the flow may be aligned in the direction of the μ^* gradient (e.g., ultrafiltration) or perpendicular to the μ^* gradient (e.g., chromatography)

$$F(=) = \text{flow and } \mu^* \text{ gradient parallel} \tag{7.7}$$

$$F(+) = \text{flow and } \mu^* \text{ gradient perpendicular} \tag{7.8}$$

The combination of the fundamental arrangements of chemical potential profiles and flow structure outlined above yields nine possible categories of separations. These are shown in Table 7.1.

Nearly all separation methods fall clearly into one of the nine basic categories. The way in which specific separation methods relate to one another when ordered into these categories is shown in Table 7.2.

Examination of the groupings of Table 7.2 verifies most of the more or less traditional associations between separation methods: chromatography with countercurrent distribution and distillation, electrophoresis with rate-zonal sedimentation, isoelectric focusing with isopycnic sedimentation, re-

TABLE 7.1 Nine Basic Categories of Separations

Continuous μ^* Profile	Discontinuous μ^* Profile	Combination μ^* Profile
Sc	Sd	Scd
$F(=)c$	$F(=)d$	$F(=)cd$
$F(+)c$	$F(+)d$	$F(+)cd$

TABLE 7.2 Grouping of Separation Methods in the Nine Basic Categories

Flow Condition	c (continuous μ^* profile)	d (discontinuous μ^* profile)	cd (combination μ^* profile)
S	Electrophoresis	Simple:	Electrodeposition
	Isoelectric	Extraction	Electrostatic
	focusing	Adsorption	precipitation
	Isotachophoresis	Crystallization	Electrolytic
	Rate-zonal	Distillation	refining
	sedimentation	Evaporation	Electrodialysis
	Isopycnic	Sublimation	Equilibrium
	sedimentation	Ion exchange	sedimentation
		Dialysis	
$F(=)$	Elutriation	Filtration	Electrofiltration
	Countercurrent	Ultrafiltration	
	electrophoresis	Reverse osmosis	
		Pressure dialysis	
		Zone melting	
$F(+)$	Hyperlayer	Chromatography	Field-flow
	field-flow	Countercurrent	fractionation
	fractionation	distribution	Thermogravitational
		Fractional	separation
		distillation	Electrodecantation
		Foam fractionation	
		Multistage two-phase	
		processes	

verse osmosis with ultrafiltration, and extraction with precipitation. However, the table also suggests some new associations: extraction with dialysis and zone melting with the filtration methods. A little thought about the matter confirms that these latter associations are valid. In both extraction and dialysis, the space of the system is partitioned into two regions separated by an interface which selectively transmits certain components. The interface, from a fundamental point of view, is simply a chemical potential barrier. For example, the pores of a membrane filter or other porous matrix (such as the support medium in exclusion chromatography) become a barrier rejecting large solute molecules because these molecules do not fit comfortably in the pores (see Chapter 2). The resulting entropy loss associated with the poor fit creates a thermodynamic (chemical potential) barrier that resists penetration [13].

Zone melting and the filtration methods are closely related because each has a selective interface sweeping through a solution, piling up solute in exponential (polarization) layers in front. The common mathematics shared by these methods was emphasized in Section 6.2.

We note that some close relationships exist between methods in different groups as well. Elutriation [$F(=)c$] is, in a fundamental sense, much like isoelectric focusing (Sc). The reason is that flow velocity v and field-induced velocity U have identical transport functions for a single component, a fact reflected in Eqs. 3.23, 3.30, 3.33, and so on. Of course, flow offers no selectivity between components. Thus, we can understand the similarities of chromatography and electrophoresis (Gaussian zones evolving in space) and also their differences (the nonselectivity of flow in chromatography requires a second phase to install selectivity). A more detailed comparison of flow and field effects appears in Section 7.4.

It must be emphasized that the nine basic categories appearing in Tables 7.1 and 7.2 constitute only a coarse grouping and that there is additional clustering within each family. This situation is not unlike the periodic system where different elements in a family may exist as metals or nonmetals, gases or solids, and so on. Separation methods, for example, subdivide into steady-state methods (which includes equilibrium) and transient or transport methods. However, the ability to generate effective steady-state separations or transport separations hinges on the underlying structure of flow and chemical potential, as noted before. Flow and chemical potential are the most fundamental variables of separation and should be sorted into proper classes and subclasses before trying to fully characterize steady-state and transport methods. The matter has not been worked out in detail and is too complex to pursue at length here. However, a few other important considerations are noted below.

In our earlier classification [1], separations were grouped into six categories rather than the nine appearing in Tables 7.1 and 7.2. Whereas the combined continuous-discontinuous (cd) profiles of μ^* were omitted as a basic class before, they are included here for completeness. The choice in the level of detail is, indeed, somewhat arbitrary. For example, we could expand (rather than contract) the classes of μ^* (or effective μ^*) profiles by including profiles having continuous (c) properties but possessing a distinct minimum. Such an expanded set of profiles is illustrated in Figure 7.2. Profiles with a minimum (the two members to the right in Figure 7.2) are

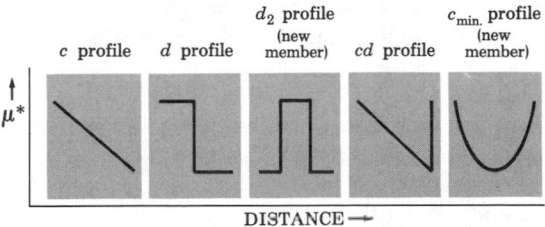

Figure 7.2. Expanded set of μ^* profiles. The profile to the far right is a new member originally included in the c class to the far left. Both classes with minima are conducive to steady-state separations. The d_2 profile, also a new member, has a tendency to maintain nonsteady-state concentration distributions on the two sides of the μ^* barrier.

ideal for steady-state separations because solute tends to collect in the minimum and approach a stationary distribution centered about that point. Thus, the minimum in cd systems leads to such steady-state methods as electrostatic precipitation and equilibrium sedimentation, both in the Scd class. In field-flow fractionation and related methods in the $F(+)cd$ category, a steady-state condition is reached along the field axis by virtue of the minimum in μ^*; however, a steady-state condition is not established along the perpendicular flow axis.

Associated with the new c_{min} class of profiles, we would find steady-state methods such as isoelectric focusing, isopycnic sedimentation, and some related methods that have been proposed [14]. The unique ability of these methods to separate mixtures into many zones or bands is related to the fact that the minimum for each component is found at a different point in space.

We could also expand the d class of profiles to include profiles with multiple discontinuities. Thus a simple membrane, with a discontinuity on each side, could be distinguished by the symbol d_2, as opposed to d for a single-interface system such as solvent extraction (see Figure 7.2). The d_2 class is more suited for nonsteady-state applications. This is particularly so if the d_2 profile consists of a μ^* barrier so high (relative to thermal energy $\mathcal{R}T$) that species cannot cross in the time of an experimental run. Thus a membrane used in dialysis can retain macromolecules in one compartment (on one side of the membrane or μ^* barrier) despite equal equilibrium concentrations on both sides.

Although it is obvious from the above discussions that there are good arguments for including profiles such as c_{min} and d_2 (or more generally d_n with $n = 2, 3, 4$, etc.) as basic classes, we choose here to keep c_{min} as a subclass of the c profiles and d_2 as a subclass of d for simplicity.

We should note that almost every system, even those in the c class, contains a discontinuity in μ^* located at the walls, electrodes, and other nonpermeable barriers that define the physical limits of the system. However, although walls represent an abrupt μ^* barrier, we do not automatically put separation in the d class because of them. If they are passive container walls simply holding the system in place, or if they are electrodes which charged particles never touch (as in electrophoresis), a d classification is not justified. However, if the walls play a major part in structuring the separation (such as membrane filters or the outer wall of a centrifuge cell which structures the solute concentration profiles in equilibrium sedimentation), those walls convey a d classification to the method.

Generally the μ^* profile is fixed in space, but in some systems μ^* is time dependent as well as position dependent. Electrical separation systems are particularly susceptible to time dependence because the ongoing transport of charged species can influence the potential distribution. Thus the μ^* minima used to trap charged components in isoelectric focusing tend to drift with time [15]. Isotachophoresis is perhaps most unique in that the continuous μ^* profile consists of straight line segments of different slopes. Each segment corresponds to a component; as the components move along the migration path, so do the μ^* segments.

The present classificatory system does not recognize as a primary class those separations depending on selective chemical reactions or electronic excitations (e.g., of isotopes). The reactions do not, in fact, constitute separation but are preparation steps designed to convert one or several components into a different chemical form more susceptible to selective displacement. (A summary of chemical conversion strategies is summarized in ref. [16].) The separation follows the preparation and is usually achieved by conventional means [1, 16].

7.4 FLOW-ASSISTED SEPARATION

We have noted that flow is an integral part of many separation methods. The role of flow in achieving separation is discussed in the remaining sections of this chapter.

We use the word "flow" to describe all bulk fluid displacements, including direct mechanical displacement and the flow process itself. These are both nonselective forms of displacement initially described in Chapter 1. The treatment of flow in this section expands on a published approach interpreting the role of flow in separation processes [1].

A perspective on flow can be gained by examining the transport equations. Here, flow velocity v (actually a vector with a direction as well as a magnitude) adds to field-induced velocity U (also a vector) to determine the rate of change of concentration dc/dt at a given point in separation space

$$\frac{dc}{dt} = -(v + U)\frac{dc}{dx} + D_T \frac{d^2c}{dx^2} \tag{7.9}$$

This expression of transport is provided by Eq. 3.30 (where D has been replaced by D_T for generality). It demonstrates that v and U have equivalent—mathematically indistinguishable—roles in component transport and thus equivalent roles in the evolution of component concentration profiles. A decrease in either v or U can be exactly offset by a gain in the other one of the pair, since only their sum, $v + U$, enters the equation.

While v and U have equivalent transport roles in a mathematical sense, they do not display equivalent characteristics in a physical sense. They are subject to different manipulations and limitations such that they complement one another. Their complementary nature allows maximum control of the overall velocity sum, $W = v + U$, and it plays a major role in optimizing separations.

Of greatest importance, only relative velocity U is selective between components, as emphasized in Chapter 1. Velocity U can also be arranged to have discontinuities (at interfaces) while this is difficult for v. The flow represented by v, on the other hand, is a far more powerful and universal agent for displacement than the motion associated with U. Under normal circumstances, v can easily be raised orders of magnitude above U (e.g., 10^2 mm/s versus 10^{-2} mm/s) and extended over great lengths (through mile-long columns in gas chromatography as an example).

While flow can play a highly constructive role in separation, it does not always do so. Flow displacement occurs under a wide variety of circumstances in separation processes. Its role is accordingly multifaceted. Below we distinguish several categories of major flow effects. Several effects can, of course, arise simultaneously. Only the final category, however, justifies classification as a flow (F) separation system (as opposed to a static (S) system) as will be explained.

7.5 TYPES OF FLOW

Flow in separation systems can be divided into the following four categories:

Parasitic Flow Flow often assumes a destructive role, despite efforts to prevent it. Convective currents frequently lead to the remixing of separated components. Electroosmotic flow (see Section 4.9) is destructive in some (not all) electrophoretic systems and is difficult to eliminate because of the ubiquitous presence of surface charges. These unintended and generally unproductive forms of flow can be termed *parasitic flow*. The following discussion serves to distinguish parasitic flows from nonparasitic flow processes.

Anywhere a chemical potential increment or gradient exists, an elementary separation step can occur. Anywhere random flow currents exist, separation is dissipated. Thus random flow currents are parasitic in regions where incremental chemical potential is used for separation. These currents should thus be eliminated, insofar as possible, in regions where electrical, sedimentation, and other continuous (c) fields are generating separations. Likewise, they should not be allowed to transport matter over discontinuous (d) separative interfaces such as phase boundaries or membrane surfaces. However, they are nonparasitic in bulk phases (removed from the separative interface) where only diffusion occurs. Here, in fact, they aid diffusion and speed the approach to equilibrium. This positive role is recognized in the following category of flow.

Mixing Flow Flow is often used to aid transport and thus hasten separation. Most often such flow is induced by some stirring motion and is used to augment diffusion within one or both phases of a two-phase separation system. (Recall from Section 5.3 that random displacements, whether induced by flow or not, act as an effective diffusion process.) The mixing of phases in simple one-stage extraction systems is an example. The rocking-stirring motion used in the Craig device for countercurrent distribution (CCD) is another example [10]. In a similar vein, turbulent motion can be induced to speed separation in chromatographic systems [17]. The circulation of liquid through a thin-cell ultrafiltration device also aids separation by effectively mixing the polarization layer (Section 6.2) with bulk solution [18].

Passive Flow   This category applies to those cases in which flow has no fundamental role in the separation process itself but in which flow is used to aid sample recovery, and so on. The bulk displacement (removal) of samples by means of a fraction collector provides an example. Another is found in continuous (deflection or curtain) electrophoresis [19] where flow displacement along an axis perpendicular to the field makes continuous collection possible (see Figure 7.3). However, flow does not aid the evolution of the separation along the principal separative axis represented by the direction of the electrical field. In fact the flow process somewhat degrades the resolution. The lateral (bulk) displacement of solute bands in continuous rotating chromatography systems [20], as described in Section 6.4, fall in the same category.

Integral Flow   In many systems flow plays an integral role in structuring the separation process. The essential character of the separation method therefore hinges on the existence, orientation, and nature of flow. Most commonly the flow—which is inherently nonselective—serves to magnify a local enrichment induced by the selective forces underlying μ^*.

The term "flow system," designated in our classification scheme by the letter F, is reserved for those methods employing integral flow because in no other case does flow alter the basic nature of separation. Two distinct classes of integral flow can be identified, $F(=)$ and $F(+)$, as noted in our classification of Section 7.3. These classes are discussed separately below.

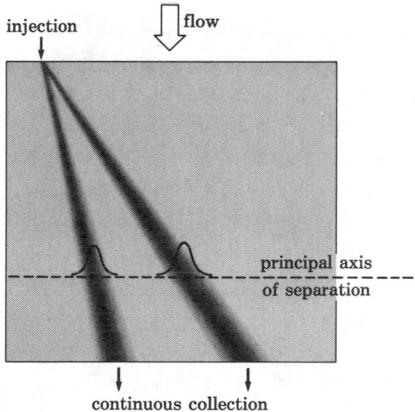

Figure 7.3. In continuous deflection electrophoresis, flow down a paper sheet or between closely spaced plates carries different solutes, which are gradually separated by electrophoresis along axis x, to different collection ports. However, the flow is classified as "passive" because it does not enhance the separation occurring along the electrical-field axis. The role of flow is to aid continuous sample collection, not to fundamentally alter the separation process.

7.6   FLOW PARALLEL TO FIELD OR INTERFACE GRADIENT, $F(=)$

Parallel flow is found in ultrafiltration and related filtration techniques, pressure dialysis, zone melting, and elutriation. In parallel flow methods— including the $F(=)c$, $F(=)d$, and $F(=)cd$ approaches—flow acts along the same axis as the selective displacement driven by μ^*. Flow thus brings the solute mixture into (or maintains its position relative to) a chemical potential gradient $[F(=)c]$ or barrier $[F(=)d]$. For instance, in ultrafiltration, $F(=)d$, flow forces the solute mixture into intimate contact with a selective barrier where separation proceeds. Without flow, selective permeation would occur only by weak diffusional transport and would be relatively slow, much like dialysis. Additionally, a fraction of permeating species would remain behind to contaminate the nonpermeating species because there would be no force to drive the former completely through the membrane barrier. Flow solves this problem by evacuating all permeable solute through the membrane.

Flow plays a similar constructive role in other filtration systems and in zone refining, all being $F(=)d$ systems. These are discussed further in Chapter 10.

7.7   FLOW PERPENDICULAR TO FIELD OR INTERFACE GRADIENT, $F(+)$

Perpendicular flow occurs in chromatography, countercurrent distribution, field-flow fractionation, and related methods. Below we explain the basic mechanism by which flow assumes its vital role in these separation techniques.

The chemical potential increments of fields and interfaces provide the selective influence generally needed for separation. The selectivity exhibits itself in the form of unequal concentration distributions along the axis of the field or across the interface. However, separation along this coordinate is often subdued by the intrinsic limitations of two-phase systems or by the inherent weakness of certain fields applied to particular classes of molecules (see Chapter 8).

Fluid flow that is perpendicular to the axis of selective displacement is capable of moving solute from one patch of interface to another downstream, or from one region of a weak field to an equivalent region downstream. In so doing, it creates a new and more extended coordinate for solute displacement and separation. The coordinate can be of great length (for example, the length of a chromatographic column) because of the enormous relative displacing power of flow. Furthermore, the displacement becomes selective because flow carries some volume elements more rapidly than others; these elements will be relatively enriched or depleted in different solutes by the primary selective forces acting perpendicular to flow.

Flow not only transports solutes into new regions of space, but it evacuates (by means of the term $-v\,dc/dx$ in Eq. 7.9) solute from old space. Evacuation makes that space or that part of the flow/separation coordinate available for other solutes carried at a different velocity [1]. Thus zones are formed and kept narrow, making multicomponent separations possible.

The evacuation of solute by flow is analogous to its evacuation by fields. Evacuation also plays a critical (but usually different) role in $F(=)$ systems, as pointed out in the previous section.

In summary, flow acting perpendicular to a field or interface gradient (i) creates a new separation axis of arbitrary length, (ii) displaces solutes selectively along that length by acting on the selective local displacements caused by the field or interface, and (iii) forces solute to evacuate large parts of the new separation axis, thus keeping solutes in zones and making multicomponent separations possible. These features combine to provide the most versatile analytical separation tool known—chromatography. In gas-liquid chromatography (GLC), for example, flow converts a static system of two phases (the stationary and mobile phases) with a maximum peak capacity of two into a powerful method capable of separating over 100 peaks.

A few less common perpendicular flow techniques are not used to generate multicomponent separations, although they are inherently capable of doing so. These techniques include the thermogravitational column [21] and electrodecantation [22]. Here the extended path serves to enhance the selectivity generated by the field and to allow the collection of two enriched or separated fractions at the two extremes of the flow coordinate.

REFERENCES

[1] J. C. Giddings, *Sep. Sci. Technol.*, **13**, 3 (1978).

[2] R. R. Sokal, *Science*, **185**, 1115 (1974); P. H. A. Sneath and R. R. Sokal, *Numerical Taxonomy*, Freeman, San Francisco, CA, 1973.

[3] B. L. Karger, L. R. Snyder, and C. Horvath, *An Introduction to Separation Science*, Wiley, New York, 1973.

[4] H. H. Strain, T. R. Sato, and J. Engelke, *Anal. Chem.*, **26**, 90 (1954).

[5] P. R. Rony, *Chem. Eng. Prog., Symp. Ser.*, **68**, 89 (1972).

[6] H. L. Lee, E. N. Lightfoot, J. F. G. Reis, and M. D. Waissbluth, in N. Li, Ed., *Recent Developments in Separation Science*, Vol. III, Part A, CRC Press, Cleveland, 1977, pp. 1–70.

[7] T. R. C. Boyde, *Sep. Sci. Technol.*, **14**, 79 (1979).

[8] H. G. Cassidy, *J. Chem. Educ.*, **27**, 241 (1950).

[9] C. J. King, *Separation Processes*, McGraw-Hill, New York, 1971.

[10] C. J. O. R. Morris and P. Morris, *Separation Methods in Biochemistry*, 2nd ed., Pitman, London, 1976.

[11] W. G. Pfann, *Sep. Sci.*, **1**, 1 (1966).

[12] H. R. C. Pratt, *Countercurrent Separation Processes*, Elsevier, Amsterdam, 1967.

[13] J. C. Giddings, E. Kucera, C. P. Russell, and M. N. Myers, *J. Phys. Chem.*, **72**, 4397 (1968).

[14] J. C. Giddings and K. Dahlgren, *Sep. Sci.*, **6**, 345 (1971).

[15] P. G. Righetti, E. Gianazza, C. Gelfi, and P. K. Sinha, *Protides of the Biological Fluids*, **34**, 3 (1986).

[16] V. G. Gaikar and M. M. Sharma, *Separation and Purification Methods*, **18**, 111 (1990).

[17] J. C. Giddings, W. A. Manwaring, and M. N. Myers, *Science*, **154**, 146 (1966).

[18] J. Murkes and C. G. Carlsson, *Crossflow Filtration*, Wiley, New York, 1988.

[19] D. S. Strickler, *Sep. Sci.*, **2**, 335 (1967).

[20] P. E. Barker, in A. Zlatkis and V. Pretorius, Eds., *Preparative Gas Chromatography*, Wiley-Interscience, New York, 1971, pp. 325–394.

[21] P. G. Grodzka and B. Facemire, *Sep. Sci.*, **12**, 103 (1977).

[22] M. Bier, in M. Bier, Ed., *Electrophoresis*, Vol. 1, Academic Press, New York, 1959, pp. 263–315.

EXERCISES

7.1(**) A glass jar is partially filled with carbon tetrachloride and water, which form two immiscible phases. Air in the system constitutes another phase. A small amount of iodine is introduced into the system; the iodine distributes selectively into the carbon tetrachloride. The jar is capped. Sketch the chemical potential profile $\mu^* = \mu^0$ of the iodine along the entire axis of the jar, extending from a point just outside the lid ($x = 0$) to a point 1 mm within the glass bottom ($x = h$).

7.2(*) Suppose you stop the flow in a chromatographic column. You would still get some separation as a result of partitioning between mobile and stationary phases. With what separation category in Table 7.1 would this technique then belong?

7.3(**) Construct an expanded form of Table 7.2 based on Figure 7.2. Redistribute the methods shown in Table 7.2 among the expanded set of categories.

8

SEPARATION BY EXTERNAL FIELDS (*Sc* METHODS): ELECTROPHORESIS AND SEDIMENTATION

We turn now from general considerations governing all methods of separation to the specifics of a simple but important class: separation in a stationary one-phase system in which displacement is caused by an external field or, equivalently, by some gradient maintained by outside influences. This is category *Sc*, as explained in Chapter 7.

8.1 TYPES AND USES OF FIELDS

For operational purposes, we define a field as any external influence extending through space capable of causing the relative displacement of components with respect to their surroundings. The most prominent examples are electrical and sedimentation fields. Other fields exist, but few have been used to any significant extent in static (nonflow) separations. Some have not been used for any separation, static or flow. The most common reason for neglect is the weakness of the forces generated by most fields. We address more specifically the criteria that must be satisfied for effective separation below.

A partial compilation of field types is shown in Table 8.1. Some static (*Sc*) methods (or phenomena) associated with such fields are shown. (Processes like photophoresis—based on radiation pressure—hardly deserve to be called methods because the separation power is so weak that no useful methodology exists. (However, natural photophoresis fractionated the gases forming our primitive earth as described in Chapter 1.) For completeness, some non-*Sc* (largely based on flow) separation systems utilizing these fields are shown in the final column of the table.

TABLE 8.1 **Types of "Fields" that Cause Relative Displacement and thus (potentially at least) Some Degree of Relative Separation**[a]

Field or Gradient	Static (Sc) Separation Methods or Phenomena	Other (non-Sc) Separation Methods
1. Electrical	Electrophoresis, isoelectric focusing, isotachophoresis	Electrodialysis, electrodeposition, electrostatic precipitation, countercurrent electrophoresis, electrical FFF[b]
2. Sedimentation	Rate-zonal sedimentation, isopycnic sedimentation	Equilibrium sedimentation, centrifugal elutriation, sedimentation FFF[b]
3. Temperature gradient	Thermophoresis	Thermogravitational methods, thermal FFF[b]
4. Electrical (nonuniform)	Dielectrophoresis	
5. Magnetic (nonuniform)	Magnetophoresis	Mass spectroscopy[c], magnetic separations, magnetic FFF[a]
6. em radiation	Photophoresis	
7. Concentration gradient	Diffusophoresis	

[a]Principal separation techniques, if they exist, are listed for each field.
[b]FFF = field-flow fractionation
[c]Inertial transport term important

By and large, analytical *Sc* separations, like most analytical separations, are carried out as a zonal procedure. This procedure is generally implemented by establishing a thin sample zone at the beginning of a separation path (see Chapter 5). A field is applied that forces the components of the sample to migrate along the separation path. Migration is differential as a consequence of the differing properties of the sample components; thus separation is achieved (see Figure 8.1). However, in some methods the zonal separation is achieved not because of differential transport rates, but because the components seek different steady-state positions along the separation path. This important variant will be described in Section 8.10.

In the pages to follow we find that the resolving power of some *Sc* methods can be very high. However, for most of these methods the zones are left distributed along the separation path when the separation is completed. Without flow there is no direct means, as in chromatography, for eluting, collecting, and examining sample fractions. Sometimes flow can be added as a separate procedure to withdraw the separated components after the separation is complete, although not without a loss of resolution. Electroosmotic flow can similarly be used in some forms of electrophoresis. (Both of these flows fall in the category of passive flow—nonessential to the separation process—as described in Chapter 7.) Other methods for the collection of fractions include the cutting of the separation medium into sections from which the components are dissolved.

Normally lacking the flow necessary to carry components to a detector positioned outside the separation system, the detecting element must be moved back to the separation path. Here, by sensing at a fixed position, successive components can be observed as they migrate by. Alternately, components are detected by scanning along the separation path and recording changes in optical absorption, fluorescence, radioactivity, and so on.

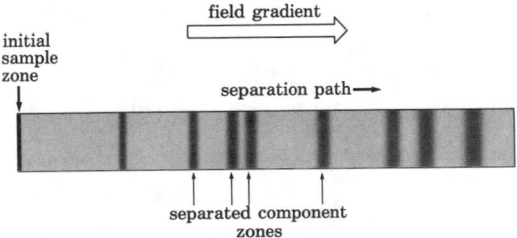

Figure 8.1. Illustration of zonal separation executed by an applied field (e.g., electrical) without flow.

8.2   TRANSPORT AND THEORETICAL PLATES

When the flow velocity v is zero, the total solute displacement velocity $W = U + v$ reduces to $W = U$. Transport therefore follows the equation

$$\frac{dc}{dt} = -U \frac{dc}{dx} + D_T \frac{d^2 c}{dx^2} \tag{8.1}$$

which comes from Eq. 3.30 or 3.34. In favorable cases zonal separation can occur by virtue of relatively small differences in U providing total diffusion coefficient D_T is small.

We note that D_T may or may not equal the molecular diffusion coefficient D, depending on the absence or presence of extraneous random dispersion mechanisms as discussed in Chapter 5. For simplicity, we relate D_T to D by

$$D_T = \theta D \tag{8.2}$$

where dispersion coefficient θ is unity when molecular diffusion acts alone and exceeds unity when other processes contribute.

We would like to know the plate height and the number of theoretical plates generated in Sc separations. Equation 5.38 gives plate height as $H = 2D_T / W$. By writing $D_T = \theta D$ and $W = U$, this becomes

$$H = \frac{2\theta D}{U} \tag{8.3}$$

Equations 3.21 and 3.19 provide fundamental expressions for D and U: $D = \mathscr{R}T/f$ and $U = -(1/f)d\mu^{\text{ext}}/dx$. When these are substituted into Eq. 8.3, we obtain

$$H = \frac{2\theta \mathscr{R}T}{-d\mu^{\text{ext}}/dx} \tag{8.4}$$

We observe that friction coefficient f (appearing both in D and U) is absent from H because it cancels out in forming the ratio D/U. Thus H is independent of f.

It was shown in Chapter 5 that the number of theoretical plates provides a direct measure of separation efficiency. Using $N = X/H$, Eq. 5.41, we have

$$N = \frac{-(d\mu^{\text{ext}}/dx)X}{2\theta \mathscr{R}T} \tag{8.5}$$

The numerator of this expression is simply the negative of the total chemical potential change $-\Delta\mu^{\text{ext}}$ gained over displacement interval X. (This is equivalent to equating the change Δy in any variable y to $(dy/dx)\Delta x$, valid when dy/dx is constant or Δx is small.) Therefore

$$N = \frac{-\Delta\mu^{\text{ext}}}{2\theta\mathscr{R}T} \tag{8.6}$$

a general equation for the efficiency of Sc methods first obtained by the author in 1969 [1].

Equation 8.6 shows N to be the ratio of two energies: $-\Delta\mu^{\text{ext}}$, which is the energy that creates and structures the separation, and thermal energy $\mathscr{R}T$ (times factor 2θ), which drives the diffusion that dissipates the separation. Intuition clearly supports Eq. 8.6 in suggesting that separation quality can be improved by increasing the structuring influence of $-\Delta\mu^{\text{ext}}$ relative to the dissipative role of $\mathscr{R}T$ and the dispersion factor θ.

We recall by reference to Eq. 5.61 that the maximum number of separable peaks (the peak capacity) is $\sim 0.25\, N^{1/2}$, which, with N from Eq. 8.6, yields

$$n_c \simeq \left(\frac{-\Delta\mu^{\text{ext}}}{32\theta\mathscr{R}T}\right)^{1/2} \tag{8.7}$$

This equation shows that n_c increases with the square root of the ratio of structuring energy $-\Delta\mu^{\text{ext}}$ to dissipating thermal energy $\mathscr{R}T$.

We noted before that some fields are not used in static (Sc) separations because the forces they generate are so weak. We see that a weak force, $-d\mu^{\text{ext}}/dx$, leads to a small structuring energy $-\Delta\mu^{ext}$, which is inadequate to provide values of N or n_c large enough for practical use.

In the following pages we describe several practical Sc techniques. We then evaluate the potential magnitude of N and n_c for these systems. This evaluation will provide a rough comparison between the Sc methods and will show their strength relative to chromatography, where N is a well-known measure of separation efficiency.

8.3 ELECTROPHORESIS: CAPILLARY, GEL, AND OTHER FORMS

In electrophoresis, an electrical field causes the differential transport of charged species. Many experimental variants exist, giving the method particular versatility in the analysis of biological materials [2–6].

Analytical electrophoresis is most commonly carried out as a zonal method (Figure 8.1). In common with other zonal methods, which are described as a class in Chapter 5, a thin band of solute is deposited at the beginning of the separation path. A voltage is then applied across that path, leading to the migration of charged species away from the starting band and along the path. Separation develops because of differences in migration velocity U. Increments in U can arise in two ways: different species have different electrical charges, which determine how vigorously they are driven by the applied voltage, and they have different degrees of frictional drag, which differentially oppose their electrophoretic motion.

The separation path in electrophoresis may follow a free electrolyte held in a tube or it may run through a granular, fibrous, or gel-like strip of supporting medium permeated with electrolyte. A major factor in the choice is the avoidance of gravity-induced convection of the type discussed in Section 4.8. This convection is amplified by the heating due to the passage of electrical current (see Section 8.5). If, for example, a region of liquid is less dense than a neighboring region due to differential heating, the lighter fluid will rise unevenly and distort the zone structure in that region. This leads to zone broadening and overlap with adjacent constituents.

Electrophoresis in a free electrolyte without a support material is most straightforward and avoids complicating interactions with the support, but it is clearly convection prone.

Several steps can be taken to combat convection in a free electrolyte. Hjerten, for example, used a horizontal tube rotating around its own axis [7]. Any small convective displacement due to gravity is exactly reversed as the tube rotates 180°. Thus the gravitational convective displacements—while not eliminated—exactly cancel one another in the course of rotation. More recent experiments aboard earth satellites are carried out under near-zero gravity (nonconvective) conditions [8, 9]. It has also been found useful to work at 4°C where water has its maximum density and where, consequently, density is least sensitive to temperature. Stabilizing density gradients have been introduced in some cases to counteract convection.

Capillary Electrophoresis

It is also possible to use extremely thin tubes so that convection is supressed by the high surface/volume ratio and the resulting high frictional drag opposing fluid motion in narrow tubes (see Section 4.8). Heat is rapidly dissipated from such tubes, making it possible to use high voltages without a significant heat buildup. The method resulting from the use of thin tubes is termed *capillary electrophoresis* (*CE*) or *capillary zone electrophoresis* (*CZE*) [6, 10–12]. (HPCZE, or high-performance CZE, provides an unnecessary proliferation of terms and will not be used here.) In 1979 Everaerts and co-workers first attempted CZE in thin Teflon tubes [10]. In 1981 Jorgenson and Lukacs developed a high resolution CZE system based on glass tubes only 75 μm (0.075 mm) inside diameter [11]. The resolving power achieved in this pioneering work is illustrated in Figure 8.2.

Interest in CZE exploded in the latter half of the 1980s [6, 12–18]. The inherent simplicity of the method and the extraordinary resolving power have drawn many adherents to this technique. The method is particularly effective with proteins, peptides, and other charged biological species; inorganic species can be separated as well.

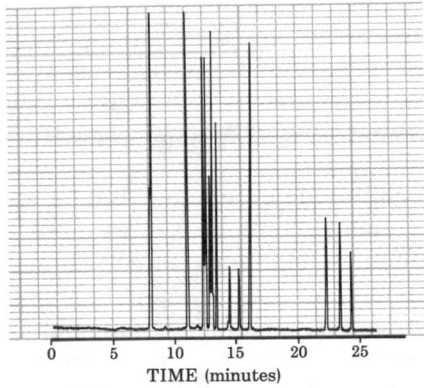

Figure 8.2. Capillary zone electrophoretic separation of dansyl amino acids. (From ref. 11. Reprinted with permission from J. W. Jorgenson and K. D. Lukacs, *Anal. Chem.*, **53**, 1298 (1981). Copyright 1981 American Chemical Society.)

A schematic of the apparatus used for CZE is shown in Figure 8.3. Here a thin capillary up to 1 m in length bridges across two buffer reservoirs. A voltage increment as high as 30,000 V is applied between the reservoirs by a power supply. A thin sample plug is introduced into one end of the capillary tube by one of several injection methods [16, 19]. The voltage then induces the migration of appropriately charged species through the capillary. The electrophoretic transport is frequently augmented by a significant level of electroosmotic flow (acting primarily as a passive flow mechanism). The electroosmotic flow is sometimes strong enough to carry species of both positive and negative charge through the capillary despite the opposite electrophoretic motion of one of the two charges.

Capillary tubes are usually made of fused silica 25–75 μm in diameter. In most buffers the wall of the capillary acquires a negative charge, leaving a cloud of positive counterions near the wall. Electroosmotic flow results when these counterions (and the surrounding fluid) are drawn to the cathode (see Section 4.9). By placing the cathode at the outlet end of the capillary tube, electroosmotic flow helps carry the species into and through the tube and detector.

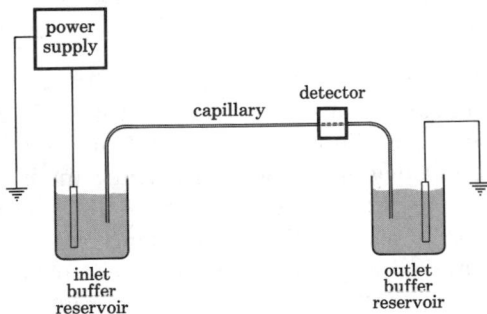

Figure 8.3. Main elements of capillary electrophoresis system.

The detector is usually built around the capillary tube which must remain unbroken so that the voltage can draw the desired components through the detection region. Detection is usually accomplished by passing a beam of light directly through the capillary tube and observing UV absorbance or fluorescence (16, 20, 21). Other detection systems are based on conductivity [22] and mass spectroscopy [23].

A number of variants of CZE have recently been introduced. Terabe has developed a technique in which micelles—microscopic soap globules—are introduced into the capillary [24]. The micelles (like most surfactants) are charged and are thus driven by electrical forces through the buffer. However, each micelle represents a miniscule domain of a second phase, relatively nonpolar compared to the surrounding water. Various species, including neutral molecules, can partition between the aqueous and micellar phases. Species that partition strongly into the micelles are carried piggyback with the micelles as they undergo electrophoretic migration; species remaining in the aqueous phase remain stationary (unless they are displaced by electroosmosis or, if charged, by their own electrophoresis). Consequently a chromatographic-like partitioning effect between two phases can lead to the differential displacement of unlike species including neutral components. The technique is as much chromatographic as electrophoretic in nature (depending on whether differential partitioning or differential mobilities are primarily responsible for separation). This hybrid technique can be termed *micellar capillary electrochromatography* (*MCE*).

More recently Karger and co-workers have introduced gel media into capillaries [25]. Gels have unique electrophoretic properties including a sieving effect (see below) that makes their use advantageous for molecular weight based protein and nucleic acid separation. When gels are incorporated within a capillary, high voltages and rapid heat dissipation lead to exceptional resolution [26].

In general CZE offers outstanding separative capabilities and it provides a more instrumental (and less manual) approach than most other electrophoretic techniques; it is consequently promising for research, routine separations, and automated analysis.

Gel Electrophoresis

As noted in Section 4.8, a support medium serves to break the liquid volume up into narrow capillary-like spaces whose high surface/volume ratio supresses convection. Support media work so effectively that they are used for most present day electrophoresis. These media consist mainly of gels; the corresponding techniques are called *gel electrophoresis* (*GE*). Other media include starch granules, membranes, paper, and so on. These materials may be used as strips or as a filler in glass or other nonconducting tubes. Historically, paper strips soaked with electrolyte dominated support-based electrophoresis at its beginning in the early 1950s [27]. Porous cellulose acetate membranes largely replaced paper within a decade. The most successful medium has been polyacrylamide gel (giving *PAGE— polyacrylamide gel electrophoresis*); agarose gel is also important. These gels, as others, consist of long molecular chains that are periodically crosslinked together. The space between the chains is occupied by the electrolyte and the charged solute.

A high degree of crosslinking in the gel reduces the pore size between polymer chains and can impede not only electroosmosis but also the motion of large solute molecules (e.g., proteins). This "sieving" effect leads to increased frictional drag. With sufficient crosslinking, the sieving effect, which depends on molecular size, becomes a selective factor in separation. Proteins, denatured and evenly charged by sodium dodecyl sulfate (SDS), can be separated according to differences in their chain length and thus molecular weight in polyacrylamide gels [2, 4]. DNA chain fragments differing by only one base can be similarly separated (see Figure 8.4), making it possible to sequence DNA [28]. Without the selective sieving effects of gels, neither DNA nor SDS treated proteins are electrophoretically separated according to molecular weight or size.

For very large biopolymer molecules (e.g., DNAs over 30×10^6 molecular weight), the polymeric chain orients with the field and snakes (reptates) through the gel. In this way the sieving effect is rendered ineffective; it can be restored by switching back and forth between fields oriented at different angles in a technique called *pulsed field gel electrophoresis*, or *PFGE* [29, 30].

Other Electrophoretic Techniques

At the opposite extreme of molecular size from DNA, small molecules and free atoms can be ionized and subjected to electrophoresis in the gas phase. The technique, named *ion mobility spectroscopy* (*IMS*) or *plasma chromatography*, would better be called *gaseous electrophoresis* for consistency. The uses and theory of this method can be found in the literature [31, 32].

A rather different form of zonal electrophoresis is isotachophoresis [3–5, 33–35]. This is a nonlinear form of electrophoresis in which the presence of one component affects the migration of another. Consequently, the linear theory of Chapter 5 leading to Gaussian zone development no longer applies. The process is understood by recognizing that the current passing through any one cross section of the electrophoretic bed must equal that passing through any other. If the conductivity is lower than average in a certain region of the bed, the voltage must be higher to maintain a constant current. If ions of several different mobilities occupy a given segment, ions of the highest mobility will naturally move ahead of those of the second highest mobility, and so on. However, faster ions cannot leave following ions behind. If they did, an ion-free gap would develop between them. This

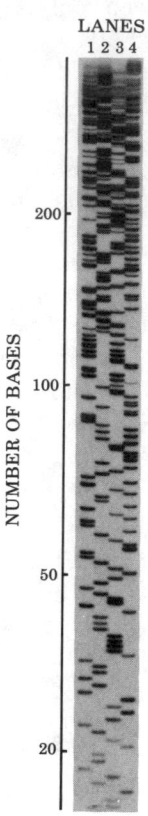

Figure 8.4. Determination of base sequence of a strand of viral (bacteriophage M13) DNA by the electrophoretic separation of negatively charged fragments derived from the DNA chain. Fragments are produced by reagents that create chains complementary to the original DNA strand acting as a template. These chains terminate at specific bases depending on the reagents. Electrophoresis of fragments is carried out from top to bottom in four lanes of a thin (0.4 mm) film of denaturing polyacrylamide gel. In lane one (left), all DNA fragments are terminated at adenine base sites by the adenine specific reagent. In lanes two, three, and four, fragments are similarly produced that terminate at guanine, cytosine, and thymine base sites, respectively. Between the four lanes, all possible fragment lengths are represented, making it possible to deduce the base sequence of the original DNA strand. Scale at left indicates number of bases in separated fragments. Run time is about 1.5 h at 1600 V. Detection of fragments is by autoradiography of gel using radioactive phosphorus (^{32}P) label. (Courtesy of Norma M. Wills and Raymond F. Gesteland, University of Utah.)

gap would have a very low conductivity and thus a high potential gradient. The high potential gradient would rapidly draw the trailing ions into the gap, thus filling it. (In most electrophoretic techniques, gaps are filled by background electrolyte.) Because of this phenomenon, gaps never form and one gets a sequence of contiguous zones following one another along at equal velocities. The concentration of a given ion type is constant across its zone. The length of the zone is proportional to the amount of that type of ion in the sample; the measurement of zone length thus provides quantitative analysis. However, the boundary between zones is not absolutely sharp because of the mixing in this boundary region caused by diffusion and related dispersion processes.

Isoelectric focusing (*IEF*) is a steady-state zonal electrophoretic technique. Using IEF, proteins can be separated according to differences in their isoelectric points. The principles of IEF are described in Sections 8.10 and 8.11.

Two-dimensional (2D) electrophoretic methods are important variants of electrophoresis. In Section 6.4 we noted that electrophoresis could combine with flow in a 2D system providing continuous (and thus preparative) separation. Fundamentally, this is simply zonal electrophoresis converted into a continuous form by nonselective flow (see Section 7.5). If we observe the separation at different positions along the flow axis (as illustrated for one position in Figure 7.3), we have essentially a series of snapshots of the zones evolving with time. Each component zone is deflected from the flow axis at a unique angle ϕ as a consequence of the evolution of the electrophoretic separation.

Many 2D planar structures have been used to implement deflection (continuous flow) electrophoresis. The primary requirement is that flow and electrophoresis be carried out simultaneously and uniformly. Hanging paper curtains soaked with electrolyte and fed a stream of electrolyte from above served admirably for this purpose when the technique was initiated in the 1950s. In recent years thin flow channels enclosed between flat plates have become important. The process is complicated by parabolic flow, which distorts and effectively broadens the electrophoretic zones. More detail is available in the cited references on electrophoresis [3–5].

Two-dimensional discrete (as opposed to continuous) electrophoresis has become a premier analytical tool for separating complex protein mixtures. Since this technique employs IEF, further discussion will be postponed to Section 8.10.

8.4 SEPARATION POWER IN ELECTROPHORESIS

Here we establish the number of theoretical plates and separable zones achievable in ideal zone electrophoresis by following the approach in ref. [1]. This analysis aids in system evaluation, comparison, and optimization.

The electrostatic force exerted on a mole of charged particles by an electrical field of strength E is

$$F = z\mathcal{F}E \tag{8.8}$$

where z is the effective charge (ionic charge minus part of the opposing double layer charge) of a single particle in proton units and $\mathcal{F}$ is the Faraday, or charge per mole of protons. The negative chemical potential drop across the separation path $-\Delta\mu^{\text{ext}}$ is the above force multiplied by the length X of migration

$$-\Delta\mu^{\text{ext}} = z\mathcal{F}EX = z\mathcal{F}V \tag{8.9}$$

in which $V = EX$ is the voltage drop along path length X. Substitution of this $-\Delta\mu^{\text{ext}}$ into Eq. 8.6 gives for the number of theoretical plates

$$N = \frac{z\mathcal{F}V}{2\theta\mathcal{R}T} \tag{8.10}$$

For an ideal process ($\theta = 1$) at $T = 290$ K this equation reduces to

$$N = 20zV \tag{8.11}$$

where we have used $\mathcal{F} = 96,500$ C/mol. Thus voltages in the range 10^2–5×10^4 V with z values in the range 1–10 elementary charges are capable of yielding 2000–10,000,000 theoretical plates, a range comparable to but extending somewhat beyond that found in chromatographic systems.

An expression for peak capacity is obtained by substituting Eq. 8.9 into Eq. 8.7

$$n_c = \left(\frac{z\mathcal{F}V}{32\theta\mathcal{R}T}\right)^{1/2} \tag{8.12}$$

With $\theta = 1$ and $T = 290$ K, this becomes

$$n_c = (1.25\ zV)^{1/2} \tag{8.13}$$

With $V = 10^2$–5×10^4 and $z = 1$–10, as above, we see that from 10 to 800 distinct peaks can theoretically be resolved in electrophoresis. (In practice the number of resolvable components is much less than n_c due to statistical peak overlap as explained in Section 6.7). This enormous resolving power is consistent with observation (see Figures 8.2 and 8.4). Theory thus provides an explanation of the unusual power of electrophoresis and an insight into the variables (z and V) that must be manipulated for increased performance.

Since a chromatographic mechanism (solute partitioning between two phases moving relative to one another) is involved in micellar capillary electrochromatography, the theory outlined above for band broadening in electrophoresis does not apply. Theoretical aspects of MCE have been developed by Terabe et al. [36] and by Davis [37].

8.5   ELECTROPHORESIS: ADDITIONAL CONSIDERATIONS

The above theory shows that separation power increases with the total system voltage V. The use of high voltage electrophoresis is designed to exploit this potential. Voltages of 30,000 V and gradients of 500 V/cm are sometimes used, as opposed to conventional gradients of 5–10 V/cm. However, the increased electrical energy expended at high voltage is converted into troublesome heat (Joule heat) through the frictional drag of the various ionic species forced along the conducting pathway.

The magnitude of the energy conversion can be calculated as follows. The amount of electrical energy converted to heat per unit time is given by the product of voltage V and current i. This heat production is spread over the volume of the conducting medium LA, where L is length and A the cross-sectional area of the medium. Thus the rate of heat input $\dot{H}$ per unit volume is

$$\dot{H} = \frac{iV}{LA} \tag{8.14}$$

Using $i = V/R$ and $R = L/\kappa A$, we get

$$\dot{H} = \frac{\kappa V^2}{L^2} \tag{8.15}$$

where κ is the conductivity (reciprocal resistivity) and R the total resistance of the system. The conductivity can be written as the sum

$$\kappa = \sum \Lambda_{mi} c_i \tag{8.16}$$

where c_i is the concentration and Λ_{mi} is the molar conductivity of electrolyte i. In dilute solutions each Λ_m value can be approximated by the infinite dilution value Λ_m^0 and can be replaced by the weighted sum of cationic and anionic conductivities at infinite dilution, λ_+^0 and λ_-^0, respectively

$$\Lambda_m^0 = \nu_+ \lambda_+^0 + \nu_- \lambda_-^0 \tag{8.17}$$

where ν_+ and ν_- are the number of cations and anions in the chemical formula (e.g., for $MgCl_2$, $\nu_+ = 1$, $\nu_- = 2$). Tables of $\lambda_+^{\,0}$ and $\lambda_-^{\,0}$ have been compiled in many places.

The above treatment shows that the energy conversion (like electrical energy losses generally incurred in conduction) increases with the square of the voltage drop V; heating effects thus escalate rapidly for increasing Vs. Heat production can be reduced by lowering κ, as shown by Eq. 8.15. A smaller κ can be realized by lowering each electrolyte concentration c_i, Eq. 8.16, and by choosing electrolytes with low ionic conductances λ_+ and λ_-, Eq. 8.17. Nonetheless, heat production remains a serious problem in high voltage electrophoresis. Excess heat will increase the tendency for convection, decompose heat-sensitive molecules, cause the evaporation of volatile components and of water from open slabs or sheets, and increase migration rate in a rather nonuniform manner. Clearly, heat must be removed as rapidly as possible.

To remove heat and stem its excessive buildup, the electrophoretic medium must be cooled. Paper strips, for example, are sandwiched between thin glass plates which are water cooled; gel-containing tubes or annuli can likewise be cooled by water or petroleum ether.

Any cooling process requires heat flow from the heated center to the cooled periphery of the electrophoretic medium. A temperature gradient must exist perpendicular to the electrophoretic path to drive the heat flow. Thus temperature differences are inevitable. In the heated center of the tube or strip, viscosity is reduced, as shown in Section 4.14. A lowered viscosity gives a reduced friction coefficient f according to Stokes law, Eq. 4.42. Reduced f values (amounting to 1–2% per °C) are associated with correspondingly higher migration velocities for a fixed electromotive driving force. Therefore, the charged species in the sample migrate more rapidly in the heated center than at the cooled outer edges of the electrophoretic medium. This uneven migration causes zone broadening and resolution loss [38, 39]. The zone broadening is equivalent to a large θ value in Eq. 8.2. Thus without extraordinary care in the design and use of high voltage electrophoresis systems, it is difficult to even approach the theoretical potential suggested by Eq. 8.13, which is based on the ideal θ value, $\theta = 1$.

The use of exceedingly thin capillary tubes (~ 50 μm) in capillary electrophoresis not only reduces convection but it also provides rapid heat dissipation. Using up to 30,000 V, on the order of 10^6 or more theoretical plates are realized in such tubes (see Figure 8.2).

Increased voltage (and the associated gain in electrical field strength E) increases ion migration velocity U and thus separation speed (providing θ does not increase too severely as a side effect). For many purposes, it is useful to have a convenient parameter to characterize migration velocities independent of E. The *electrophoretic mobility* μ fulfills this role; it is defined by

$$\mu = \frac{U}{E} \tag{8.18}$$

Since U equals force/friction coefficient, and the force on the charged species is given as $z\mathscr{F}E$ by Eq. 8.8, we have

$$\mu = \frac{z\mathscr{F}}{f} \tag{8.19}$$

Separation is achieved only when mobilities differ from one another; we see that different μs arise from different zs or fs, or both.

For quantitative calculations of μ, the charge z must be known. It was noted that z is the charge (in proton units) of the species undergoing migration less part of the double layer charge. This is explained as follows.

Each charged species accumulates, by electrostatic forces, a cloud of opposite charges around it [40]. The oppositely charged cloud remaining in solution is termed the *diffuse double layer* (or *Gouy–Chapman layer*). As the charged species is pulled through the medium by the electrical field, it drags along that part of the counterion cloud closest to the species. However, the cloud, because of its opposite charge, has an opposing force exerted on it by the electrical field; its opposing pull therefore slows the progress of the charged species under consideration. This retardation is called the *electrophoretic effect*. The retardation, measured by the effective decrease in z, is greatest for diffuse double layers that are thin and thus clustered tightly around the charged species where the drag effects are severe.

The characteristic thickness of the double layer is given by Debye–Hückel theory as h^{-1}, where h is

$$h = \left(\frac{8\pi e^2 s}{\epsilon k T}\right)^{1/2} \tag{8.20}$$

where e is the charge of a single proton or electron, ϵ is the dielectric constant, k is Boltzmann's constant, and s is the ionic strength

$$s = \frac{1}{2} \sum c_i z_i^2 \tag{8.21}$$

Depending on s, h^{-1} may vary from 3–1000 Å, or beyond. For example, in 10^{-2} M NaCl, $h^{-1} \simeq 30$ Å. The important point is that double layer thickness h^{-1} varies as $1/s^{1/2}$ and thus also with overall electrolyte concentration c as $1/c^{1/2}$. Consequently, retardation by the electrophoretic effect, which we indicated to be greatest for thin double layers (small h^{-1}), increases with electrolyte concentration.

It can be shown that the effective charge z is approximately related to the true ionic charge z_{ion} by [38]

$$\frac{z}{z_{\text{ion}}} = \frac{1}{1 + ha} = \frac{h^{-1}}{h^{-1} + a} \tag{8.22}$$

where a is the radius of the charged species. This equation confirms that retardation (small z) is enhanced by large h, large s, and therefore large c. The retardation is clearly highly significant for larger species (such as macromolecules and colloids) where radius a often exceeds the double layer thickness. For such species, mobility is often expressed in terms of the zeta potential [2].

For colloidal particles with $a \gg h^{-1}$, Eq. 8.22 reduces to $z/z_{ion} = h^{-1}/a$. Since the charge on a particle, z_{ion}, is proportional to its surface area, $4\pi a^2$ (if the particle is a sphere), z is proportional to radius a. Friction coefficient f is also proportional to a as shown by Stokes law, Eq. 4.42. Thus mobility μ, Eq. 8.19, is largely independent of size for most colloidal particles of like composition, making the size-based separation of colloids by electrophoresis difficult. Similarly, the mobility of uniformly charged polymer chains (such as DNA and SDS denatured proteins) is nearly independent of chain length in free solution.

Mobility is also reduced slightly by another phenomenon termed the *relaxation effect*. Here the charged species, as it is displaced by the electric field from the center of the double layer, is acted on by the opposite charge of the double layer to pull it back [38].

Electroosmotic flow, described in Section 4.9, is another complicating factor in electrophoresis. The electroosmotic flow process is often responsible for nonselective ion transport superimposed on the electrophoretic transport. When electroosmotic displacement is significant, it must be kept in mind that the distance X in the preceding plate height equations is the displacement distance due to electrophoresis alone; it does not include electroosmotic displacement. Also the voltage V must be calculated as that applied over the path of electrophoretic displacement only, not including the distance of electroosmotic displacement [41].

Other theoretical aspects of electrophoresis are discussed by Wieme [38].

8.6 SEDIMENTATION

Sedimentation is a process in which gravitational or centrifugal acceleration is used to force particles through a fluid. The displacement process requires that the particles have a density different from that of the fluid. Differential displacement, leading to separation, arises because particles have different densities, different masses, and/or different friction coefficients, which may partially originate in different particle shape.

For sufficiently large particles, gravity is an adequate driving force for sedimentation. Gravitational sedimentation is a common phenomenon; suspended solids dropping through flowing water over time have filled entire valleys with sediments. Not only are the solid materials separated from the water in such a process (solid-liquid separation), but granules of different sizes are separated from one another, deposited in a sequence from large to small as the water slows down because of a gradient loss. In a similar way, the atmosphere is constantly purified of injected dust and particulate material by gravitational sedimentation. Many industrial processes also depend upon the settling power of gravity.

The sedimentation of particles in the colloidal and macromolecular range generally requires the more powerful forces of a centrifuge [42, 43]. A schematic diagram of a multicomponent separation induced by centrifugal sedimentation is shown in Figure 8.5. The separation of a number of components is typical of the analytical use of centrifugation. We should note that most centrifugation—industrial or lab scale—is used for crude solid-liquid separation. Here one simply spins all particulate material above a certain particle size to the outer wall where it condenses as a precipitate (or *pellet*) and can be collected. Such procedures are not only commonly used to concentrate valuable solid products but also to purify liquids, for example, contaminated water.

The greatest hurdle to overcome in analytical centrifugation, as in electrophoresis, is the tendency for convection. In electrophoresis, convective effects are multiplied by the heat produced. In sedimentation, the convection problem is amplified because regions with any density differences, no matter how small, are subjected to differential centrifugal forces (leading to differential convective motion) up to 10^5 times stronger than those generated by gravity. In nearly all forms of analytical centrifugation, a density gradient must be established to prevent this convection. To install a density gradient, a density-altering component must be added to the separation medium at unequal concentration levels so that the density increases as one proceeds out along radial coordinate r.

Many different mechanical arrangements have been used in the construction of rotors for centrifugation. For analytical centrifugation, these rotors must be designed to overcome the problem of convection and to make it possible to detect the component zones which are normally hard to track

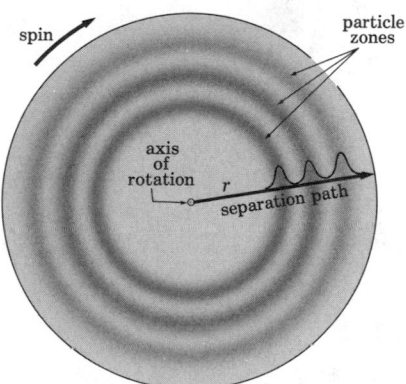

Figure 8.5. Schematic diagram of separation by centrifugal sedimentation.

because they are held within the bulky, spinning rotational apparatus. Two approaches are particularly useful.

In the disc centrifuge, the separation occurs in a cavity between two transparent plastic discs [44]. A mild density gradient is set up in the cavity. A thin optical beam can be transmitted through the discs during rotation, leading to sample detection. An example of the separation of colloidal particles by disc centrifugation is shown in Figure 8.6. Disc centrifuges have been used up to about 10^4 rpm, which gives a force equivalent to approximately 10^4 gravities.

A series of zonal rotors, some of which can be used for analytical centrifugation, have been developed by Norman Anderson [45]. These rotors utilize a cylindrical or bowl-shaped cavity. The rotors are generally made of titanium whose high strength-weight ratio allows operation in some cases up to 100,000 rpm. In a zonal rotor, special connecting seals are used to introduce a liquid with a preformed density gradient into the centrifuge during slow rotation. A band of sample material is then introduced at the inner edge (meniscus) of the spinning liquid. Once the gradient material and the sample are in place, the rotation rate is greatly increased to carry out the sedimentation process. When separation is complete, the rotor is slowed and the contained liquid with its separated bands is slowly withdrawn from the rotor so that the separated components can be detected and, if necessary, collected.

Sedimentation in an ultracentrifuge is a powerful means for separating biological particles and large macromolecules. Sedimentation is the only static field (*Sc*) method other than electrophoresis (in all of its various forms) commonly used for multicomponent separations.

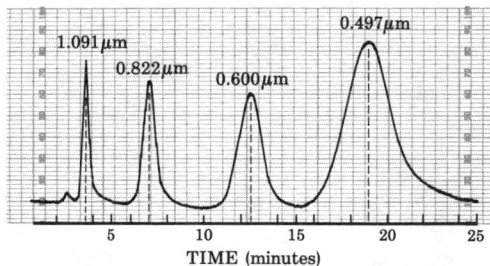

Figure 8.6. Separation of polystyrene latex beads of four different diameters (indicated in the figure) by a disc centrifuge operated at 3586 rpm. (From ref. 44. Reprinted with permission from R. M. Holsworth, T. Provder, and J. J. Stansbrey, in T. Provder, Ed., *Particle Size Distribution*, ACS Symposium Series No. 332, American Chemical Society, Washington, DC, 1987, Chapter 13. Copyright 1987 American Chemical Society.)

8.7 SEPARATION POWER IN SEDIMENTATION

The sedimentation force acting on a mole of particles of molecular weight M is

$$F = M(1 - \bar{v}\rho)G \tag{8.23}$$

where $\bar{v}$ is the partial specific volume of the particles, ρ is the solvent density (which enters in relationship to the buoyancy force), and G is the acceleration $\omega^2 r$ existing at radial coordinate r at a rotation rate of ω radians/s. Quantity $\bar{v}$ is defined as the rate of increase in solution volume V with respect to added mass m of particles

$$\bar{v} = (\partial v / \partial m)_{T,p} \tag{8.24}$$

The value of $\bar{v}$ can often be approximated as reciprocal particle density, ρ_s^{-1}.

The negative chemical potential increment associated with migration from radius r_1 to r_2 is the force integrated over the r_1 to r_2 interval. This becomes

$$-\Delta\mu^{\text{ext}} = M(1 - \bar{v}\rho)\omega^2 \int_{r_1}^{r_2} r \, dr = M(1 - \bar{v}\rho) \frac{\omega^2(r_2^2 - r_1^2)}{2} \tag{8.25}$$

The liquid density ρ in the above equations may vary substantially over the separation path due to the density gradients, in which case ρ should technically be included in the integration, or assigned a properly weighted average value.

The number of theoretical plates can be obtained by substituting the above $-\Delta\mu^{\text{ext}}$ into Eq. 8.6

$$N = M(1 - \bar{v}\rho) \frac{\omega^2(r_2^2 - r_1^2)}{4\theta \mathcal{R} T} \tag{8.26}$$

If we replace $\bar{v}$ by $1/\rho_s$ where ρ_s is solute density and $r_2^2 - r_1^2$ by $2\bar{r}\Delta r$ where $\bar{r}$ is the average radial coordinate $(r_1 + r_2)/2$ and where Δr is the displacement distance $r_2 - r_1$, we have

$$N = M\left(\frac{\rho_s - \rho}{\rho_s}\right) \frac{\omega^2 \bar{r}\Delta r}{2\theta \mathcal{R} T} \tag{8.27}$$

These equations show that N increases in direct proportion to molecular weight M, suggesting that sedimentation becomes effective only at high M values [1]. The following calculations illustrate this point.

Some modern high-speed centrifuges of $\sim 10^5$ gravities yield $\omega^2 \bar{r}\Delta r$ of $\sim 10^9$ cm^2/s^2, while $1 - \bar{v}\rho$ is sometimes as large as 0.5. At $T = 300$ K with $\theta = 1$, N becomes

$$N \sim 10^{-2} M \tag{8.28}$$

Thus, a reasonable N for analytical separation, about 1000, can only be achieved for a molecular weight of $\sim 10^5$ or higher, and then only at these very high rotation rates. Consequently, sedimentation methods are not particularly powerful for mixtures of proteins, many of which have $M < 10^5$. Most synthetic polymers, for which M is typically $\sim 10^5$, are also marginally separated by sedimentation. However, sedimentation should be highly effective for virus (and larger) size particles, for which M is typically $\sim 10^7$. In this case, $N \sim 10^5$ and the potential effectiveness of separation should begin to rival that of electrophoresis. Unfortunately, the high resolving power thus predicted has not been realized experimentally.

While there are many practical problems in realizing the theoretical power of sedimentation, major advances have accompanied the development of the disc centrifuge and of the zonal rotor [43–45].

8.8   THERMAL DIFFUSION: OVERLOOKED OPPORTUNITY?

Thermal diffusion is a process in which solute is driven through solvent by the action of a temperature gradient rather than by a concentration (or chemical potential) gradient [46]. It is a natural outgrowth of the laws of irreversible thermodynamics (Section 3.2) in which all driving forces are expected to be associated with some transport of matter.

We have observed that the direct separation of components in static systems is commonly carried out with both electrical and sedimentation fields. By contrast, we rarely hear of direct (*Sc*) separations using thermal diffusion. At present, thermal diffusion is nearly always employed with flow in such a manner that the flow multiplies the separability [46]. It is reasonable to inquire if an opportunity has been overlooked in the neglect of static separations by thermal diffusion. In obtaining an answer to this question, we develop an approach that can be used for judging the potential effectiveness of other uncommon gradients and fields in direct *Sc* separations as well.

Since thermal diffusion is a nonisothermal process and thus cannot be considered as driven by chemical potential gradients, we must go directly to the solute flux equations to understand the capacity of thermal diffusion for separation. The basic law expressing the flux density caused by thermal diffusion [46–48] is

$$J = \frac{\alpha c D}{T} \frac{dT}{dx} \tag{8.29}$$

where α is the *thermal diffusion factor*, a dimensionless constant expressing the magnitude of the thermal diffusion effect. Values of α are distributed roughly as shown in Table 8.2.

TABLE 8.2 Range of Values of Thermal Diffusion Factor α

Class of Substances	Range of α Values
Light gases and isotopes	0–0.5
Common liquids	0–5
Polymers	0–300

Inasmuch as mean molecular velocity is given by $U = J/c$, Eq. 3.14, the velocity induced by thermal diffusion is simply

$$U = \frac{\alpha D}{T} \frac{dT}{dx} \tag{8.30}$$

Plate height H is related to U by Eq. 8.3, $H = 2\theta D/U$, which, when combined with Eq. 8.30, gives

$$H = \frac{2\theta T}{(\alpha \, dT/dx)} \tag{8.31}$$

The number of plates, $N = X/H$, becomes

$$N = \frac{\alpha X}{2\theta T} \frac{dT}{dx} \tag{8.32}$$

Inasmuch as $X \, dT/dx$ equals temperature drop ΔT over the path of separation, N is simply

$$N = \frac{\alpha}{2\theta} \frac{\Delta T}{T} \tag{8.33}$$

The maximum magnitude of N can be estimated by assuming $\theta = 1$ and a very large relative temperature increment $\Delta T/T = 1/3$, corresponding to a ΔT of about 100°C at room temperature. We get

$$N_{\text{max}} \sim \frac{\alpha}{6} \tag{8.34}$$

Considering the magnitude of the α values in Table 8.2, we see that N_{max} fails to reach unity for ordinary gases and liquids but may climb to 50 in the most favorable cases for polymers. Inasmuch as peak capacity is $\sim 0.25 \, N^{1/2}$, Eq. 5.61, 50 plates yields only ~ 2 peaks. This number is expected only under favorable circumstances, assuming everything at its maximum.

Thermal diffusion, then, simply fails to act on molecules with sufficient vigor to generate, by itself, good separations. However, thermal diffusion can be augmented by flow (in an $(F+)cd$ system) to produce strong separation effects. Such favorable separations occur in both Clusius–Dickel (thermogravitational) columns and thermal field-flow fractionation (Chapter 9).

8.9   OPTIMIZATION IN STATIC FIELD (*Sc*) SYSTEMS

Both resolution and peak capacity, with a given relative velocity difference $\Delta U / U$, increase with plate number N, as noted in Sections 5.7 and 5.8. Thus optimization of parameters to maximize both R_s and n_c requires the maximization of N. Equation 8.6 shows how this is to be done: The chemical potential drop must be as large as possible while θ must be minimal. In electrophoresis, the former involves increasing the effective charge or zeta potential of the solute and increasing the total voltage drop. In sedimentation, the centrifugal acceleration must be large and the effective mass high. In both cases θ is minimized by utilizing homogeneous media and by stabilizing the system against undesirable convective currents.

Because the transport of large molecules is inherently sluggish (due to a large friction coefficient f, Section 4.13), the speed at which macromolecules are separated by electrophoresis and sedimentation becomes an important consideration in optimization. Since N measures the efficiency of separation, the ratio $N/$time measures the time-based rate of generation of efficiency [1]. We clearly wish to maximize the N/t ratio.

From Eq. 8.5 we get

$$\frac{N}{t} = \frac{-(d\mu^{\text{ext}}/dx)}{2\theta\mathscr{R}T}\frac{X}{t} \tag{8.35}$$

Quantity X/t is simply migration velocity U. If U is replaced by $-(1/f)\,d\mu^{\text{ext}}/dx$, as suggested by Eq. 3.19, we find

$$\frac{N}{t} = \frac{(d\mu^{\text{ext}}/dx)^2}{2\theta\mathscr{R}Tf} \tag{8.36}$$

This expression can be used as a basis for maximizing separation speed.

We note that N/t depends on friction coefficient f while N itself, Eq. 8.6, does not. Thus, separation speed N/t can be maximized by any factor that reduces f, most notably by viscosity reduction. (The importance of f and the means for reducing it were discussed generally in Sections 4.12 through 4.14. The elevation of temperature to the maximum degree compatible with solute integrity is clearly the most effective way to reduce f and thus increase N/t in *Sc* systems.) Plate number N, by contrast, is independent of f and thus of viscosity: a viscous medium retards the migration and the resultant splitting apart of component zones, but it retards zone diffusion to an identical degree. Separation at high viscosity thus proceeds slowly, but for a given $-\Delta\mu^{ext}$ it produces the same level of resolution as would be realized at any other viscosity [1].

The above conclusions are relevant in comparing gel electrophoresis and electrophoresis in free solution. The polymer strands of the gel network are necessary to produce a selective sieving effect, but they increase f. This, by itself, has no adverse effect on efficiency N, but by reducing mobility (Eq. 8.19), it reduces separation speed as expressed by Eq. 8.36. This loss of speed can be offset by increasing the temperature, thus decreasing the local viscosity of the solution between the polymer strands and correspondingly decreasing f.

Another difference in the strategy for maximizing N as opposed to that for maximizing N/t arises as follows: N is increased by the total chemical potential drop $-\Delta\mu^{ext}$, while N/t is independent of $-\Delta\mu^{ext}$ but increases with the square of the gradient $d\mu^{ext}/dx$, equivalent to the molar force squared. Thus, using electrophoresis as an example, efficiency N is maximized by increasing voltage V (to which $-\Delta\mu^{ext}$ is proportional) as far as possible, and speed N/t is maximized by pushing the voltage gradient (electric field strength) to a maximum. These steps must be taken without creating significant convective currents or else increases in θ will offset any gains made [1].

8.10 STEADY-STATE VARIANTS OF Sc METHODS: ISOELECTRIC FOCUSING AND ISOPYCNIC SEDIMENTATION

As noted in Section 7.2, some chemical potential (μ^*) profiles of continuous (c) form have minima, the shape of which is shown on the far right side of Figure 7.2. The uptrending slope of the μ^* profile on either side of the minimum gives rise to a force focused toward the minimum; this force in turn leads to a velocity component directed toward the minimum from both sides. It can be shown that for small displacements from the minimum, the force and resultant velocity U are approximately proportional to the distance y from the minimum [49]

$$U = -ay \tag{8.37}$$

This expression is equivalent to Eq. 6.14 with $n = 1$. Since the force (and consequently the velocity) is focused toward $y = 0$, a solute will tend to seek out and settle about $y \sim 0$, forming a steady-state concentration profile centered on that point. If the position of the minimum for each solute component is different, the components will separate into distinct zones, each centered around its unique focusing position. Rather spectacular separations can result.

Components cannot be focused around different points by a single gradient. Like the transient field-driven (*Sc*) approaches, the steady-state methods require a basic driving field, which we call the *primary gradient*. However, another gradient or force must generally be superimposed on the primary gradient in order to modulate its force and create focusing conditions [49]. The superimposed gradient, which may not by itself cause transport, is termed a *secondary gradient* [50].

In *isoelectric focusing* (IEF), μ^* minima are formed by superimposing a pH gradient (the secondary gradient) on an electrical field (the primary gradient). The pH gradient, through shifts in acid-base equilibrium, alters the effective charge on each species as a function of position. Each species is drawn by electrical forces toward the position of its *isoelectric point*—the position of zero charge (see Figure 8.7). If the species should somehow move from one side to the opposite side of its isoelectric position (perhaps by diffusion), it would acquire a charge of opposite sign and reverse its direction of migration; once again it would head for its isoelectric point. Consequently, all motion (except diffusion) is focused toward the isoelectric point, which explains the name "isoelectric focusing" [35, 51].

The key step in the development of isoelectric focusing was the invention of a method for setting up stable pH gradients. Svensson (whose name was later changed to Rilbe) first recognized that such a gradient would be established by a mixture of ampholytes in solution, each with a different isoelectric point [52]. With several preparations of such ampholytes available commercially, IEF can be carried out in various support media and in a free electrolyte with the usual precautions to avoid convection. However, Righetti and co-workers have improved the stability and resolution of IEF by superimposing an immobilized pH gradient (with ampholyte species bonded to a gel matrix) on that established by ampholytes in solution [53, 54]. With such gradients they have resolved proteins whose isoelectric points differ by only 0.003 pH units. For example, they have resolved native hemoglobin from a mutant form in which just one of the 574 amino acids

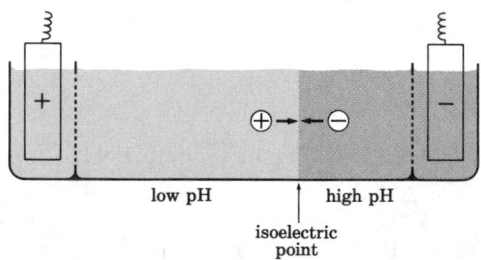

low pH   high pH

isoelectric
point

Figure 8.7. The "focusing" of isoelectric focusing is a consequence of the opposite charge assumed by a species on the two sides of its isoelectric point in a pH gradient. Different species with different isoelectric points focus at different positions, thus becoming separated.

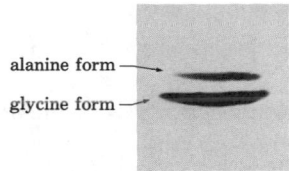

Figure 8.8. Resolution by isoelectric focusing in an immobilized pH gradient (pH range 7.35–7.55 over 20 cm path) of two human fetal hemoglobins, one with glycine (lower) instead of alanine (upper) in position 136 of the gamma chains. The isoelectric points of the hemoglobins differ by only 0.003 pH units. (Courtesy of P. G. Righetti, University of Milano.)

(alanine) in this protein has been replaced by another (glycine). The separation is illustrated in Figure 8.8 [55]. The run time was 6 h.

IEF is highly effective in separating proteins (or other components differing in isoelectric points) as described above. However, its most spectacular application is realized in combination with gel electrophoresis (GE). These two high resolution steps, each with a peak capacity n_c of the order of 10^2, can be applied successively and at right angles to generate a two-dimensional (2D) separation with $n_c \sim 10^4$, as explained in Section 6.5. The *IEF × GE* 2D method has been used to isolate proteins, hundreds at a time, from biological fluids and tissues [56]. Proteins specific to various diseases have been identified, showing that IEF × GE is promising for medical diagnoses [57, 58]. An example of IEF × GE is shown in Figure 8.9, which displays the distribution of spots of the numerous proteins extracted from a few milligrams of human heart tissue. (The observed streaking, common in this method, is likely due to low protein solubility and overloading [59].) Separation along the IEF axis is based on differences in isoelectric points

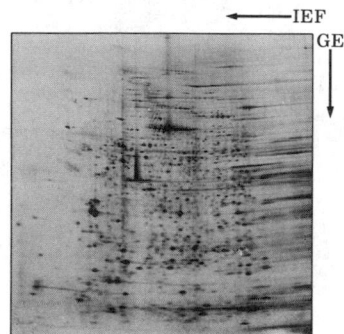

Figure 8.9. Resolution of proteins from heart muscle by 2D electrophoresis: isoelectric focusing (IEF) × gel electrophoresis (GE). Sample size, 1 mg wet weight; silver stain. (Courtesy of E. Jellum, University of Oslo.)

and along the GE axis is based on differences in molecular weights, as explained above for the individual IEF and GE techniques.

Isopycnic sedimentation [42, 43] occurs when a density gradient (the secondary gradient) is superimposed on a sedimentation field (the primary gradient). The density gradient modulates the density difference between the component and the background liquid, controlling the direction of migration. Any time component particles are more dense than the liquid, migration is outward from the center of rotation; in places where particles are less dense, buoyancy forces push them inward. Component transport is analogous to that in isoelectric focusing, with all motion focused toward the point of neutral buoyancy for the component in question. The separation spectrum is therefore based on density differences rather than on unequal isoelectric points.

We found earlier that transient (nonsteady-state) sedimentation required density gradients to stabilize against convection. Isopycnic sedimentation relies on density gradients not only for anticonvective purposes but also as the secondary gradient needed to establish steady-state conditions. The difference in the two cases is found in the magnitude of the density gradient and in the degree to which components are allowed to approach their steady-state condition. The equipment is similar; the zonal rotor developed by Anderson is used for isopycnic as well as transient zonal separations [45].

An integrated treatment of isoelectric focusing and isopycnic sedimentation was presented in 1971 [50]; the following is based on that treatment.

8.11   SEPARATION POWER OF STEADY-STATE METHODS

In isoelectric focusing, as in electrophoresis, the force acting on a species having z elementary charges is $z\mathscr{F}E$, Eq. 8.8. Charge z, of course, is zero at the isoelectric point, equivalent to the focusing point, $y = 0$. At positions close to $y = 0$, z increases as follows

$$z = \frac{dz}{dy}\, y = \frac{dz}{d(pH)}\, \frac{d(pH)}{dy}\, y \tag{8.38}$$

(This equation is simply the first nonzero term in a Taylor's expansion of z about $y = 0$.) The far right term in this expression is written to account for the fact that z is a function of pH. When the z of Eq. 8.38 is substituted into Eq. 8.8, we get

$$\text{force} = \mathscr{F}E\, \frac{dz}{d(pH)}\, \frac{d(pH)}{dy}\, y \tag{8.39}$$

Velocity U is simply the force divided by friction coefficient f

$$U = \frac{\mathscr{F}E}{f} \frac{dz}{d(pH)} \frac{d(pH)}{dy} y \tag{8.40}$$

The proportionality of U to y is in accord with Eq. 8.37, $U = -ay$. The direct comparison of Eqs. 8.40 and 8.37 yields parameter a

$$a = \frac{-\mathscr{F}E}{f} \frac{dz}{d(pH)} \frac{d(pH)}{dy} \tag{8.41}$$

It was shown in Section 6.1 that a component focused at $y = 0$ by the velocity expression $W = U = -ay$ would form a Gaussian zone centered at $y = 0$. Equation 6.18 shows that the σ of the zone is $(D_T/a)^{1/2}$. In the absence of convection, D_T can be replaced by molecular diffusion coefficient D, yielding

$$\sigma = \left(\frac{D}{a}\right)^{1/2} \tag{8.42}$$

When this expression is combined with Eq. 8.41 we get

$$\sigma = \left[\frac{Df/\mathscr{F}E}{-\dfrac{dz}{d(pH)} \dfrac{d(pH)}{dy}}\right]^{1/2} \tag{8.43}$$

Writing $D = \mathscr{R}T/f$, Eq. 3.21, we obtain

$$\sigma = \left[\frac{\mathscr{R}T/\mathscr{F}E}{-\dfrac{dz}{d(pH)} \dfrac{d(pH)}{dy}}\right]^{1/2} \tag{8.44}$$

an expression showing how σ and zone width (4σ) depends on the pH gradient and other parameters of the system.

Although the component zones are Gaussian, plate height expressions are meaningless because the zones (and the resulting separation structure) do not evolve continuously as a result of steady migration. However, resolution and peak capacity are still valid indices of separation, as discussed below.

To calculate resolution, we need the distance ΔX between the centers of the two zones of interest as well as mean zone width 4σ. The distance ΔX can be written as [50]

$$\Delta X = \frac{\Delta(pI)}{d(pH)/dx} \tag{8.45}$$

where $\Delta(pI)$ is the increment in the pH between the isoelectric points of the two species. The substitution of Eqs. 8.45 and 8.44 into the basic resolution expression $R_s = \Delta X/4\sigma$, Eq. 5.52, yields

$$R_s = \frac{\Delta(pI)}{4} \left(\frac{\mathscr{F}E}{\mathscr{R}T} \frac{-dz/d(pH)}{d(pH)/dx} \right)^{1/2} \tag{8.46}$$

where we have used $d(pH)/dy = d(pH)/dx$. Equation 8.46 shows that resolution increases with the square root of electrical field strength E. It also increases with the increment in isoelectric points and with the average rate at which the absolute value of the charge of the species increases with pH, $dz/d(pH)$. However, R_s decreases with increasing pH gradient, $d(pH)/dx$.

Clearly the pH gradient, the electrical field strength, and a number of other parameters can be controlled in a calculable way to achieve desired resolution levels. The excellence of resolution between components having almost identical isoelectric points was illustrated in Figure 8.8.

In the above equations $dz/d(pH)$ can be replaced by $(f/\mathscr{F})d\mu/d(pH)$, as is clear from Eq. 8.19. The equations then assume a form like those first derived by Svensson [52]. This substitution exchanges the readily visualized parameter z for the directly measurable parameter μ.

A treatment paralleling the above can be given for isopycnic sedimentation [50]. We begin with the force expression of Eq. 8.23, force $= M(1 - \bar{v}\rho)G = M\bar{v}[(1/\bar{v}) - \rho]G$. A gradient exists in density ρ and is represented by $d\rho/dy$. Quantity $(1/\bar{v}) - \rho$ is zero $(\rho = 1/\bar{v})$ at the isopycnic point, $y = 0$, and increases on either side of that point in proportion to y

$$\left(\frac{1}{\bar{v}} \right) - \rho = -\left(\frac{d\rho}{dy} \right)y \tag{8.47}$$

an expression replacing Eq. 8.38. From this point the derivation is identical to that used above for isoelectric focusing and produces the following expressions to replace Eqs. 8.39, 8.40, 8.41, 8.44, 8.45, and 8.46, respectively

$$\text{force} = -M\bar{v}G\left(\frac{d\rho}{dy} \right)y \tag{8.48}$$

$$U = -\frac{M\bar{v}G}{f} \left(\frac{d\rho}{dy} \right)y \tag{8.49}$$

$$a = \frac{M\bar{v}G}{f} \frac{d\rho}{dy} \tag{8.50}$$

$$\sigma = \left(\frac{\mathscr{R}T/M\bar{v}G}{d\rho/dy} \right)^{1/2} \tag{8.51}$$

$$\Delta X = \frac{\Delta \rho_s}{d\rho/dx} \tag{8.52}$$

$$R_s = \frac{\Delta \rho_s}{4} \left(\frac{M\bar{v}G}{\mathscr{R}T(d\rho/dx)} \right)^{1/2} \tag{8.53}$$

The final expression, Eq. 8.53, shows how the resolution of two species varies with their average molecular weight M, the difference $\Delta \rho_s$ in their density, the field strength G, and the density gradient $d\rho/dx$. Like Eq. 8.46 for isoelectric focusing, Eq. 8.53 can be used to quantitatively desribe the parameters necessary for resolution in isopycnic sedimentation [50].

Peak capacity n_c, like resolution, can be calculated in a straightforward manner for both isoelectric focusing and isopycnic sedimentation. However, it is most interesting to cast n_c in a form such that these steady-state methods can be related to the analogous transient methods, namely, zone electrophoresis and rate-zonal sedimentation. To do this we substitute σ from Eq. 8.42 into the basic peak capacity expression $n_c = L/4\sigma$, Eq. 5.59. Then using $D = \mathscr{R}T/f$, we get

$$n_c = \left(\frac{afL^2}{2} \right)^{1/2} \left(\frac{1}{8\mathscr{R}T} \right)^{1/2} \tag{8.54}$$

To understand the meaning of the first term, $(afL^2/2)^{1/2}$, we note that the force exerted by the primary electrical or sedimentation field on a mole of solute particles is fU or, with Eq. 8.37

$$\text{force} = -afy \tag{8.55}$$

The maximum energy (or chemical potential) change possible for any solute occurs if the force acts over the entire separation path L, which gives

$$\Delta \mu_{max}^{ext} = \int_0^L (\text{force}) \, dy = \frac{-afL^2}{2} \tag{8.56}$$

Although this expression stretches the credibility of the assumed linearity between force and coordinate y, it suggests that the first term in Eq. 8.54 is approximately $(-\Delta \mu_{max}^{ext})^{1/2}$, giving for peak capacity

$$n_c \sim \left(\frac{-\Delta \mu_{max}^{ext}}{8\mathscr{R}T} \right)^{1/2} \tag{8.57}$$

an equation identical in form to Eq. 8.7 for electrophoresis and rate-zonal sedimentation. (The twofold numerical difference in the two expressions is not regarded as significant.) The near identity of Eqs. 8.7 and 8.57 shows that steady-state and transient methods using an electrical or sedimentation field have essentially the same overall capability for separation. However, the sequence and grouping of peaks, plus experimental factors, may make one approach preferable to another in specific cases.

Additional details on the comparison of the steady-state and transient methods, and on the optimization of separation speed as well as resolution, can be found in the original publication [50].

REFERENCES

[1] J. C. Giddings, Sep. Sci., **4**, 181 (1969).

[2] C. J. van Oss, *Separation and Purification Methods*, **8**, 119 (1979).

[3] C. J. O. R. Morris and P. Morris, *Separation Methods in Biochemistry*, 2nd ed., Pitman, London, 1976.

[4] Z. Deyl, Ed., *Electrophoresis. Part A: Techniques*, Elsevier, Amsterdam, 1979.

[5] C. F. Simpson and M. Whittaker, Eds., *Electrophoretic Techniques*, Academic Press, London, 1983.

[6] B. L. Karger, Ed., *J. Chromatogr.*, **480**, 1–435 (1989). (Special Symposium issue devoted to capillary electrophoresis.)

[7] J. Hjerten, *Chromatogr. Rev.*, **2**, 122 (1967).

[8] F. J. Micale, J. W. Vanderhoff, and R. S. Snyder, *Separation and Purification Methods*, **5**, 361 (1976).

[9] R. S. Snyder, P. H. Rhodes, T. Y. Miller, F. J. Micale, R. V. Mann, and G. V. F. Seaman, *Sep. Sci. Technol.*, **21**, 157 (1986).

[10] F. E. P. Mikkers, F. M. Everaerts, and Th. P. E. M. Verheggen, *J. Chromatogr.*, **169**, 11 (1979).

[11] J. W. Jorgenson and K. D. Lukacs, *Anal. Chem.*, **53**, 1298 (1981).

[12] J. W. Jorgenson, *Anal. Chem.*, **58**, 743A (1986).

[13] Y. Walbroehl and J. W. Jorgenson, *Anal. Chem.*, **58**, 479 (1986).

[14] W. Th. Kok and G. J. M. Bruin, *Eur. Chromatogr. News*, **2**, 22 (1988).

[15] A. G. Ewing, R. A. Wallingford, and T. M. Olefirowicz, *Anal. Chem.*, **61**, 292A (1989).

[16] R. A. Wallingford and A. G. Ewing, in J. C. Giddings, E. Grushka, and P. R. Brown, Eds., *Advances in Chromatography*, Vol. 29, Marcel Dekker, New York, 1989, Chapter 1.

[17] J. H. Knox and K. A. McCormack, *J. Liq. Chromatogr.*, **12**, 2435 (1989).

[18] M. Novotny, *J. Microcolumn Separations*, **2**, 1 (1990).

[19] J. W. Jorgenson, D. J. Rose, and R. T. Kennedy, *Am. Lab.*, **20**, 32 (1988).

[20] S. Wu and N. J. Dovichi, *J. Chromatogr.*, **480**, 141 (1989).

[21] L. Gross and E. S. Yeung, *J. Chromatogr.*, **480**, 169 (1989).

[22] X. Huang, T. J. Pang, M. J. Gordon, and R. N. Zare, *Anal. Chem.*, **59**, 2747 (1987).

[23] R. D. Smith, C. J. Barinaga, and H. R. Udseth, *Anal. Chem.*, **60**, 1948 (1988).

[24] S. Terabe, K. Otsuka, and T. Ando, *Anal. Chem.*, **57**, 834 (1985).

[25] A. S. Cohen and B. L. Karger, *J. Chromatogr.*, **397**, 409 (1987).

[26] A. Guttman, A. S. Cohen, D. N. Heiger, and B. L. Karger, *Anal. Chem.*, **62**, 137 (1990).

[27] R. J. Block, E. L. Durrum, and G. Zweig, *A Manual of Paper Chromatography and Paper Electrophoresis*, 2nd ed., Academic Press, New York, 1958.

[28] G. L. Trainor, *Anal. Chem.*, **62**, 418 (1990).

[29] D. C. Schwartz and C. R. Cantor, *Cell*, **37**, 67 (1984).

[30] E. Lai, B. W. Birren, S. M. Clark, M. I. Simon, and L. Hood, *Biotechniques*, **7**, 34 (1989).

[31] T. W. Carr, Ed., *Plasma Chromatography*, Plenum, New York, 1984.

[32] S. Rokushika, H. Hotano, M. A. Baim, and H. H. Hill, *Anal. Chem.*, **57**, 1902 (1985).

[33] F. M. Everaerts, J. L. Beckers, and Th. P. E. M. Verheggen, *Isotachophoresis, Theory, Instrumentation, and Applications*, Elsevier, Amsterdam, 1976.

[34] W. Thormann, *Sep. Sci. Technol.*, **19**, 455 (1984).

[35] O. Vesterberg, *J. Chromatogr.*, **480**, 3 (1989).

[36] S. Terabe, K. Otsuka, and T. Ando, *Anal. Chem.*, **61**, 251 (1989).

[37] J. M. Davis, *Anal. Chem.*, **61**, 2455 (1989).

[38] R. J. Wieme, in E. Heftmann, Ed., *Chromatography*, Van Nostrand-Reinhold, New York, 1975, pp. 228–281.

[39] E. Grushka, R. M. McCormick, and J. J. Kirkland, *Anal. Chem.*, **61**, 241 (1989).

[40] S. Ross and I. D. Morrison, *Colloidal Systems and Interfaces*, Part III, Wiley, New York, 1988.

[41] J. C. Giddings, *J. Chromatogr.*, **480**, 21 (1989).

[42] P. Sheeler, *Centrifugation in Biology & Medical Science*, Wiley, New York, 1981.

[43] D. Rickwood, Ed., *Centrifugation*, 2nd ed., IRL Press, Oxford, 1984.

[44] R. M. Holsworth, T. Provder and J. J. Stansbrey, in T. Provder, Ed., *Particle Size Distribution*, ACS Symposium Series 332, ACS, Washington, DC, 1987, Chapter 13.

[45] N. G. Anderson, Ed., *Zonal Centrifugation*, National Cancer Institute Monograph No. 21, 1966.

[46] P. G. Grodzka and B. Facemire, *Sep. Sci. Technol.*, **12**, 103 (1977).

[47] H. J. V. Tyrrell, *Diffusion and Heat Flow in Liquids*, Butterworth, London, 1961.

[48] M. E. Hovingh, G. H. Thompson, and J. C. Giddings, *Anal. Chem.*, **42**, 195 (1970).

[49] J. C. Giddings, *Sep. Sci. Technol.*, **21**, 831 (1986).

[50] J. C. Giddings and K. Dahlgren, *Sep. Sci.*, **6**, 345 (1971).

[51] N. Catsimpoolas, Ed., *Isoelectric Focusing*, Academic Press, New York, 1976.

[52] H. Svensson, *Acta Chem. Scand.*, **15**, 325 (1961).

[53] P. G. Righetti, E. Gianazza, and C. Gelfi, *Separation and Purification Methods*, **16**, 105 (1987).

[54] P. K. Sinha and P. G. Righetti, *J. Biochem. Biophys. Methods*, **12**, 289 (1986).

[55] P. G. Righetti and G. Cossu, personal communication, 1987.

[56] N. G. Anderson and N. L. Anderson, *Anal. Biochem.*, **85**, 331 (1978).

[57] E. Jellum, *J. Chromatogr.*, **239**, 29 (1982).

[58] K. E. Willard, *Clinical Chem.*, **28**, 1031 (1982).

[59] N. G. Anderson, personal communication, February 1990.

EXERCISES

8.1(*)   The *electrophoretic mobility*, μ, is defined as the velocity U per unit of electrical field strength, E. Show that $\mu/D = ze/kT$.

8.2(*)   The *sedimentation coefficient*, s, is defined as the velocity induced per unit of acceleration, $s = U/\omega^2 r$. Show that $s/D = (\rho_0 - \rho)\bar{V}/\mathscr{R}T$ (the Svedberg equation), where $\bar{V}$ is molar volume.

8.3(**)   A spherical molecule of diameter $40.0\,\text{Å}$ in a dilute aqueous solution at 20°C ($\eta = 0.0100$ poise) carries an effective charge of 3 units ($z = 3$). What is the magnitude of its velocity in an electrical field of strength 50.0 V/cm? What is its electrophoretic mobility?

8.4(*)   What is the velocity of the molecule described in the preceding problem when in a centrifugal field of $75{,}000 \times g$? Assume its density to be $1.40\,\text{g/cm}^3$. What is its sedimentation coefficient?

8.5(**)   Sedimentation, used in conjunction with diffusion measurements, has been used for the determination of the molecular weights of macromolecules. At 20°C in a dilute aqueous solution, the sedimentation coefficient of hemoglobin, density $1.33\,\text{g/cm}^3$, is 4.3×10^{-13} s. Its diffusion coefficient is $6.9 \times 10^{-7}\,\text{cm}^2/\text{s}$. Calculate the molecular weight. How is this result affected by the hydration shell?

8.6(***)   The lactate and acetate ions have one unit of charge and the respective electrophoretic mobilities of 13.1×10^{-5} and $16.6 \times 10^{-5}\,\text{cm}^2/\text{V s}$ in 0.1 M alkaline solution at 20°C. How far through such a solution must they travel before unit resolution is achieved using an electrophoretic field strength of 20 V/cm? How long does it take and how many theoretical plates are needed? Assume ideal electrophoresis ($\theta = 1$).

8.7(**)   Two ions each with $z = 2$ but with slightly different friction coefficients (averaging 5×10^{15} g/s) require 25,000 plates for separation. What voltage drop is needed? How long does the separation take when executed over a 10.0 cm path length? 50.0 cm? Assume $\theta = 1$ and $T = 290$ K. (Watch your units!)

8.8(**) Working on a planet in another galaxy, where laboratory temperatures are 200 K, a scientist discovers a new force field selective for isotopes whose separation promises to fuel a bomb to blow an adjacent planet to smithereens. The maximum force exerted by this field on the two isotopes is 5.00×10^{10} dyn/mol. What is the lowest plate height the scientist can hope to achieve in an Sc-class separation? In this field the velocity difference is 2% of the mean (which is remarkably selective for isotopes). What length separation tube is needed to separate the isotopes at unit resolution? (Remember, lofty goals are worth the trouble.)

8.9(*) In a classical experiment, Meselson and Stahl (Proc. Natl. Acad. Sci. U.S.A., **44**, 671 (1958)) attempted to resolve labeled (N^{14}) and unlabeled *E. coli* DNA, which differ in density by 0.014 g/cm^3, the density and molecular weight of the lighter unlabeled DNA being 1.710 g/cm^3 and 9.3×10^6, respectively. The density gradient, created in a CsCl solution, was approximately 0.080 g/cm^4. The acceleration was $140,000 \times g$ and the temperature was 27°C. Calculate the resolution predicted for these DNAs (the observed resolution was 1.5).

8.10(*) Separation by density-gradient (isopycnic) centrifugation is attempted using a 1.00 cm long cell (extending from 6 to 7 cm from the axis) with a CsCl concentration gradient. At 44,700 rpm the concentrated solution at the outer end of the cell has a density of 1.790 g/cm^3, compared with 1.671 g/cm^3 at the inner end. The gradient may be assumed linear. Suppose you have two DNA molecules of average molecular weight 14.2×10^6 with respective densities of 1.710 and 1.732 g/cm^3 at 20°C. What is the average zone width 4σ in distance units? What is the distance between zone centers? Based on these two numbers, calculate the expected resolution.

8.11(*) For the protein ovalbumin, the effective charge z decreases 4.5 units between pH = 4.3 and 4.8. The isoelectric point is 4.6. For a pH gradient of 0.050 pH units/cm, a field of 25 V/cm, and a temperature of 4°C, calculate the peak width 4σ (in distance units) for the zone. Calculate the width 4σ in pH units.

8.12(**) For a general treatment of isoelectric focusing, the restoring force F acting toward the isoelectric point at $y = 0$ cannot always be assumed proportional to y, as indicated by Eq. 8.39. All that we really know in the most general case is that F is some function of y, $F(y)$. If we expand $F(y)$ in a Taylor's series around $y = 0$, we end up with $F(y) = a_0 + a_1 y + a_2 y^2 + \cdots$. The first term is zero because F vanishes at $y = 0$. By neglecting all terms of third-order and higher, and by assuming $D_T = D$, show that the concentration profile is given by

$$\frac{c}{c_0} = \exp\left[\frac{(a_1 y^2/2) + (a_2 y^3/3)}{\mathscr{R} T}\right]$$

which approaches the Gaussian profile suggested for isoelectric focusing only where $a_1 y \gg a_2 y^2$.

9

SEPARATION USING PERPENDICULAR FLOW, $F(+)$: FIELD-FLOW FRACTIONATION, CHROMATOGRAPHY, AND RELATED METHODS

The basis for effective $F(+)$ separation was discussed in Section 7.7. It was pointed out that slight enrichment in the direction of a field or across an interface could be converted into an effective (sometimes spectacular) separation along a flow axis perpendicular to the axis of enrichment. The magnification of enrichment by flow is sufficiently large that many components can be separated in a single run. This is best illustrated by chromatography, the most important analytical separation method now in use. Another $F(+)$ approach of analytical importance is field-flow fractionation, a relatively new family of techniques applicable to macromolecules, colloids, and related materials.

Although $F(+)$ separation methods are powerful, they are relatively complicated in physicochemical detail. In this chapter we will provide a general framework for $F(+)$ methodology, outline the principles and applications of field-flow fractionation, and introduce the theoretical basis of chromatography. Chromatography will be treated in greater detail in Chapters 10–12.

9.1 ROLE OF CHEMICAL POTENTIAL PROFILE

In comparing separation techniques, we generally find a striking difference in methods based on continuous (c) chemical potential profiles and those involving discontinuous (d or cd) profiles. There is, for example, a glaring contrast in instrumentation, applications, experimental techniques, and the capability for multicomponent separations between the two basic static systems, Sc (e.g., electrophoresis) and Sd (e.g., extraction). Similarly, there are large differences in the two parallel-flow categories $F(=)c$ (e.g., elutriation) and $F(=)d$ (ultrafiltration). However, the normally sharp contrast between c- and d-based (or cd-based) systems is somewhat moderated in the perpendicular-flow case treated in this chapter. Multicomponent separations, for example, can be carried out both by chromatography $[F(+)d]$ and field-flow fractionation $[F(+)cd]$ using similar flow-based instrumentation. There is much difference in detail and capability between these methods but they are closely related both fundamentally and experimentally. The close relationship is due to the fact that the flow axis—now the axis of separation—differs little in intrinsic form between $F(+)d$ and $F(+)cd$ methods. Given this common axis of separation, the nature of the (perpendicular) enrichment axis is less relevant to the general form of the separation (although it is immensely important to the details of separation). Thus a common theoretical foundation, given below, underlies both.

9.2 SEPARATION POWER IN $F(+)$ SYSTEMS

We expand on the separation power exhibited along the flow axis. We noted earlier that flow is a powerful transport mechanism, capable of carrying components rapidly over considerable distances. Flow is also capable of keeping the components in fairly compact zones by virtue of its power to evacuate component material from one region as it carries it into another. These capabilities, as noted in Section 7.7, stem from the flow transport term, $-v \, dc/dx$, in the basic transport equation, Eq. 3.30. When acting alone, this term gives

$$\frac{dc}{dt} = -v \, \frac{dc}{dx} \tag{9.1}$$

which shows that the rate of change in concentration due to flow transport is proportional to flow velocity v; the extent of the transport is proportional to the length L of the transport path. Both v and L can be large in flow systems, giving considerable separation power.

The above idea takes more concrete form when expressed in terms of the number N of theoretical plates. When migration distance X is replaced by total path length L and the general velocity term W replaced by v, Eq. 5.43 takes the form

$$N = \frac{vL}{2D_T} \tag{9.2}$$

which confirms that separation power generally increases with both v and L, both terms capable of reaching high values in flow systems.

Equation 9.2 is useful conceptually in showing the power of flow expressed through high vs and Ls. The downside of increased flow is that it usually generates high D_T. Thus the benefit of a high v in Eq. 9.2 is often cancelled by a large D_T term.

The origin of D_T is interesting. We have already noted that $F(+)$ systems are selective for different solutes because the enrichment processes acting in a direction perpendicular to flow work in concert with (and require) nonuniform flow. However, it is the nonuniformity in flow that tends to increase D_T in Eq. 9.2, thus reducing separative efficiency. The positive and negative sides of nonuniform flow require that optimization be undertaken carefully. More details are given below.

9.3 ZONE MIGRATION

Whether fields (c or cd profiles) or interfaces (d profiles) establish selective concentration changes in different microscopic regions, separation in $F(+)$ methods takes place by subjecting solute accumulated in each small region to a *different* flow velocity. (This requires nonuniform flow as alluded to above.) In chromatography, for example, solutes are distributed selectively between stationary and mobile phases or regions; separation occurs because one region moves (by means of flow) relative to another. In general there can be many microscopic regions holding solute in $F(+)$ methods, each subject to a unique velocity (including zero velocity). These different regions of unique velocity can be termed velocity states [1]. Solute molecules can pass freely (and usually rapidly) between the regions or velocity states by diffusion or other mechanisms.

The net rate of displacement of a component downstream depends on how it is distributed among the regions. Species distributed mainly in high velocity regions will advance more rapidly than species favoring more stationary regions. This is the basis of all $F(+)$ separation. The quantitative basis of this observation follows.

If we imagine slicing across the separation path at right angles to flow, our cut will intercept a representative population of velocity states (see Figure 9.1). If in any small element i of the exposed cross section the concentration is c_i and the velocity is v_i, the average velocity $\mathcal{V}$ of solute crossing that plane is simply

$$\mathcal{V} = \frac{\sum c_i v_i}{\sum c_i} \tag{9.3}$$

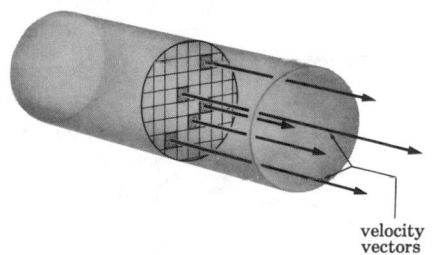

velocity
vectors

Figure 9.1. Division of cross section of flow system into small elements, each with a given velocity and concentration. Some of the velocity vectors are shown.

The same velocity will be found at all other cross sections along the separation path if (i) the medium is homogeneous, (ii) the flowrate is constant up and down the flow path, and (iii) the microscopic concentrations c_i are distributed in the same relative (not necessarily absolute) amounts. The latter condition—that the relative concentration distribution remains constant—will be met if the concentrations are fixed by equilibrium. In the most important cases they are.

Equation 9.3 confirms that velocity $\mathscr{V}$ will be greatest for species which heavily populate (i.e., have high c_i in) regions where the v_i are large. Thus the different component distributions, leading to different displacement velocities, are responsible for separation.

The summations in Eq. 9.3 can be replaced by integrals

$$\mathscr{V} = \frac{\int cv \, dA}{\int c \, dA} \tag{9.4}$$

where the integrals extend over *all* elements (flowing and stationary) of cross-sectional area A.

The downstream motion of component bands in $F(+)$ methods is characterized by the *retention ratio* R, a dimensionless parameter defined as

$$R = \frac{\mathscr{V}}{\langle v \rangle} \tag{9.5}$$

We see that R specifies $\mathscr{V}$ relative to the average cross-sectional fluid flow velocity $\langle v \rangle$. Since the flow velocities at all points in the cross section are scaled to one another and to $\langle v \rangle$ (see Section 4.4), the velocity $\mathscr{V}$ of a component carried by the sum of these microscopic flows is also proportional to $\langle v \rangle$. Thus the ratio $R = \mathscr{V}/\langle v \rangle$ is constant, independent of flowrate.

A band of solute distributed evenly over the flow stream—not concentrated more in one region than another as needed for separation—would travel with the flow stream, that is, it would be displaced at a velocity $\mathscr{V}$ exactly equal to $\langle v \rangle$. For such a *nonretained* component, $R = 1$. Most solutes in practical $F(+)$ systems concentrate in low-velocity regions, thus causing "retention" relative to the overall flow stream. The retention is reflected in the inequalities $\mathscr{V} < \langle v \rangle$ and $R < 1$.

The cross-sectional average of v is expressed by

$$\langle v \rangle = \frac{1}{n} \sum_{i=1}^{n} v_i \tag{9.6}$$

or in integral form

$$\langle v \rangle = \frac{1}{A_f} \int v \, dA \tag{9.7}$$

where n is the number of velocity states and A_f is the cross-sectional area occupied by the fluid.

We now obtain a generalized equation for R by substituting Eqs. 9.4 and 9.7 into Eq. 9.5, which gives

$$R = \frac{A_f \int cv \, dA}{\int c \, dA \int v \, dA} \tag{9.8}$$

9.4 ZONE SPREADING

We reemphasize that a basic feature of $F(+)$ methods is that different parts of any given component zone are carried at different velocities along the flow axis by the nonuniform flow-induced displacement. The unequal displacement velocities are necessary for separation but they have a fault; they break the zone apart along the flow axis, causing zone broadening and thus interfering with the effective resolution of components. This was noted in Section 9.2; it is further illustrated in Figure 9.2. Clearly, means must be found to minimize this effect, although we recognize it cannot be fully eliminated because nonuniform displacement is needed for separation.

Our remedy is based on the observation that if component molecules diffuse or otherwise transfer rapidly between fast and slow velocity states, zone broadening will be reduced. The reduction occurs because rapid diffusion will quickly shuttle molecules between high and low velocity states, thus preventing them from getting very far behind or ahead of neighboring molecules occupying different states.

A rapid diffusional exchange is encouraged by reducing the dimensions

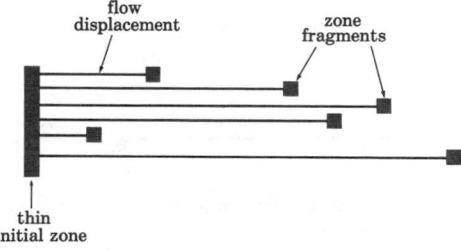

Figure 9.2. The breakup of a thin initial zone by different flow displacement rates is shown for methods with perpendicular flow, $F(+)$. The consequent zone broadening can only be contained by rapid solute exchange between the different velocity states.

over which diffusion occurs. This reduction is a major consideration in all important analytical $F(+)$ systems. Among other things, the general need to reduce diffusional dimensions explains why the particles used to pack columns for high-performance liquid chromatography (HPLC) are so small (often 5 μm or less in diameter).

We additionally note that for effective $F(+)$ operation—for an ability to separate multicomponent mixtures—the exchange must occur so rapidly that each component is virtually in a state of equilibrium between the velocity states. Thus near-equilibrium distributions can be assumed in most cases.

A simplistic but informative theory of zone broadening can be developed using the random-walk model of Section 5.3. (For background see refs. [2] and [3].) Rather than attempt to account for the multiple transfers between all possible velocity states, we assume that there is one velocity v_2 representing the fast-moving regions and a lower velocity v_1 representing the sluggish (or even stagnant) regions (see Figure 9.3). Solute will partition between the two velocity states and it will thereby acquire an overall mean displacement velocity of $\mathcal{V}$ intermediate between v_2 and v_1.

We focus first on velocity state v_2. Any molecule caught up in this high-velocity state will remain there for some average equilibration time t_{eq} before being lost or exchanged to state v_1 by diffusion or some equivalent mechanism. During occupancy of the high-velocity state, the molecule will outpace the average molecule in the zone by the velocity $v_2 - \mathcal{V}$. Since a molecule persists for an average time t_{eq} at velocity v_2, it gains a distance

$$l = (v_2 - \mathcal{V})t_{eq} = \Delta\mathcal{V}t_{eq} \tag{9.9}$$

with respect to the zone center.

The transfer of molecules back and forth between velocity states is a two-way random process; thus step distance l may be positive or negative. The random sequence of positive and negative steps constitutes a random walk (see Section 5.3). The total number n of such steps, forward and backward, is the total time (retention time) of the process t_r over exchange time t_{eq}

$$n = \frac{t_r}{t_{eq}} = \frac{L}{\mathcal{V}t_{eq}} \tag{9.10}$$

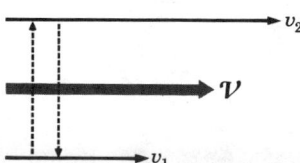

Figure 9.3. Simple model of $F(+)$ system. Solute transfers continuously (dashed arrows) between high-velocity state v_2 and low-velocity state v_1, acquiring a mean velocity $\mathcal{V}$.

where the right-hand expression comes from replacing t_r by displacement (column) length L over mean component velocity $\mathcal{V}$.

The substitution of Eqs. 9.9 and 9.10 into Eq. 5.24a, $\sigma^2 = l^2 n$, gives the following zone variance

$$\sigma^2 = \left(\frac{\Delta \mathcal{V}}{\mathcal{V}}\right)^2 \mathcal{V} t_{eq} L \tag{9.11}$$

The plate height, $H = \sigma^2/L$ (see Eq. 5.40), is thus

$$H = \left(\frac{\Delta \mathcal{V}}{\mathcal{V}}\right)^2 \mathcal{V} t_{eq} \tag{9.12}$$

We note that $\Delta \mathcal{V}/\mathcal{V}$ is a measure of the nonuniformity of flow experienced by a component in the system. It is a dimensionless constant reflecting the geometry of the flow space. It does not change with flowrate because, as noted in Section 4.4, all velocities in a given cross section subject to laminar flow maintain the same proportion to one another. Thus when flowrate increases, all individual velocities increase proportionately, as does $\mathcal{V}$ and $\Delta \mathcal{V}$. The ratio $\Delta \mathcal{V}/\mathcal{V}$ remains constant.

The broadening of zones by molecules caught in different flow lines is the most fundamental mechanism of band dispersion in $F(+)$ methods. In some cases ordinary diffusion up and down the separation path also contributes measurably to zone broadening. In the simplest case ordinary diffusion contributes a variance of $\sigma^2 = 2Dt$ and thus a plate height, σ^2/L, of

$$H_D = \frac{2Dt}{L} \tag{9.13}$$

The ratio L/t equals zone velocity $\mathcal{V} = R \langle v \rangle$. Thus

$$H_D = \frac{2D}{R\langle v \rangle} = \frac{B}{\langle v \rangle} \tag{9.14}$$

which shows that H_D is inversely proportional to flow velocity $\langle v \rangle$. In heterogeneous systems like chromatography, solute occupies two (mobile and stationary) phases and experiences two different diffusion coefficients. Such complications will be treated in Section 11.3.

9.5   OPTIMIZATION

Although Eq. 9.12 is approximate, it contains a great deal of information bearing on the optimization of $F(+)$ systems. We discuss some very general strategies for optimization below [3].

We note first that the plate height increases strongly with the relative dispersion $\Delta \mathcal{V}/\mathcal{V}$ in downstream velocities. It would be tempting to conclude that we should eliminate all dispersion in flow velocity, but we again remember that separation requires differential flow. A reduction in $\Delta \mathcal{V}/\mathcal{V}$ may therefore mean a reduction in separation power. However there are some types of flow dispersion that contribute to plate height but not to selectivity. For example in chromatography, flow inequalities between different microscopic flow channels contribute to band broadening without aiding separation. (Separation requires only a difference in velocity between mobile and stationary phases; it is not aided by additional velocity differentiation in the mobile phase.) These interchannel effects can be reduced by increasing packing uniformity (see Section 11.5). They can also be reduced by using electroosmotic flow (Section 4.9) in place of pressure driven flow. In general, all nonessential components of velocity differences should be minimized.

Given a certain relative dispersion in flow velocities, Eq. 9.12 shows that the plate height increases with the mean velocity $\mathcal{V}$ of the solute. Clearly then, H can be reduced by decreasing the flowrate. However a reduction in flow will lead to longer retention times and slower separation. Therefore the choice of flow velocity requires a compromise between separation speed and efficiency. This subject has been treated in some detail for chromatography [2]; we will say more about it in Chapter 12.

Perhaps the most important term in Eq. 9.12 is the time t_{eq} required for the transfer or equilibration of solute molecules between high- and low-velocity states. Time t_{eq} may be controlled by diffusion or, if the separation tube is packed, by a combination of diffusion and hydrodynamic factors. If the transfer between velocity states requires the crossing of an interface, t_{eq} can depend upon interfacial kinetics.

In the simplest case, t_{eq} is determined by the rate of diffusion over some characteristic distance d between velocity states; we can estimate t_{eq} as

$$t_{eq} \sim \frac{d^2}{2D_T} \tag{9.15}$$

When molecular diffusion acts alone to transfer solute, we have $D_T = D$, which leads to

$$t_{eq} \sim \frac{d^2}{2D} \tag{9.16}$$

Equation 9.12 then assumes the form

$$H \simeq \left(\frac{\Delta \mathcal{V}}{\mathcal{V}}\right)^2 \mathcal{V} \frac{d^2}{2D} \tag{9.17}$$

Equation 9.17 shows that H is normally smallest for systems with the highest diffusivity D. A factor of greater importance, however, is d^2. Rapid equilibration requires that the distance d between the high- and low-velocity states be minimal. In chromatography, d can often be taken as a simple multiple of the particle diameter, which means the particle diameter must be small. This simple conclusion is the basis of the "high-performance" in HPLC.

In summary, the requirements of selectivity show that the relative velocity difference $\Delta \mathcal{V}/\mathcal{V}$ must remain substantial, but Eq. 9.17 tells us that the velocity increments must occur over a short distance d. As the distance d between velocity extremes decreases, the shear rate and the pressure gradient needed to generate high shear increases. Thus a trend to increasing pressure drops in $F(+)$ systems is a natural outgrowth of these basic considerations. Again the most direct confirmation of theory is HPLC, where pressures of several hundred atmospheres are common.

In some cases the speed of separation becomes crucial, so we change our focus, as in Section 8.9, to that of maximizing the number of plates per unit time N/t. From Eq. 9.17, $N = X/H$ is

$$N = \frac{2(X/\mathcal{V})D}{(\Delta \mathcal{V}/\mathcal{V})^2 d^2} \tag{9.18}$$

Since $X/\mathcal{V}$ is the time t expended in reaching point X, we have

$$\frac{N}{t} \simeq \frac{2D}{(\Delta \mathcal{V}/\mathcal{V})^2 d^2} \tag{9.19}$$

Quantity N/t is increased in much the same way as H is decreased: by finding ways to reduce d and to increase D. However H is proportional to zone velocity $\mathcal{V}$ (which we are therefore advised to reduce) whereas N/t is independent of $\mathcal{V}$. (If we add in the effects of ordinary longitudinal diffusion, we find N/t largest at high velocities.)

While the random walk model employed here is widely applicable to $F(+)$ methods, it fails if the molecules do not transfer rapidly between velocity states, equivalent to many random steps. Such a limitation applies to electrodecantation (noted below), where the distances are too great for rapid diffusional exchange. The random walk model is most meaningful for zonal separation methods such as chromatography and field-flow fractionation.

9.6 EQUILIBRIUM AND NONEQUILIBRIUM

We have emphasized that $F(+)$ separations depend on velocity differences in the system and the differential partitioning of species between the resulting velocity states. We have also shown that the exchange of molecules between velocity states must be fast for effective separation, a condition that leads to a virtual state of equilibrium. However, a deeper look shows us that there are subtle departures (both positive and negative) from equilibrium over most of a component zone. These small departures have big effects.

If flow were to cease entirely, we could reasonably expect sample components to reach equilibrium rapidly between different velocity states. However with flow, the molecules carried into any cross section of the system will arrive from upstream where the concentration is higher or lower than the existing level. The concentration change due to inflow will unbalance any previously established equilibrium. The solute will respond by repartitioning between velocity states (usually by diffusion), but even as this proceeds the unbalancing effect of flow continues. With ongoing flow, equilibrium remains just out of reach [2].

An exception occurs at the peak maximum. Here the concentration profile is momentarily flat so that incoming solute does not change the existing concentration level. Thus equilibrium is established. Consequently the peak center will move at a rate governed by equilibrium, and the migration equations (such as Eq. 9.3) based on equilibrium concentrations are valid.

The front and back parts of migrating zones are a different matter. At the leading edge, the arriving solute comes from richer regions located upstream toward the zone center. The fastest streamlines bring in the richest solute. Thus the concentration c_f in the fast streams is perturbed in the direction of a higher-than-equilibrium concentration, $c_f > c_f^*$. The excessive c_f values lead to a higher-than-equilibrium rate of migration at the front of the zone. The front, therefore, moves faster than the zone center. Likewise at the rear of the zone where $c_f < c_f^*$, solute falls behind that at the zone center. Thus front and rear regions of the zone gradually pull away from each other and cause the zone to broaden.

The above band-broadening mechanism assumes a more concrete form when it is described specifically for chromatography in Section 10.6. When this mechanism is expressed mathematically it becomes the nonequilibrium theory, an important tool describing zone evolution in chromatography (Section 10.6) and field-flow fractionation [2, 3].

Clearly, departures from equilibrium—along with the resultant zone spreading—will decrease as means are found to speed up equilibrium between velocity states. One measure of equilibration time is the time defined in Section 9.4 as t_{eq}, equivalent to the transfer or exchange time between fast- and slow-velocity states. Time t_{eq} must always be minimized; this conclusion is seen to follow from either random-walk theory or nonequilibrium theory. These two theories simply represent alternate conceptual approaches to the same band-broadening phenomenon. Thus the plate height from Eqs. 9.12 and 9.17 may be considered to represent simultaneously both nonequilibrium processes and random-walk effects.

9.7   OVERVIEW OF $F(+)cd$ SEPARATIONS

In this section we briefly survey the $F(+)cd$ methods. In Table 7.2 we listed among $F(+)cd$ methods the techniques of *field-flow fractionation*, *thermog-*

ravitational separation, and *electrodecantation*. In all of these methods, a field or gradient is applied perpendicular to the axis of a tube or flow chamber as shown schematically in Figure 9.4. The field interacts with the components and pushes them toward one wall or boundary of the chamber as illustrated in the figure. A steady-state condition may be reached; in this case the concentration profile generally assumes an exponential form as shown in Section 6.2. The thickness of the steady-state layer, of course, will vary from component to component.

Flow is then instituted along the axis of the chamber and a differential flow pattern develops. This differential flow carries each component along in some uneven pattern. Different components distributed over the cross section with different concentration profiles will be carried downstream unequally because of the nonuniformity of flow. The net result is that the different components will be displaced at different average velocities in the tube and separation will be realized.

The flow pattern itself can vary depending on the source of the flow. Ordinary pump-driven flow will create a parabolic (bullet-shaped) profile in narrow tubes, with a high velocity in the center and low velocities at the tube edges (see Section 4.4). Convective flow, on the other hand, is often countercurrent, with dense fluid sinking and less dense fluid rising (Section 4.8).

Convective flow is used in both the thermogravitational (Clusius–Dickel) column and in electrodecantation. In the thermogravitational system, one wall of the channel is heated or, alternatively, a hot wire is placed along the axis of the channel. The fluid at the cold surface then tends to sink relative to that at the hot surface. Simultaneously, thermal diffusion (Section 8.8) causes different levels of enrichment in the hot and cold regions of the channel. The enriched solutes then move up and down the channel at a rate depending upon their distribution between hot and cold regions. In binary

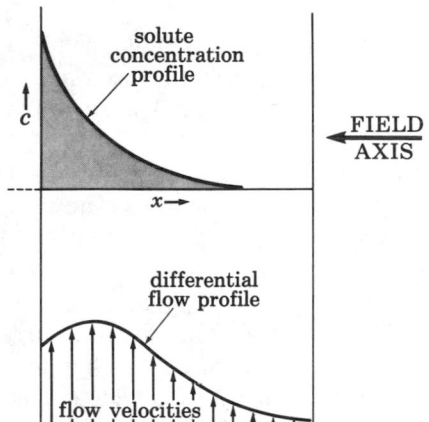

Figure 9.4. Schematic illustration of a $F(+)cd$ separation system with arbitrary flow profile.

(carrier-free) solute mixtures, one component tends to move up and the other to move down the channel. This method has been especially prominent in the separation of isotopes. Further details are available in reference (4).

Electrodecantation works in a similar (but less effective) way. In this method, charged components (e.g., proteins) are drawn toward a wall or barrier by an electrical field. If the component mixture is denser than the solvent, it will sink as a mass along the wall and can be collected at the bottom. A steady-state distribution of solute along the electrical field axis is rarely approached. (The flow in electrodecantation has a very limited role in governing separation; flow mainly aids sample recovery after an essentially static (*Scd*) separation. Thus one can argue that the flow should be categorized as passive, Section 7.5, and the system should not be grouped with the flow methods.) Further details on electrodecantation are provided in reference [5].

The perpendicular flow methods have a number of important similarities and differences [6]. Most significantly, both thermogravitational separation and electrodecantation are limited to the separation of only two fractions (i.e., peak capacity equals two). These methods are thus restricted to preparative applications. Field-flow fractionation is the only method in the $F(+)cd$ class designed specifically for multicomponent analytical separations. We describe this method in greater detail below.

9.8   FIELD-FLOW FRACTIONATION

Field-flow fractionation (*FFF*) is a family of separation methods first conceived in the 1960s [7]. The various subtechniques of FFF are best suited for the separation of macromolecules, colloids, and particles, including biological components ranging in size from proteins to living cells, environmental colloids and particles, and industrial polymers, powders, latexes, and emulsions.

The separation process in FFF is carried out in a thin ribbon-like channel. The channel is cut from a spacer and sandwiched between two walls as shown in Figure 9.5. An enlarged edge view of the channel is shown in Figure 9.6.

A narrow band of sample material is injected at the head of the channel. A field or gradient is then applied across the face of the channel as shown in the figures. In *normal operation* (variants will be described in Section 9.11) the field causes the components to migrate to one wall, termed the *accumulation wall*. Each component quickly reaches (in a process termed *relaxation*) a steady-state distribution close to that wall. The distribution is exponential as described by Eq. 6.19 and illustrated in Figure 9.6. The mean thickness of the component layer so formed is given by Eq. 6.20, $\ell = D/|W|$, where W (specifically its component U) is the field-induced velocity

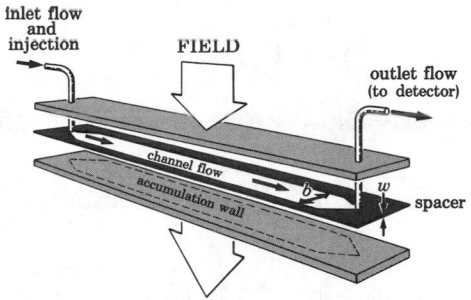

Figure 9.5. Diagram showing FFF flow channel cut from a thin spacer and sandwiched between two walls. The field is applied perpendicular to the flow.

directed at the wall. Since diffusion coefficient D and perpendicular velocity W (and U) are different from one component to another, mean layer thickness ℓ varies from one component to another in the FFF channel.

After the components have reached their steady-state distributions, flow is started through the channel. The flow profile across the channel is parabolic as shown in Figure 9.6. With parabolic flow, the flow velocity decreases upon approaching the accumulation wall and becomes zero at the wall. The components compressed most tightly against the accumulation wall—those having the smallest ℓ values—occupy the slowest flow regions adjacent to the wall and therefore are carried downstream most slowly of all the components. This results in separation as illustrated in Figure 9.7. (The mechanism of separation is further elaborated in refs. [8–12].)

The components separated by FFF are washed out of the channel and into a detector and/or fraction collection device. The detectors are mainly those used in liquid chromatography where light absorption, refractive index changes, and so on, are measured as components flush through a small flow cell following their elution.

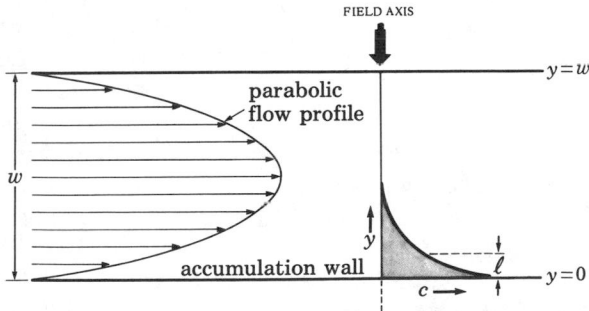

Figure 9.6. Schematic illustration of flow and exponential concentration profile across the narrow gap between walls in an FFF channel.

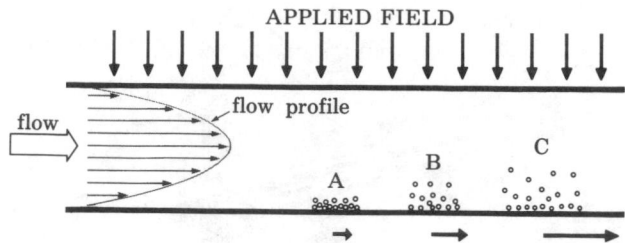

Figure 9.7. Separation mechanism of FFF. Three components, shown here as A, B, C, are compressed to different levels against the wall by the field or gradient. The distribution of each component is exponential as shown in Figure 9.6. Component A lies closest to the wall where the downstream fluid motion approaches zero. Consequently A is displaced very slowly downstream, as indicated by the short arrow below. Component B, protruding into a more rapid flow region, is carried more rapidly downstream. Component C, least compacted because of a weaker interaction with the field, is displaced most rapidly by flow, as indicated by the long arrow.

The flow channel in FFF is typically 25–100 cm long and 1–3 cm in breadth. The thin dimension (thickness) of its ribbon-like structure is generally 50–500 μm, or 0.05–0.5 mm. The channel is open and unobstructed—it contains no packing material. There is no need for packing to support a stationary phase (as in chromatography) because there is no stationary phase: retention is induced by the external field.

Since the geometry and flow profile (parabolic) of FFF channels is well-defined, rather exact theoretical predictions of retention and plate height can often be made. The correlation of these experimental parameters with various physicochemical properties of component species (such as molecular weight, charge, and size) is therefore possible (11, 13).

Retention in FFF is controlled by an externally applied field; this field can be varied quickly and flexibly to adapt to different kinds of samples and to institute various time-based programs (14). This gives the FFF system versatility in handling a wide range of species and in optimizing resolution and speed.

Relative to chromatography, the absence of a high interfacial area between stationary and mobile phases gives FFF an advantage in dealing with macromolecules and particles, since these species are normally quite interactive (e.g., adsorptive or reactive) with interfaces. On the other hand most fields are not as selective as interfaces, nor are they strong enough to adequately retain low molecular weight materials. In most respects FFF and chromatography are complementary: the former is highly effective for macromolecules and particles largely out of the range of chromatography and the latter is superior for low-molecular-weight substances. However, FFF, by far the most recent entry (dating from 1966 [7]) among major families of analytical separation techniques, has not yet had its potential fully defined.

Among the various fields and gradients used to drive species to the accumulation wall, four specific kinds have been most developed. Each of these fields, utilized as above, yields a different *subtechnique* of FFF. (Other subtechniques are described in Section 9.11.) Each subtechnique has its own advantages and disadvantages and is applicable to samples of different types having different molecular weight ranges. By choosing the appropriate field and thus the FFF subtechnique corresponding to that field, nearly any kind of macromolecular or colloidal material is subject to separation.

One of the subtechniques of FFF employs an electrical field; this subtechnique is known as *electrical FFF* (*ElFFF*). The principles are simple. Species with the highest electrical charge are driven most forcefully toward the wall and form the most compact layers (e.g., zone A of Figure 9.7). These species, forced to reside in the low-flow region near the wall, migrate more slowly than the low-charge species (such as B and C in Figure 9.7) and thus have longer retention times. Thus retention and separation are based on electrical charge. This is somewhat different from electrophoresis where separation is based on electrophoretic mobility, determined by a combination of the charge and the friction coefficient (Eq. 8.19). Another difference is that ElFFF is intrinsically an elution method, like chromatography. With elution techniques, components are flushed from the channel and are more readily subjected to post-channel detection and collection.

Electrical fields have the advantage, pointed out in Chapter 8, of being exceedingly powerful, which means they can induce the migration of virtually all charged species (small ions and macroions) toward the accumulation wall. However the implementation of ElFFF has been difficult because of the need to avoid the generation of electrolysis products in the channel. Thus despite its promise, electrical FFF has had limited applications, mainly to the separation of protein mixtures [15].

Sedimentation FFF (*SdFFF*) is far more advanced. In SdFFF the channel is coiled to fit inside a centrifuge basket, as shown in Figure 9.8. The centrifuge has special seals to introduce the carrier stream from outside and to withdraw the sample upon elution from the channel. The level of retention is determined by the sedimentation force exerted on the particles. Species with the highest mass and density are subject to the strongest driving forces and are retained most. Consequently small (low-mass) species are eluted first, large ones later. The selectivity between particles of different size in SdFFF is relatively high.

As pointed out in Chapter 8, the forces of centrifugation are too weak to influence the distribution of small molecules. The molecular weight M of species must be $\sim 10^6$ in order to generate the necessary force in SdFFF. However for $M > 10^6$, there are many important separation problems involving polymers, biological macromolecules (such as DNAs), subcellular particles, emulsions, and a great variety of natural and industrial colloids. SdFFF has been applied to many such systems [10–12, 16]. An example of the separation of colloidal polystyrene latex microspheres is shown in Figure 9.9.

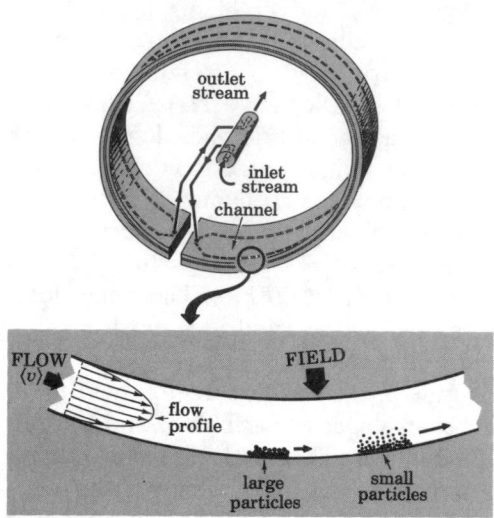

Figure 9.8. Schematic of apparatus for sedimentation FFF.

Thermal FFF (ThFFF) is a subtechnique that depends on the existence of a temperature gradient between the walls of the FFF channel. The temperature gradient is formed between two heavy copper bars clamped (as suggested by Figure 9.5) over a plastic spacer containing the channel. The copper bar at the top is heated and the bottom one is cooled. The resulting temperature difference (typically 10–100 K) induces the phenomenon of thermal diffusion (see Sections 3.2 and 8.8). Thermal diffusion by itself does not generate strong forces (Section 8.8), but when combined with perpendicular flow in an FFF system it becomes highly effective for resolving

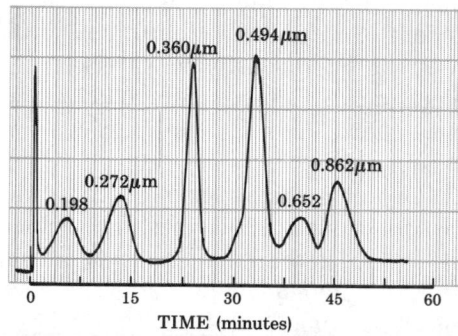

Figure 9.9. Separation of polystyrene latex beads of indicated diameters in the colloidal size range by SdFFF programmed from 1500 to 75 rpm at 2 mL/min flowrate. (Courtesy of Bhajendra N. Barman, FFFractionation. Inc.)

synthetic polymers in organic solvents (17). ThFFF is particularly effective for the separation of synthetic polymers with $M \sim 10^3$–10^7.

In the fourth subtechnique, *flow FFF* (*FlFFF*), an external field, as such, is not used. Its place is taken by a slow *transverse flow* of the carrier liquid. In the usual case carrier permeates into the channel through the top wall (a layer of porous frit), moves slowly across the thin channel space, and seeps out of a membrane-frit bilayer constituting the bottom (accumulation) wall. This slow transverse flow is superimposed on the much faster down-channel flow. We emphasized in Section 7.4 that flow provides a transport mechanism much like that of an external field; hence the substitution of transverse flow for a transverse (perpendicular) field is feasible. However this transverse flow—*crossflow* as we call it—is not by itself selective (see Section 7.4); different particle types are all transported toward the accumulation wall at the same rate. Nonetheless the thickness of the steady-state layer of particles formed at the accumulation wall is variable, determined by a combination of the crossflow transport which forms the layer and by diffusion which breaks it down. Since diffusion coefficients vary from species to species, exponential distributions of different thicknesses are formed, leading to normal FFF separation.

Flow FFF has been applied to a variety of species including proteins, water-soluble synthetic polymers, organic-soluble synthetic polymers, and various colloidal particles. The fractionation of proteins is illustrated in Figure 9.10.

Flow FFF is the most versatile FFF subtechnique (and one of the most versatile of all separation techniques) because it requires nothing more than the interaction of the transverse stream with a suspended or dissolved component. This interaction is made up of the drag forces which induce

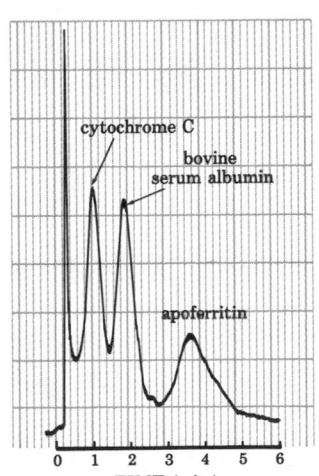

Figure 9.10. Rapid separation of three proteins by flow FFF at a channel flowrate of 8.0 mL/min and a crossflow rate of 6.8 mL/min. (Courtesy of Min-Kuang Liu, University of Utah.)

entrained species to follow flow displacements. This mechanism of transport is universal; all species, large or small, are forced into transport by the motion of surrounding fluid. Thus flow FFF, which utilizes this phenomenon, approaches the status of a universal separation technique [18].

Other aspects of FFF operation and applications have been described in various survey articles [8–14] and in one book [19].

9.9   FACTORS CONTROLLING RETENTION IN FFF

The zone velocity for a specific type of particle or macromolecule in FFF can be formulated in terms of Eq. 9.4. The use of this equation requires knowing the profiles of both concentration c and flow velocity v over the channel cross section. However c and v vary only with transverse coordinate y. Velocity v depends only on y (except near the edges) because of the thinness and uniformity of FFF channels. Concentration varies only with y because the field is applied along y. Furthermore area element dA can be replaced by $b\,dy$, where b is the channel breadth (see Figure 4.2 or 9.5 for explanation).

Specifically, the concentration distribution in the channel, as we noted earlier, is exponential in coordinate y; its form is given by Eq. 6.19, that is

$$\frac{c}{c_0} = \exp\left(-\frac{y}{\ell}\right) \tag{9.20}$$

where y is the distance out into the channel measured from the accumulation wall. The mean layer thickness ℓ, also discussed in the last section, was first given by Eq. 6.20, repeated here for clarity

$$\ell = \frac{D}{|W|} \tag{9.21}$$

The flow velocity profile needed here is that for thin rectangular channels as developed in Section 4.4. Equation 4.17 gives

$$v = 6\langle v \rangle \left(\frac{y}{w} - \frac{y^2}{w^2}\right) \tag{9.22}$$

The substitution of Eqs. 9.22 and 9.20 into 9.4, along with the use of $dA = b\,dy$, gives

$$\mathcal{V} = \frac{6\langle v \rangle \int_0^w \exp\left(-\frac{y}{\ell}\right)\left(\frac{y}{w} - \frac{y^2}{w^2}\right) dy}{\int_0^w \exp\left(-\frac{y}{\ell}\right) dy} \tag{9.23}$$

where the integrals extend from the lower (accumulation) wall at $y = 0$ to the upper wall at $y = w$. From this expression the retention ratio R, equal to $\mathcal{V}/\langle v \rangle$ (Eq. 9.5), can be obtained as

$$R = \frac{\dfrac{6}{w} \displaystyle\int_0^w \exp\left(-\frac{y}{\ell}\right) y \, dy - \dfrac{6}{w^2} \displaystyle\int_0^w \exp\left(-\frac{y}{\ell}\right) y^2 \, dy}{\displaystyle\int_0^w \exp\left(-\frac{y}{\ell}\right) dy} = \frac{A - B}{C} \tag{9.24}$$

The evaluation of these integrals is straightforward, requiring integration by parts. We find

$$A = -6\ell[\exp(-\lambda^{-1})(\lambda + 1) - \lambda] \tag{9.25}$$

$$B = -6\ell[\exp(-\lambda^{-1}) + 2\lambda \exp(-\lambda^{-1})(\lambda + 1) - 2\lambda^2] \tag{9.26}$$

$$C = -\ell[\exp(-\lambda^{-1}) - 1] \tag{9.27}$$

where we have defined *retention parameter* λ as

$$\lambda = \frac{\ell}{w} \tag{9.28}$$

We see that λ is a dimensionless form of ℓ, representing the ratio of ℓ to channel thickness w.

The substitution of the A, B, and C expressions back into Eq. 9.24, followed by rearrangement, yields

$$R = 6\lambda\left[\coth\left(\frac{1}{2\lambda}\right) - 2\lambda\right] \tag{9.29}$$

The dependence of R on λ as specified by this equation is shown in Figure 9.11.

For the most practical FFF operation we strive (for reasons we shall see) for small λ and ℓ values, in which case Eq. 9.29 can be shown to approach the limit

$$R = 6\lambda \tag{9.30}$$

The simple linear form of Eq. 9.30 (corresponding to the linear region on the left-hand side of Figure 9.11) is convenient for relating observed retention to the physicochemical parameters making up λ.

We now examine λ more closely. From Eqs. 9.21 and 9.28 we obtain

$$\lambda = \frac{D}{|W|w} \tag{9.31}$$

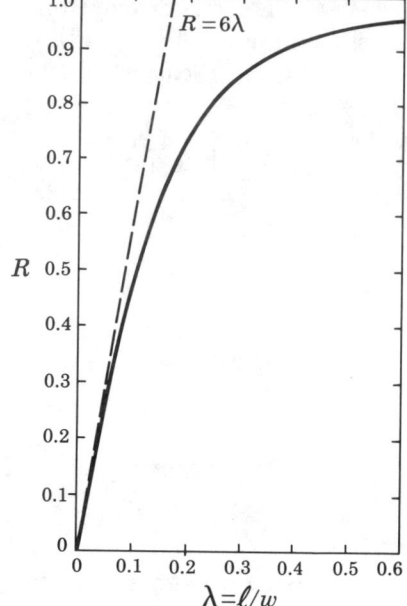

Figure 9.11. Variation of retention ratio R with λ according to Eqs. 9.29 (solid line) and 9.30 (broken line).

where W is the lateral velocity induced by the applied driving force, equal to the sum $U + v$. For the subtechnique of flow FFF, v is finite and U is zero, giving $W = v$ and $\lambda = D/|v|w$. For the other FFF subtechniques where lateral flow is absent we have

$$W = U = \frac{F}{f} \tag{9.32}$$

that is, U equals driving force F divided by friction coefficient f, as shown by Eq. 3.19. If we write $D = \mathscr{R}T/f$, Eq. 3.21, Eq. 9.31 becomes

$$\lambda = \frac{\mathscr{R}T}{|F|w} \tag{9.33}$$

Thus retention parameter λ can be considered as a ratio of two energies: dissipative energy $\mathscr{R}T$ and the energy $|F|w$ expended in driving components across channel thickness w. We have shown before (see Section 8.2) that energy ratios of this kind are crucial to separation.

The application of the above development can be illustrated for sedimentation FFF. In this subtechnique the channel, fitted in a circular configuration at the inner circumference of a centrifuge basket (Figure 9.8), is subjected to a lateral sedimentation force due to spin. All particles denser than the carrier are spun to the outside wall, which becomes the accumulation wall along which the particles migrate. (Less dense particles drift to the inside wall and undergo FFF separation there.)

To characterize migration we need λ, which requires that we substitute the sedimentation force into Eq. 9.33. We use Eq. 8.23 for force. With sample density ρ_s replacing $1/\bar{v}$ and $\omega^2 r$ used for acceleration G, this force equals

$$F = M[1 - (\rho/\rho_s)]\omega^2 r \tag{9.34}$$

Thus λ becomes

$$\lambda = \frac{\mathscr{R}T}{Mw|1 - (\rho/\rho_s)|\omega^2 r} \tag{9.35}$$

With this λ substituted into Eq. 9.29 or 9.30, we get retention ratio R. Clearly, R varies strongly with molecular weight M because λ depends directly on M. In the simple case in which λ is small and Eq. 9.30 can be used for R, we obtain

$$R = \frac{6\mathscr{R}T}{M|1 - (\rho/\rho_s)|\omega^2 r} \tag{9.36}$$

which shows that zone migration velocity $R\langle v \rangle$ is inversely proportional to M. We thus expect to get an effective separation based on differences in molecular weight or particle mass (for particles of equal density). This is borne out by Figure 9.9, which shows the separation of colloidal particles of different diameters and thus of different masses.

The preceding equations yield experimental R values in terms of the particle characteristics M and ρ_s. If M (or ρ_s) is not known, it can be obtained from the above equations, given an experimental value for R. Thus sedimentation FFF can be used to measure the mass and density of colloidal particles, normally a difficult task. This approach has been used on viruses, liposomes, and other important colloids. The methodology has been discussed fully in various papers [9–12, 20–23].

An approach similar to that used above for sedimentation FFF can be applied to the other FFF subtechniques. This approach yields

$$\lambda = \frac{D}{\mu E w} \qquad \text{(electrical FFF)} \tag{9.37}$$

$$\lambda = \frac{DV^0}{\dot{V}_c w^2} \qquad \text{(flow FFF)} \tag{9.38}$$

$$\lambda = \frac{T}{\alpha w(dT/dy)} \qquad \text{(thermal FFF)} \tag{9.39}$$

where μ is the electrophoretic mobility, E the electrical field strength, V^0 the channel void volume, $\dot{V}_c$ the volumetric cross-flowrate, and α the thermal diffusion factor.

When the above λs are substituted into Eq. 9.29 or 9.30, we again end up with a linkage between experimental retention measurements (as reflected in R) and various electrical (μ), diffusive (D), and thermal (α) properties of the components of interest. It is clear that species are separated on the basis of differences in these properties. For a particular sample we can, in theory, choose the property with the largest differences (the highest selectivity) between species, then apply the corresponding FFF subtechnique to that sample.

We note that all λs are inversely proportional to the strength of the field or gradient employed: $\omega^2 r$, E, $\dot{V}_c$, and dT/dy. Thus the λs and the resulting R values can be readily manipulated by changes in field strength. This adaptability makes FFF applicable to many different sample types over a wide molecular weight range. It also makes possible *field programming*, in which the field strength is gradually reduced in the course of a run to handle a wide variety of components within that run [14].

9.10 PLATE HEIGHT AND OPTIMIZATION IN FFF

Equation 9.17 can be used to obtain a plate height equation for field-flow fractionation. (The longitudinal diffusion term, Eq. 9.14, is generally negligible in FFF.) We assume that the relative flow dispersion $\Delta \mathcal{V}/\mathcal{V}$ is approximately unity. Most diffusion processes are limited to paths not much larger than ℓ; thus we can write $d \sim 2\ell$. With these assumptions we get the equation [24]

$$H = \frac{q\mathcal{V}\ell^2}{D} \tag{9.40}$$

where the coefficient q is ~ 2. However, a theory based on nonequilibrium considerations, more rigorous than the random walk model, leads to an exact result, which for $\ell \ll w$ is $q = 4$.

Equation 9.40 can be expressed in terms of λ by writing $\ell = \lambda w$ (from Eq. 9.28), $\mathcal{V} = R\langle v \rangle$ (Eq. 9.5), and $R = 6\lambda$ (Eq. 9.30). We get

$$H = \chi \frac{w^2 \langle v \rangle}{D} \tag{9.41}$$

where we find $\chi = 6q\lambda^3$, which equals $24\lambda^3$ for small λ and ℓ values [24].

The last two equations are important for the optimization of FFF. Equation 9.41 shows that plate height H is reduced by decreasing the values of $\langle v \rangle$ and w and by increasing D. The latter is achieved by decreasing the carrier viscosity by working closer to the boiling point as described in Chapter 4. Most importantly, Eq. 9.40 shows that H is directly proportional to the square of layer thickness ℓ. In seeking high resolution, therefore, it is imperative to keep ℓ small. A reduction in ℓ, however, leads to reduced λ

and R values. The reduction in R gives slower zones and thus appears to reduce separation speed. This can be offset by an increased flow velocity as shown below.

To pursue fast separation we can maximize N/t, which we obtain directly from Eq. 9.19. However we follow here an alternate optimization scheme that is more like that generally applied to chromatography (Section 14.5). Following this approach, we examine the time $L/\mathcal{V}$ needed to bring the last component through the channel. If the separation requires N plates, then the channel length must be $L = NH$, in which case separation time is

$$t = \frac{L}{\mathcal{V}} = N \frac{H}{\mathcal{V}} \tag{9.42}$$

The $H/\mathcal{V}$ ratio is obtained from Eq. 9.40, leading to

$$t = N \frac{4\ell^2}{D} \tag{9.43}$$

Thus separation time is reduced by minimizing ℓ and increasing D. It is also reduced by working with the least possible number N of plates. We minimize N by choosing the most selective FFF system; this often proves to be sedimentation FFF.

We see that all optimization goals depend strongly on reducing ℓ. Experimentally, ℓ values down to 1 μm and less have been realized. Such thin solute layers yield high-speed separations, often requiring only a few minutes for completion. However with small ℓ, particle-wall interactions increase and in some cases lead to departures from theory [25].

9.11 VARIANTS OF FIELD-FLOW FRACTIONATION

FFF is a very broad family of methods. The theory laid out above describes only one category (or operating mode) called *normal FFF*, which by itself consists of a number of subtechniques as described in Section 9.8. In normal FFF the concentration distribution along y is exponential as a consequence of the balance between field-induced displacements and diffusion. However under other circumstances the concentration distribution of a species may be different, altering the characteristics of the system and invalidating the foregoing exponential-based theory. For example the concentration distribution may approach a δ-function along y, centered either near or away from the wall, or the distribution may be time dependent. Each of these distributions (produced by special arrangements of transverse forces) can be combined with a variety of fields to produce a plethora of subtechniques. Some of the many possible subtechniques are shown in Table 9.1. Only a few of these, mainly in the normal FFF category, have been carefully tested and evaluated.

TABLE 9.1 A Partial Compilation of Possible Subtechniques of FFF, Each Arising as a Unique Combination of Driving Force (or field) and Operating Mode

Operating Mode	Transverse Driving Force[a]					
	Sd	El	Th	Fl	Mg	Dl
Normal (Nl)	SdFFF	ElFFF	ThFFF	FlFFF	MgFFF	DlFFF
Steric (St)	Sd/StFFF	El/StFFF	Th/StFFF	Fl/StFFF	Mg/StFFF	Dl/StFFF
Hyperlayer (Hy)	Sd/HyFFF	El/HyFFF	Th/HyFFF	Fl/HyFFF	Mg/HyFFF	Dl/HyFFF
Cyclical field (Cy)	Sd/CyFFF	El/CyFFF	Th/CyFFF	Fl/CyFFF	Mg/CyFFF	Dl/CyFFF
Second. equil. (Sy)	Sd/SyFFF	El/SyFFF	Th/SyFFF	Fl/SyFFF	Mg/SyFFF	Dl/SyFFF
Chrom. hybrid (Ch)	Sd/ChFFF	El/ChFFF	Th/ChFFF	Fl/ChFFF	Mg/ChFFF	Dl/ChFFF

[a]Sd = sedimentation, El = electrical, Th = thermal, Fl = flow, Mg = magnetic, Dl = dielectrical.

The large number of potential FFF subtechniques creates a nomenclature problem. Each subtechnique is characterized by a specific driving force or field and by an operating mode corresponding to the type of concentration distribution. The operating mode is most fundamental and should be designated by the immediate prefix to FFF: *StFFF* is *steric FFF*, *HyFFF* is *hyperlayer FFF*, and so on. The designation of field type precedes the above: for example *Sd/StFFF* is *sedimentation/steric FFF*. Sometimes the "Nl" for normal FFF is omitted for historical reasons: *SdFFF* is the same as *Sd/NlFFF*. However sedimentation FFF should not be abbreviated as "SFFF" because the single letter "S" could stand for either "sedimentation" or "steric" [26].

The most prominent variation from normal FFF is an operating mode known as *steric FFF*. Steric FFF, which can be used with any of the fields or gradients described previously, is applicable to large particles (usually over 1 μm diameter) that are driven forcefully toward the wall and that have negligible diffusion [27]. Thus particle motion toward the wall is halted, not by diffusion, but by the particle's approach to physical contact with the wall. The distribution of particles relative to the wall is then determined by particle size. Thus separation is based on size. The largest particles, which because of their size protrude furthest into the flowstream, emerge first in steric FFF. Smaller particles follow. This elution order is inverted relative to that observed for normal FFF where smaller particles emerge first.

Steric FFF (mainly in the form *Sd/StFFF*) has been applied to the separation of a number of large-particle samples, including chromatographic support material, fine-ground coal, and human and animal cells. Separation can be accomplished rapidly despite the normal sluggishness of large-particle transport (28). An example is shown in Figure 9.12.

We note that in steric FFF the centers of equal-sized particles, unaffected by Brownian motion, will settle into the same plane, thus effectively forming

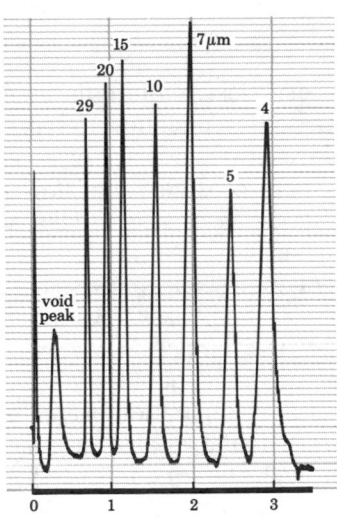

Figure 9.12. Rapid separation of large (>1 μm) polystyrene latex microspheres of different diameters by steric FFF. (Courtesy of Myeong Hee Moon, University of Utah).

a δ-function distribution located approximately one particle radius away from the wall. However it has been observed that the particle centers move further away from the wall as the flow velocity increases. This has been attributed to *hydrodynamic lift forces* [28, 29]. When carried to an extreme by rapid flow, the particles are elevated to a level substantially above the wall. The plane of particle centers then forms a "hyperlayer" and the methodology becomes *hyperlayer FFF*. This particular form of hyperlayer FFF separates larger particles very rapidly because of the high flowrates [26, 30].

Hyperlayer FFF is, by itself, a broad methodological class with many subtechniques possible through various combinations of transverse forces and gradients. (Some are shown in the third row of Table 9.1.) Only a few of the hyperlayer subtechniques have been experimentally realized.

Hyperlayer FFF is, in a sense, the ultimate $F(+)$ separation method. Ideally each particle type in hyperlayer FFF is forced into a very thin layer (approaching a δ-function) at its own characteristic elevation. Here it occupies a thin fluid layer flowing at a fixed velocity. Thus the variation $\Delta \mathcal{V}$ in velocity for particles of a given type is very small and the ratio $\Delta \mathcal{V}/\mathcal{V}$ in Eq. 9.17 is almost negligible. By contrast species undergoing chromatographic or normal FFF migration are smeared over a group of velocity states and $\Delta \mathcal{V}/\mathcal{V}$ is of order unity. The difference is that hyperlayer FFF is the only methodology for which each distinct species is narrowly confined to its own unique velocity state. Very strong forces are required for such confinement and effective hyperlayer FFF has so far only been developed for larger particles [30–32]. It is yet to be determined how far this category of FFF methods will go in solving practical separation problems.

9.12   OVERVIEW OF $F(+)d$ SEPARATIONS

The $F(+)d$ methods, like techniques in the $F(+)cd$ class, require differential flow for separation. However the nature of the differential flow differs in the two cases. Since the $F(+)d$ methods induce enrichment *only* at an interface between two phases, the sole requirement of differential flow is that one phase assumes motion relative to another phase. It is usually an easy matter to instigate the relative motion of phases.

A few simple differences in the properties of immiscible phases make possible their relative displacement. Most simply, if the phases have different densities they will automatically acquire a relative motion in a gravitational field. Thus in adsorptive bubble separation methods, bubbles injected into a column of liquid rise toward the upper surface. Separation occurs by combining the relative enrichment of components at the bubble interface with the continuous displacement of bubbles through the liquid [33–35].

Fractional distillation works similarly. There is a division into vapor and liquid phases, the first enriched in solute(s) of relatively high vapor pressure and the second enriched in solutes of low vapor pressure. This enrichment is amplified by the relative displacement of phases. Again the relative motion occurs by virtue of the differences in densities of the phases coupled with the action of gravity. The downward flow (reflux) of the liquid stream is a direct consequence of gravity and is responsible for the accumulation of high-boiling components at the bottom of a distillation column. The motion of the vapor, countercurrent to the liquid stream, sweeps low boilers to the top [36].

Countercurrent distribution (CCD), as it is practiced in the so-called Craig machine, utilizes gravity in combination with machine activity to give relative motion to two immiscible liquid phases. After the two phases are stirred together by a rocking motion of the machine (to exchange solute), the phases are given time to settle. The upper phase is then drained off into a subsequent tube. Thus, gravity (combined with density differences) is responsible for the stratification of the two liquids in the tube and for the drainage process where additional relative motion of phases is realized. Other forms of CCD are described in the literature [37].

There is another common mechanism for achieving the relative flow of phases. This mechanism, used in chromatography, takes advantage of the fact that phases differ in their ability to wet and adhere to various solid surfaces. Thus if one phase (say a liquid phase in gas-liquid chromatography) is drawn into the fine capillaries of a porous support by surface forces, its flow will be rendered negligible, even with a pressure gradient, by the flow resistance of the large internal surface and/or the adhesion of the liquid to the surface. In some cases one phase is chemically bonded to the surface (Chapter 10). The phase adhering in one of these ways to the particles or column walls is termed the *stationary phase*.

The phase left in the large interstices between support particles is unanchored and can be driven by an applied pressure gradient. It is necessary, of course, that the viscosity of this *mobile phase* be relatively low so that flow can occur without excessive pressure drops.

We note that electroosmotic flow, originating at any charged interface (i.e., almost any real interface), is an effective mechanism for relative phase displacement. Its use in chromatography has been championed by Knox and co-workers [38].

The singular importance of chromatography in analytical separations is well known. However, dynamically chromatography is one of the most complicated separation methods [2]. An abbreviated treatment of the principles of chromatography, developed below, will integrate chromatography with other $F(+)$ separations and will, at the same time, set the stage for the more detailed treatment provided in the remaining three chapters of this book.

9.13   CHROMATOGRAPHY

A solute undergoing chromatographic migration partitions between the stationary and mobile phases, a process driven by thermodynamic equilibrium. At equilibrium (established fully only at the zone center), the concentration in the stationary phase (c_s) relative to that in the mobile phase (c_m) is given by the thermodynamic distribution constant K, as shown by comparing Eqs. 2.18 and 2.19. Thus

$$\left(\frac{c_s}{c_m}\right)_{eq} = K \tag{9.44}$$

where K and the factors affecting it were discussed in Chapter 2.

The amount of solute accumulating in the stationary phase relative to that in the mobile phase is termed the capacity factor k'; it is the product of K and of the ratio of the volume V_s of stationary phase to the volume V_m of mobile phase in the column

$$k' = \frac{V_s}{V_m}\left(\frac{c_s}{c_m}\right)_{eq} = \frac{V_s K}{V_m} \tag{9.45}$$

By the same reasoning, the fraction R' of the total solute found in the mobile phase is simply

$$R' = \left(\frac{V_m c_m}{V_m c_m + V_s c_s}\right)_{eq} = \frac{V_m}{V_m + K V_s} = \frac{1}{1 + k'} \tag{9.46}$$

A solute that is "inert" to partitioning in the stationary phase, and therefore occupies only the mobile phase ($R' = 1$), will constitute a nonretained solute as described in Section 9.3. This solute will be swept directly through the column at velocity $\langle v \rangle$ with the mobile-phase flow. A partitioning solute, for which fraction $R' < 1$ occupies the mobile phase, will travel at only a fraction R' of the velocity $\langle v \rangle$ of the inert peak. Therefore its velocity will be

$$\mathscr{V} = R'\langle v \rangle \tag{9.47}$$

We see that this is equivalent to Eq. 9.5, which shows that the fractional velocity, defined as $R = \mathscr{V}/\langle v \rangle$, is the same as the fractional occupancy R' of the mobile phase. Thus

$$R' = R \tag{9.48}$$

Equation 9.47 can be obtained directly from Eq. 9.4. Each integral in the latter is broken into two integrals, one extending (as indicated by label m) over the cross-sectional area A_m of the mobile phase and the other (label s) over stationary phase area A_s. This gives

$$\mathcal{V} = \frac{\int_m c_m v \, dA + \int_s c_s v \, dA}{\int_m c_m \, dA + \int_s c_s \, dA} \tag{9.49}$$

where the upper right-hand integral is zero because local flow velocity v is zero everywhere in the stationary phase. If we assume equilibrium partitioning between phases then mobile and stationary phase concentrations c_m and c_s will be constant within the phases and can be removed from the integrals. Thus

$$\mathcal{V} = \frac{c_m \int_m v \, dA}{i_m \int_m dA + c_s \int_s dA} = \frac{c_m \langle v \rangle A_m}{c_m A_m + c_s A_s} \tag{9.50}$$

Multiplying numerator and denominator by column length L and recognizing that $A_m L$ and $A_s L$ are the respective volumes V_m and V_s of mobile and stationary phases in the column, we have

$$\mathcal{V} = \langle v \rangle \frac{c_m V_m}{c_m V_m + c_s V_s} = \langle v \rangle \frac{V_m}{V_m + (c_s/c_m)V_s} \tag{9.51}$$

or with the equilibrium expression, $c_s/c_m = K$, we obtain

$$\mathcal{V} = \frac{V_m \langle v \rangle}{V_m + KV_s} = R \langle v \rangle \tag{9.52}$$

which is the same as Eq. 9.47 and which (with $R' = R$) is also given by Eq. 9.46.

Returning to the inert or nonretained solute traveling at velocity $\langle v \rangle$, we note that such a solute will emerge after one unit of the mobile phase volume V_m has been swept through the column. A retained solute traveling at the lesser velocity $R \langle v \rangle$ will require a greater volume of mobile phase, $V_r = V_m/R$, to wash it through the column. Combining this with Eq. 9.46 (recognizing that $R' = R$), we get

$$V_r = V_m + KV_s \tag{9.53}$$

where the volume V_r is termed the *retention volume*.

The V_r expression of Eq. 9.53 is strictly valid only for columns without gradients along their length. The correction for the nonuniformity arising from gas compressibility in gas chromatography will be derived in Section 10.4.

The simplest picture of chromatographic zone broadening comes from the random walk model. The random walk model for chromatography will be developed at some length in Chapter 11; here we present a synopsis.

We imagine molecules hopping erratically between mobile and stationary phases. The hops represent random steps forward and backward with respect to the zone center. Equation 9.12 shows that plate height H for such a process is proportional to transfer or equilibration time t_{eq}, in agreement with our conclusions based on nonequilibrium considerations.

In most forms of gas chromatography (GC) and in some forms of liquid chromatography (LC) at high flowrates, the slow (and therefore controlling) step for solute transfer is diffusion through a stationary film or droplet of depth d. Under these circumstances, Eq. 9.17 applies. Assuming $\Delta \mathcal{V}/\mathcal{V} \sim 1$, and using $\mathcal{V} = R \langle v \rangle$, we get

$$H \sim \frac{1}{2} R \frac{d^2 \langle v \rangle}{D_s} \tag{9.54}$$

where D_s is the diffusion coefficient for solute in the stationary phase. A more exact analysis by nonequilibrium theory yields [2]

$$H = qR(1 - R) \frac{d^2 \langle v \rangle}{D_s} \tag{9.55}$$

where q is a configuration factor dependent on the exact geometry of the stationary pool. Clearly the depth d of the pool must be kept minimal for optimum performance (see Chapter 12).

Transfer through the mobile phase is more complicated because diffusion and flow both act to shuttle molecules to and from the stationary phase (see Chapter 11). The effective coefficient for transfer is $D_T = D_{err} + D$, Eq. 5.34, where D_{err} arises from random flow currents. Equation 9.17 assumes the form

$$H \sim \frac{d_p^2 \langle v \rangle}{D_{err} + D_m} \tag{9.56}$$

where we have assumed $\Delta \mathcal{V}/\mathcal{V} \sim 1$, $d \sim d_p$ (the support particle diameter), and $\mathcal{V} \sim \langle v \rangle$, the average velocity applicable to the mobile phase where the transfer is taking place. The diffusion coefficient for solute in the mobile phase is designated by D_m.

Equation 9.56 is a form of the coupling equation for mobile-phase plate height, so named because of the coupling of flow (through term D_{err}) and diffusion (term D_m) in transporting solute between phases [2]. The equation illustrates the importance of reducing particle diameter d_p in chromatography, a point emphasized several times earlier. However Eq. 9.56 has some less obvious properties that help explain the behavior and optimization of LC systems. For example, D_{err} is proportional to $\langle v \rangle$. (This can be deduced from Eq. 5.26 in which $\dot{n}$ is clearly proportional to $\langle v \rangle$.) Consequently at high velocities, $D_{err} \gg D$ and the mobile phase H becomes independent of $\langle v \rangle$.

A more detailed treatment of chromatography (see Chapter 11) suggests a plate height equation of the form [2]

$$H = \frac{B}{\langle v \rangle} + qR(1 - R)\frac{d^2\langle v \rangle}{D_s} + \sum \left[\frac{1}{(1/2\lambda_i d_p) + (D_m/\omega_i d_p^2 \langle v \rangle)} \right]$$
$$(9.57)$$

where B is a coefficient (see Eq. 9.14) representing longitudinal diffusion and where the λ_i and ω_i values are constants. The summation refers to different transport paths over which solute must be shuttled to aid equilibrium in the mobile phase.

The role of Eq. 9.57 and similar equations in the description and manipulation of modern chromatography is very broad, but the details are beyond the scope of this chapter (see Chapter 12). The equation can be used to optimize, design, and understand the behavior of both GC and LC systems. Because the equation is written to be valid for all chromatographic forms, its most important function, from the point of view of this chapter, is to show that a common theoretical framework stretches across all chromatographic methods. This conclusion is a microcosm of the theme of this book, that all separation methods are dominated by common basic features.

REFERENCES

[1] J. C. Giddings, *J. Chem. Educ.*, **44**, 704 (1967).

[2] J. C. Giddings, *Dynamics of Chromatography, Part I: Principles and Theory*, Marcel Dekker, New York, 1965.

[3] J. C. Giddings, in I. M. Kolthoff and P. J. Elving, Eds., *Treatise on Analytical Chemistry*, Part I, Vol. 5, Wiley, New York, 1982, Chapter 3.

[4] P. G. Grodzka and B. Facemire, *Sep. Sci. Technol.*, **12**, 103 (1977).

[5] M. Bier, in M. Bier, Ed., *Electrophoresis*, Vol. 1, Academic, New York, 1959, pp. 270–315.

[6] J. C. Giddings, M. N. Myers, and K. D. Caldwell, *Sep. Sci. Technol.*, **16**, 549 (1981).

[7] J. C. Giddings, *Sep. Sci.*, **1**, 123 (1966).

[8] J. C. Giddings, M. N. Myers, K. D. Caldwell, S. R. Fisher, in D. Glick, Ed., *Methods of Biochemical Analysis*, Vol. 26, Wiley, New York, 1980, pp. 79–136.

[9] J. C. Giddings, *Sep. Sci. Technol.*, **19**, 831 (1984).

[10] K. D. Caldwell, *Anal. Chem.*, **60**, 959A (1988).

[11] J. C. Giddings, *C&E News.* **66**, 34 (1988).

[12] L. F. Kesner and J. C. Giddings, in P. R. Brown and R. A. Hartwick, Eds., *High Performance Liquid Chromatography*, Wiley, New York, 1989, Chapter 15.

[13] J. C. Giddings, G. Karaiskakis, K. D. Caldwell, and M. N. Myers, *J. Colloid Interface Sci.*, **92**, 66 (1983).

[14] J. C. Giddings and K. D. Caldwell, *Anal. Chem.*, **56**, 2093 (1984).

[15] K. D. Caldwell, L. F. Kesner, M. N. Myers, and J. C. Giddings, *Science*, **176**, 296 (1972).

[16] J. J. Kirkland and W. W. Yau, *Science*, **218**, 121 (1982).

[17] J. C. Giddings, in B. J. Hunt and S. Holding, Eds., *Size Exclusion Chromatography*, Blackie and Son, Glasgow, 1989, Chapter 8.

[18] J. C. Giddings, F. J. Yang, and M. N. Myers, *Science*, **193**, 1244 (1976).

[19] J. Janča, *Field-Flow Fractionation: Analysis of Macromolecules and Particles*, Marcel Dekker, New York, 1988.

[20] J. C. Giddings, K. D. Caldwell and H. K. Jones, in T. Provder, Ed., *Particle Size Distribution: Assessment and Characterization*, ACS Symposium Series No. 332, ACS, Washington, DC, 1987, Chapter 15.

[21] R. Beckett, G. Nicholson, B. T. Hart, M. E. Hansen, and J. C. Giddings, *Wat. Res.*, **22**, 1535 (1988).

[22] H. K. Jones and J. C. Giddings, *Anal. Chem.*, **61**, 741 (1989).

[23] J. C. Giddings, B. N. Barman, and H. Li, *J. Colloid Interface Sci.*, **132**, 554 (1989).

[24] J. C. Giddings, *J. Chem. Educ.*, **50**, 667 (1973).

[25] M. E. Hansen and J. C. Giddings, *Anal. Chem.*, **61**, 811 (1989).

[26] J. C. Giddings, X. Chen, K.-G. Wahlund, and M. N. Myers, *Anal. Chem.*, **59**, 1957 (1987).

[27] J. C. Giddings and M. N. Myers, *Sep. Sci. Technol.*, **13**, 637 (1978).

[28] T. Koch and J. C. Giddings, *Anal. Chem.*, **58**, 994 (1986).

[29] K. D. Caldwell, T. T. Nguyen, M. N. Myers, and J. C. Giddings, *Sep. Sci. Technol.*, **14**, 935 (1979).

[30] X. Chen, K.-G. Wahlund, and J. C. Giddings, *Anal. Chem.*, **60**, 362 (1988).

[31] S. K. Ratanathanawongs and J. C. Giddings, *J. Chromatogr.*, **467**, 341 (1989).

[32] B. N. Barman, M. N. Myers, and J. C. Giddings, *Powder Technol.*, **59**, 53 (1989).

[33] B. L. Karger, L. R. Snyder, and C. Horvath, *An Introduction to Separation Science*, Wiley, New York, 1973.

[34] A. N. Clark and D. J. Wilson, *Foam Flotation: Theory and Application*, Marcel Dekker, New York, 1983.

[35] R. B. Grieves, in P. J. Elving and I. M. Kolthoff, Eds., *Treatise on Analytical Chemistry*, Part I, Vol. 5, Wiley, New York, 1981, Chapter 9.

[36] A. Rose and E. Rose, in P. J. Elving and I. M. Kolthoff, Eds., *Treatise on Analytical Chemistry*, Part I, Vol. 5, Wiley, New York, 1982, Chapter 10.

[37] H. L. Rothbart and R. A. Barford, in P. J. Elving and I. M. Kolthoff, Eds., *Treatise on Analytical Chemistry*, Part I, Vol. 5, Wiley, New York, 1982, Chapter 12.

[38] J. H. Knox and I. H. Grant, *Chromatographia*, **24**, 135 (1987).

EXERCISES

9.1(**) Use a mathematical (not physical) approach to determine R for a component swept through an empty tube of square cross section when the component's concentration (but not the flow velocity) is uniform over that cross section.

9.2(*) A component with $D = 0.80 \times 10^{-5}$ cm^2/s is carried at 1.6 mm/s down the flow axis of a $F(+)$ system in which $d = 50\ \mu$m. Assume $\Delta \mathcal{V}/\mathcal{V} = 1$. Estimate H and N/t. Repeat the calculation for $d = 5.0\ \mu$m. Note the strong effect of d on your result.

9.3(***) Show that the substitution of the integrated forms of A, B, and C into Eq. 9.24 yields Eq. 9.29.

9.4(**) Prove that the retention ratio R in FFF, given by Eq. 9.29, approaches $R = 6\lambda$ (Eq. 9.30) as $\lambda \rightarrow 0$.

9.5(*) The sedimentation coefficient s, often used to characterize biological macromolecules and particles, is equal to the ratio of the velocity $|U|$ induced by a sedimentation field to the strength G of that field measured as acceleration. Show that s can be related to the retention parameter λ in sedimentation FFF by $s = D/\lambda Gw$.

9.6(**) For sedimentation FFF, we consider retention in a $w = 0.254$ mm channel held at 293 K in a high-speed centrifuge rotating at $1.00 \times 10^5 \times g$. Assume that the carrier density is 1.00 g/mL and that the particle density is 1.50 g/mL. Calculate ℓ, λ, and R for molecular weight $M = 10^6$. Repeat the calculation for $M = 10^5$ and $M = 10^7$.

9.7(**) In the text it is stated that a molecular weight M of $\sim 10^6$ is necessary for effective operation in sedimentation FFF. We can examine this matter by further considering the system described in 9.6 above. If the two molecules ($M = 10^6$, $M = 10^7$) are spherical globules, estimate the plate height and the number of plates in a channel 50.0 cm long. The flow velocity $\langle v \rangle$ is 1.25 cm/s.

9.8(**) When transversely oriented focusing forces of the type used in isoelectric focusing or isopycnic sedimentation are used to confine different components to different thin bands or laminae in an FFF channel, we have a variant of FFF called hyperlayer FFF. Suppose component A is focused at $y = 15.2\ \mu$m and B at $y = 127\ \mu$m (0.005 in). You may assume that the thin bands form a δ-function distribution in coordinate y and that channel thickness $w = 254\ \mu$m. What is the R value for each of the two components?

9.9(**)   Imagine a system resembling capillary chromatography except for the presence of three phases: (i) a stationary phase (composed of a gummy material) applied to the inside wall of the tube, (ii) a flowing annulus of liquid inside the stationary phase, and (iii) a core consisting of another nonmiscible fluid. For simplicity assume that the three phases all occupy an equal volume and that phases 2 and 3 both flow uniformly at 2.5 and 4.0 mm/sec. Furthermore, assume that for a particular solute, values for two of the distribution constants are $K_{21} = (c_2/c_1)_{eq} = 0.76$ and $K_{32} = (c_3/c_2)_{eq} = 1.6$. Calculate the velocity of the solute band down the tube.

10

CHROMATOGRAPHY: A DEEPER LOOK

We have singled out chromatography for a more probing analysis. The reasons for this focus are twofold. First of all, chromatography, because of its outstanding separation power and versatility, is the most widely used of all separation methods. Second, the underlying processes governing chromatographic performance are among the most complicated of those controlling any separation method. These processes require a considerable depth of understanding on the part of anyone seeking to maximize and extend chromatographic capabilities.

Despite the mechanistic complications, chromatography is intrinsically simple in operation. A piece of tubing packed with a granular sorptive material and carrying a stream of gas or liquid through its length will convert a droplet of a complex mixture, injected at one end, into dozens of molecular constituents, which emerge individually from the other end. With such directness and simplicity, chromatography has a universal clientele. It serves the life scientist in attempting to unravel the complex chemical basis of life processes; the industrial chemist in determining the composition of plastics, drugs, pesticides, food, petroleum, and so on; the environmental/ analytical chemist in tracing contaminants through air and water; the nuclear chemist in the separation of transuranium elements and some isotopes; the organic and physical chemist in studies of reaction mechanisms and rates as revealed by the analysis of reaction products; and scientists from literally dozens of other disciplines requiring the analysis of complex multicomponent materials.

While a basic level of separation can be realized fairly easily in chromatography, the complex chemical mixtures confronting many scientists require a higher level of sophistication. The challenge in chromatography is not simply in making it work, but in making it work well; it must work exceptionally well to deal with truly complex mixtures. Great strides have been made in the past few decades, largely following the guidelines of theory. Much work is still needed because our world is such a complex place that some of its common mixtures (e.g., a drop of petroleum or the contents of a bacterial cell) still defy any more than fragmentary resolution.

Chromatography was developed near the beginning of this century (1903) by the Russian botanist Mikhail Tswett, generally considered the father of chromatography. Tswett was interested in separating some important plant pigments (green chlorophylls and yellow carotenoids). He deposited a mixture of these pigments (dissolved in a small amount of solvent, mainly petroleum ether) at the top of a narrow glass tube or column filled with powdered calcium carbonate. The mixture was then washed into the column with additional solvent. Because the pigments adsorbed to some extent on the solid, they did not wash freely through the column but were retarded by adsorption. Since some pigments adsorbed more strongly than others, they were retarded to a greater extent and soon fell behind the weakly adsorbing pigments. Thus there developed a differential motion (or migration) down the column, each species moving at a rate determined by its tendency to adsorb. The components soon separated into individual zones (see Figure 10.1), each located at a different point along the column's length and each visible because of its color. The colored bands led Tswett to employ the name chromatography (Greek: color writing).

A more complete historical perspective on this and subsequent developments can be found in a book edited by Ettre and Zlatkis [1].

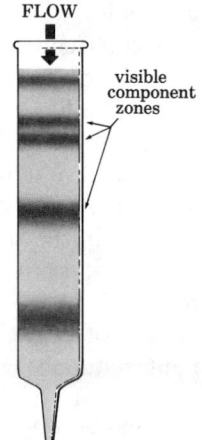

Figure 10.1. Schematic of Tswett's pioneering chromatography: The separation of visible components into separate zones in a glass-walled chromatographic column.

10.1 SCOPE OF CHROMATOGRAPHY

The Tswett procedure is only one of hundreds that has now inherited the name "chromatography." All these methods depend upon the basic chromatographic mechanism: a differential retardation caused by unequal degrees of sorption of substances washed along in a stream of fluid. However there are numerous variations in the procedure as suggested below.

Tswett's fluid was a liquid; hence this method was a form of *liquid chromatography* or *LC*. His liquid or *mobile phase* was nonpolar; hence *normal-phase LC*. His retardant (*stationary phase*) was an adsorbent; thus *adsorption chromatography*. However, the nonpolar liquid that washed Tswett's components through the column can be exchanged for a more polar one, including water (*reversed-phase liquid chromatography, RPLC*), or can even be exchanged for a gas if the components are volatile (*gas chromatography* or *GC*). The adsorbent can be a different chemical type or, in its place, there can be small "droplets" of some liquid held motionless in the pores of granular particles, in which case retardation is caused by differential absorption or solution (*partition chromatography*) rather than adsorption or attachment to a surface.

Additional variations entail using columns of different sizes, made of different materials, and operated at different pressures and temperatures. For narrow-bore columns the walls can be coated with a sorbent layer and the packing removed (*open tubular* or *capillary chromatography*). Alternately one can do away with the column entirely, using an open granular bed (*thin-layer chromatography, TLC*) or a simple sheet of paper (*paper chromatography, PC*) into which liquids flow by capillary action (Section 4.7).

These and many other variations (see below) make it possible to find a chromatographic system suitable for application to most complex mixtures. The species to be separated may be large or small, polar or nonpolar, isomeric or homologous, molecular or ionic, volatile or nonvolatile, and, of course, colored and thus visible (as with Tswett's work) or, more commonly, invisible, requiring a sensitive detector based on UV adsorption, selective ionization, and so on.

From the above discussion it is apparent that the names used to describe different chromatographic techniques are related to some important feature of that technique. Gas and liquid chromatography (GC and LC) are the two major categories of chromatography. Clearly, *gas* and *liquid* refer to the mobile phase. However, *adsorption* in adsorption chromatography refers to the mechanism of retardation. The modifier *paper* in paper chromatography applies when a sheet of paper is used in place of a column. These adjectives are quite descriptive, which helps in remembering them, but they apply to a hodgepodge of different and often unrelated features (mobile phase, retarding mechanism, migration medium). There are still other descriptive names used in chromatography that refer to the shape of the incoming solute profile, the linearity of the sorption isotherm, variations in temperature, and so on.

Categories of Variations

The scope of chromatography and the relationship among techniques can be stated most clearly by categorizing the major variations and exploring their function and limits. While not all variations are in widespread use, some breadth is useful for perspective and for suggesting solutions to future problems. By and large we follow IUPAC terminology [2]; more recent compilations of terms have been published [3, 4].

Type of Mobile Phase   The mobile phase carrying solutes through the column must be some kind of fluid since it must flow. As noted earlier, the fluid is commonly a liquid (giving LC) or a gas (GC). Increasingly, supercritical fluids or SFs (high density gases above their critical temperature as described in Section 2.5) are being used as mobile phases, giving SFC.

GC, LC, and SFC are similar in principle and have the same theoretical basis [5]. However, gases and liquids differ widely in physical constants important to separations, especially viscosity and diffusivity. The key role of viscosity in regulating separative transport (including diffusion) and pressure drop was outlined in Chapter 4 and will be reiterated for chromatography in Chapter 12. We observed that the viscosities of most gases were two orders of magnitude lower ($\sim 10^{-4}$ poise, Table 4.2) than those of common liquids ($\sim 10^{-2}$ poise, Table 4.1), making gases more favorable for transport. SFs are intermediate in viscosity and diffusivity [6]. Thus based on transport speed, all other things being equal, the order of preference for mobile phases is gas $>$SF$>$ liquid.

All other factors are not equal. Gases, being noninteractive with solutes, contribute neither selectivity nor solvent power (the ability to dissolve solutes) to the chromatographic process (see Section 2.5). GC can therefore be applied only to volatile solutes. Liquids are the most versatile and generally the most effective mobile phases based on their solvent properties. SFs are again intermediate.

The above discussion shows that gases, liquids, and SFs have their own unique strengths and weaknesses as mobile phases. The choice among them depends on the nature of the sample material, the requirements of the separation, and various other factors.

Mechanism of Retention   The retardation of solute molecules occurs by several different retention mechanisms that depend on the different forms of equilibrium realized between the mobile phase and a plethora of stationary phases of different form and function. Different stationary phases have different effects on equilibrium according to the intermolecular forces involved, in general accordance with the guidelines of Chapter 2.

If retardation is caused by adsorption on granular solids or other fixed surfaces, the technique is called adsorption chromatography, as noted above. If the solid surfaces merely act as a scaffold to hold an absorbing liquid (which may be of a chosen polarity) in place—perhaps within the pores of solid particles—we have partition chromatography. Many cases lie between these extremes; this occurs whenever the solid acts as a support for liquid but retains some adsorptive activity [7]. A special case exists with *chemically-bonded phases* (*CBPs*), which usually consist of a one-molecule thick layer of hydrocarbon (often C_{18}) chemically bonded to the solid surface. (These nonpolar phases are frequently used for nonpolar solutes in RPLC.) Partitioning into such a thin layer is affected by the nearby surface, especially since the configuration and motion of each attached molecule is restricted by its fixed anchor to the surface [8].

The names used for adsorption and partition chromatography in GC are *gas-solid chromatography* (*GSC*) and *gas-liquid chromatography* (*GLC*), respectively. With liquid as the mobile phase (LC), the equivalent terminology is *liquid-solid* and *liquid-liquid chromatography*, or *LSC* and *LLC* respectively.

A different mechanism of retention is found in *ion-exchange chromatography*, *IEC* (or simply *ion chromatography*). The ion exchange materials (often in the form of small beads) have ionic groups that can exchange positions with ions in the mobile phase, thus immobilizing the latter. Hence ionic solutes are retarded by the ion-exchange process [9, 10].

Size exclusion chromatography (*SEC*) differs significantly from all the above because the solutes—usually macromolecules—are *repelled* and thus partially excluded from the stationary phase rather than being differentially *attracted* to the immobilized phase. Exclusion (repulsion) is induced by a porous support in which the pores are so small that macromolecules fit in awkwardly, losing entropy [11, 12] (see Section 2.7). Large macromolecules fit worst of all, lose the most entropy, and are thus excluded most (see Section 2.7). This mechanism is applicable both in aqueous solutions (where the technique is sometimes called *gel filtration chromatography* or *GFC*) and in nonaqueous mobile phases (often called *gel permeation chromatography* or *GPC*).

In *hydrophobic interaction chromatography* (*HIC*), nonpolar components are selectively expelled from an aqueous mobile phase due to the cohesive forces induced in water by hydrogen bonding. These forces can be modulated by the concentration of dissolved salts. The nonpolar species adsorb on or partition into a nonpolar stationary phase largely as a result of these forces. The nonpolar phase is often a bonded phase as noted above.

A more complete discussion of retention mechanisms can be found in two recent books [13, 14].

Incoming Solute Concentration Profile Most commonly, the solute mixture is introduced at the head of the column as a narrow zone, giving *zonal chromatography*. Differential migration induced by flow of the mobile phase leads to discrete and well-separated zones of approximately Gaussian shape as was discussed in Chapter 5. It is possible, however, using a technique called *frontal analysis*, to introduce the solute mixture into the column as a continuous stream or "train" mixed with the entering mobile phase. The train, of course, is made up of a number of subtrains, one for each component. When the front of the train encounters the sorptive material, each of its subtrains is slowed to a different extent due to chromatographic retardation. Thus we end up with a set of subtrains projecting themselves to different distances into the column. This is illustrated in Figure 10.2 and compared with the more usual zonal method. Notice that frontal analysis does not lead to a clear separation of solutes; considerable overlap is inherent in the method. It does, however, break up the front of the train into subfronts. The concentration jump at each subfront can be monitored by a detector, thus indicating, for purposes of analysis, the presence of a discrete solute in the original mixture.

Zonal chromatography is used predominantly in both gas and liquid chromatography. The separated zones are swept (eluted) from the column into a detector where they are sensed and recorded as the sequential peaks of a *chromatogram*. Typical chromatograms are shown in Figure 10.3.

Linearity of the Isotherm The sorption isotherm (where "sorption" is the general term for both adsorption and absorption) may be either linear or nonlinear. A linear isotherm implies that the amount of nonsorbed solute is proportional to the amount sorbed, that is, the amounts of nonsorbed and sorbed solute are in a constant ratio to one another. Linearity holds only for very dilute solutions in adsorption chromatography, but generally applies (to a good approximation) over a fairly large concentration range in partition chromatography. For analytical chromatography there are advantages (i.e., constant peak positions) in keeping the isotherm linear (*linear chromatography*); hence solute concentrations are generally kept low. At higher

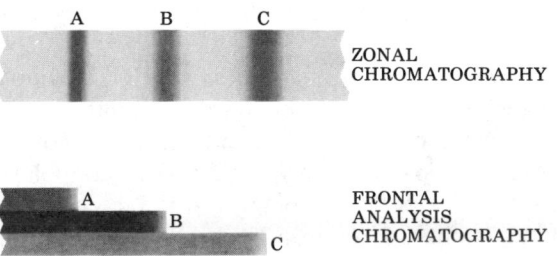

Figure 10.2. A comparison of zonal chromatography with frontal analysis.

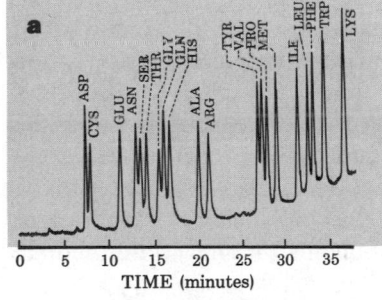

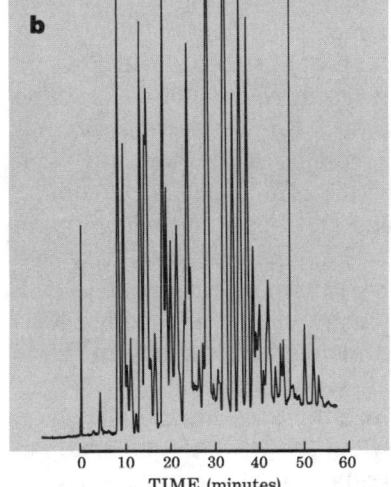

Figure 10.3. Examples of chromatograms. (*a*) The separation of 20 essential amino acids (in derivatized form) by RPLC using 1 mm inside diameter by 150 mm long microbore column packed with 4 μm silica support (with C_{18} CBP) particles and a water/acetonitrile mobile phase gradient. (*b*) The fractionation of gasoline using SFC with CO_2 mobile phase (temperature and pressure programmed) in 0.25 mm × 50 cm packed column with 5 μm polymeric support particles. (Courtesy of Frank J. Yang.)

concentrations a noticeable departure from linearity occurs and peak position depends on sample size. This condition is called *overloading* and results in *nonlinear chromatography*. For the high concentrations needed for preparative chromatography, nonlinear operation is inevitable; nonlinear chromatographic systems have been studied intensively by Guiochon and his co-workers (15, 16).

The Langmiur isotherm—used by Guiochon and others in the study of preparative scale chromatography—is based on the concept that adsorption nonlinearity occurs when there are so many molecules that they compete with one another for a limited number of adsorption sites. It is obvious that when two concentrated solutes are present at the same time, they will interfere with one another's adsorption. The one that adsorbs most strongly will almost totally displace the weaker adsorber. This is the basis of *displacement chromatography*, a nonlinear form developed by Tiselius in 1943 (17) and revived recently by Horvath [18].

In displacement chromatography, a band of solute is first established at the head of the column. Flow then commences using a mobile phase having a component which adsorbs more strongly than any of the solutes. The mobile phase component pushes the solutes ahead of it by virtue of displacement. The solutes themselves line up in the order of adsorption strength as each solute zone, pushed from behind by a stronger adsorbate, pushes the next weaker adsorbate ahead of it. One thus obtains the separation of solutes into compact adjoining zones. (The mechanism resembles that observed for isotachophoresis, Section 8.3.) The zones can be prevented from mutual contact and overlap (thus enhancing the separation) by purposely introducing outside *spacer* components that have intermediate adsorption strengths.

Column and Planar Procedures The sorptive medium of chromatography is ordinarily contained in a column to give it the proper structure (length and diameter), to allow the application of pressure to drive flow, and to confine the mobile phase along the desired path. All such forms are categorized as *column chromatography*. In contrast to the column techniques, the *non-column* (or *planar*) methods operate without the confinement of a column wall. These methods employ a layer of granular (*thin-layer chromatography*) or fibrous (*paper chromatography*) material. The mobile phase is restricted to flow along the layer by capillary forces (see Section 4.7) rather than by a column wall. While the noncolumn methods are simple and convenient, flexibility is lacking in controlling flow and extending bed length to optimize separation. The subject has been reviewed by Geiss [19].

Since the bed in planar methods is generally a thin two-dimensional layer, chromatographic migration can take place in either of two directions or, if desired, in one direction followed by the other. If the mobile phase is changed for the second migration in order to alter the separation parameters, we get two-dimensional chromatography. The advantage of two-dimensional separation was noted in Sections 6.4 and 6.5. We will return to two-dimensional methods shortly as a distinct category of chromatography.

Column chromatography is of two kinds. In one the column is packed full of granular sorptive material, giving a *packed column*. In the other the column is empty except for a thin layer of sorptive material (usually a liquid film or bonded layer) coated on the wall of the column. Such columns, usually of small diameter (see Figure 1.3), are called *capillary columns*. They are used mainly in GC (20) but are undergoing rapid development in LC and SFC [21–23].

Gradient Methods In the simplest case a constant environment is maintained during chromatography, consisting of a constant temperature and a steady flow using a mobile phase of fixed composition (often a pure liquid or gas). However it is often possible to improve the range of separation by slowly changing one of the variables affecting migration. In a method called *gradient elution*, used in LC, the composition of the entering mobile phase is gradually changed [24]. In *programmed temperature* methods, used primarily with GC, the entire column is slowly heated. In the less-common *chromathermography*, a temperature gradient is established over part of the column and then moved down along the column. With *programmed flow*, the flowrate is gradually increased.

The merits of the gradient methods is that a wide range of conditions can be used in a single run, during which, in theory, a window of best conditions can be found for each component. Thus high resolution conditions are established for the peaks eluting early, while later peaks are urged along by the changing environment, avoiding excessive separation time.

Dimensionality   While chromatography is normally operated as a one-dimensional (1D) system, two-dimensional (2D) operation is possible (see Sections 6.4 and 6.5). Continuous (preparative) 2D chromatography in a rotating annulus was described in Section 6.4; this approach has not yet been widely applied. Discrete (noncontinuous) 2D analytical chromatography has been widely used in paper and thin-layer chromatography. It is relatively easy to separate components first along one axis of a thin-layer bed, then independently along the second axis with a different mobile phase. Guiochon has described the extension of this technique to column chromatography [25]. The enhanced resolving power of such 2D systems were described in Section 6.5.

Dimensionality can be realized in either 2D or higher dimensional form by using *coupled column* systems. The resulting *multidimensional chromatography* provides a powerful extension of 1D chromatography (see Section 6.4 and ref. [26]).

Physical Scale   The contrast in the physical dimensions of chromatographic systems—already pointed out in Chapter 1—is growing as preparative demands push large columns to greater size [27] and analytical needs drive small columns toward microscopic dimensions [23]. Experimental methods are strongly affected by these scale factors but chromatographic principles change little with size unless linear/nonlinear differences are involved.

By way of summing up, Table 10.1 has been abstracted from the preceding discussion. Clearly the many approaches discussed above can be combined in a great number of ways to fit the needs of different separation problems. These many degrees of freedom explain, in part, the great versatility and wide scope of chromatographic methods.

It is apparent from the above that the names commonly used with different techniques, although descriptive, do not pinpoint the method explicity. For example, the term *liquid* chromatography does not tell whether the technique is adsorption or partition, zonal or frontal, gradient or nongradient. If nothing more is specified, it is usually taken for granted that the most common and/or simple choices are made: partition; zonal; nongradient. Column techniques are more common than noncolumn methods. While the lack of specificity can be annoying, the specifics usually emerge with detailed descriptions of the method.

TABLE 10.1   Major Variants of Chromatography

1. Mobile phase	Gas
	Liquid $\begin{cases} \text{normal phase} \\ \text{reversed phase} \end{cases}$
	Supercritical fluid
2. Mechanism of retention	Adsorption
	Partition
	Bonded phase
	Ion exchange
	Exclusion
	Hydrophobic
3. Incoming solute profile	Zonal
	Frontal
4. Isotherm linearity	Linear
	Nonlinear $\begin{cases} \text{displacement} \\ \text{zonal} \end{cases}$
5. Column or planar	Column $\begin{cases} \text{packed} \\ \text{capillary} \end{cases}$
	Planar $\begin{cases} \text{thin-layer} \\ \text{paper} \end{cases}$
6. Gradients	Nongradient
	Gradient $\begin{cases} \text{gradient elution} \\ \text{programmed temperature} \\ \text{chromathermography} \\ \text{programmed flow} \end{cases}$
7. Dimensionality	One-dimensional
	Two-dimensional $\begin{cases} \text{continuous} \\ \text{discrete} \end{cases}$
	Multidimensional
8. Physical scale	Microbore
	Standard
	Preparative-scale

10.2   CHROMATOGRAPHIC MIGRATION

Chromatographic separation is a result of the differential migration of components. Differential migration can be understood only if the basic migration process is well characterized. This section and several to follow will be devoted to the fundamentals of migration, emphasizing the physicochemical processes that control the motion of each solute component down the column or bed. The treatment here expands on the introduction to this subject in Section 9.3.

As a first approximation we assume that the distribution of solute between phases quickly reaches equilibrium, at least at the zone center. When one introduces a narrow sample zone at the head of the column, an equilibrium fraction $1 - R'$ of the solute enters the stationary phase, the remaining fraction R' being left in the mobile phase (see Section 9.12). The value of R' remains fixed through the run providing we have linear chromatography and no disturbing gradients. However R' is different from one solute to the next, depending on the strength of adsorption or absorption.

As pointed out earlier, zone migration is slow for strongly sorbed solutes, that is, those with a low R' value. This can be shown quantitatively as follows. Of the total solute in the zone, the fraction R' is carried along with the mobile phase at an average velocity v. (For simplicity we henceforth use v instead of $\langle v \rangle$ for the cross-sectional average velocity.) The remaining fraction, $1 - R'$, is held stationary; this fraction has zero velocity. The velocity of the zone as a whole is the average velocity of its solute: (fraction in mobile phase, R') $\times v +$ (fraction in stationary phase, $1 - R'$) $\times O = R'v$. Thus zone velocity is directly proportional to R'

$$\mathcal{V} = R'v = Rv \tag{10.1}$$

which is the same as Eq. 9.52. The second equality, $\mathcal{V} = Rv$, merely repeats the definition of R as $\mathcal{V}/v$ (see Eq. 9.5). This approach again shows the equivalence of the thermodynamic parameter R' and the relative velocity term R (Eq. 9.48). (Henceforth, both R' and R, although defined differently, will be expressed as R.) Separation achieved by making velocity $\mathcal{V}$ different for each component is seen to depend ultimately on unequal R values.

Equation 10.1 shows that R is a measure of the retardation or slowing of the zone or peak with respect to mobile phase velocity. A peak that experiences no retardation because its solute does not partition into the stationary phase ($R = 1$) is termed a *nonretained peak* or *void peak*; such a peak travels at mobile phase velocity v. Solute retained to some extent by the stationary phase migrates as a *retained peak*, for which $R < 1$. (The smaller R, the greater the retention.) Because of its key role in specifying retention, R is termed the *retention ratio*.

The capacity factor k', to be discussed shortly, is an alternate measure of retention. While k' is used more often than R in chromatography, the use of R is advantageous because (i) it is directly proportional to peak migration velocity and is thus a more direct measure of retention than k', (ii) most equations describing chromatography are simpler when expressed in terms of R rather than k', and (iii) R is a more universal measure of retention; R but not k' applies to other perpendicular flow methods such as field-flow fractionation.

The *retention time* t_r is the time needed for the center of the peak to migrate to the end of the column at distance L. Clearly

$$t_r = \frac{L}{\mathcal{V}} = \frac{L}{Rv} \tag{10.2}$$

that is, t_r is inversely proportional to peak velocity $\mathscr{V}$ and thus inversely proportional to R as well. For a nonretained peak, for which $R = 1$, the "retention" time is

$$t^0 = \frac{L}{v} \tag{10.3}$$

The *retention volume* V_r is the volume of the mobile phase, measured as it emerges at the outlet, necessary to flush the peak center to the end of the column. For a nonretained peak, traveling entirely in the mobile phase, it is necessary to disgorge all the mobile phase, occupying volume V_m in the column, to bring the peak from the beginning to the end of the column. For such a peak, the retention volume is simply V_m (Section 9.12). For a retained peak traveling only the fraction R of the velocity of the nonretained peak, the flushing time and volume are increased by the factor $1/R$, giving

$$V_r = \frac{V_m}{R} \tag{10.4}$$

Equilibrium and Retention Parameters

We have noted that R (because it is equal to R') is the equilibrium fraction of a component in the mobile phase. The equilibrium distribution of the component between phases can be described by classical thermodynamical methods. Therefore, the general discussion of phase equilibrium in Chapter 2 applies. The effect of this equilibrium on chromatographic parameters will be discussed next. A preliminary development of these concepts was presented in Chapter 9.

Consider the column interior, within which lies a volume V_s of stationary phase and V_m of mobile phase. Assume that the concentrations of solute in the stationary and mobile phases are the equilibrium values c_s^* and c_m^*. The amount of solute occupying the mobile phase is the concentration-volume product, $c_m^* V_m$; for the stationary phase, the amount is $c_s^* V_s$. Therefore the ratio of solute quantity in the stationary and mobile phases is $c_s^* V_s / c_m^* V_m$. This ratio also equals the fraction $1 - R$ of stationary-phase solute over the fraction R of mobile-phase solute. Thus

$$\frac{1 - R}{R} = \frac{c_s^* V_s}{c_m^* V_m} \tag{10.5}$$

Capacity factor k' is defined as equaling this ratio (Section 9.12)

$$k' = \frac{c_s^* V_m}{c_m^* V_m} \tag{10.6}$$

When the solute forms a dilute solution with both stationary and mobile phases, the ratio c_s^* / c_m^* equals K, the distribution coefficient or constant (see

Section 2.3). Solutes in chromatography generally form ideal or near-ideal solutions during most of their migration because they are highly dilute. (Deviations from ideality, if significant, lead to nonlinear chromatography.) Ordinarily we can substitute K for c_s^*/c_m^*, giving

$$k' = \frac{1 - R}{R} = \frac{KV_s}{V_m} \tag{10.7}$$

Chromatography, as compared to other separation methods based on phase equilibrium, stems from a notoriously heterogeneous physical system. Even in partition chromatography, where supposedly partition between bulk phases is predominant, the different kinds of high-area interfaces often lead to surface effects. In the most general case, we assume that there are several mechanisms of retention, both bulk and interfacial. In this case the numerator of the last equation must be enlarged to incorporate the other mechanisms

$$k' = \frac{1 - R}{R} = \frac{\sum K_i V_{si}}{V_m} \tag{10.8}$$

where the different V_{si} represent active interfacial areas as well as the bulk volume of one or more stationary phases.

The above equations can be rearranged to yield the retention ratio R explicitly. The last equation gives

$$R = \frac{V_m}{V_m + \sum K_i V_{si}} \tag{10.9}$$

In column chromatography the degree of retention is often measured by the retention volume V_r. The substitution of Eq. 10.9 into 10.4 yields

$$V_r = V_m + \sum_i K_i V_{si} \tag{10.10}$$

When a single stationary phase can be assumed this reduces to Eq. 9.53, that is

$$V_r = V_m + KV_s \tag{10.11}$$

In paper and thin-layer chromatography, retention is usually measured by an R_f value, similar but not equal to R. There is no simple equilibrium expression for R_f, as will be explained in Section 10.5.

In ordinary practice the summation term $\sum K_i V_{si}$ in the above equations is approximated by KV_s as in Eq. 10.11. This not only simplifies the study of retention but it also reflects the fact that in most cases only the major mechanism of sorption is known, the others being masked by its dominating effect.

We see that retention depends on one or more distribution coefficients. The precise nature of retention is revealed only by utilizing the thermodynamic parameters of equilibrium, particularly the free energy expressed as the chemical potential. Thus for each distribution coefficient we have, in accordance with Eq. 2.19

$$K = \exp\left(-\frac{\Delta\mu^0}{\mathcal{R}T}\right) \tag{10.12}$$

where $\Delta\mu^0$ is the standard chemical potential change for transporting solute from the mobile phase to the sorbed state. As noted in Chapter 2 (Eq. 2.31), the $\Delta\mu^0$ value for a particular solute i can often be approximated by a sum of group effects, $\Delta\mu_i^0 = \Sigma \Delta\mu_k^0$. Each term in the summation represents a chemical structural unit such as CH_2, NH_2, Cl, and so on.

Once thermodynamic parameters are known, R values can be obtained under a wide range of conditions. Knowledge of R values leads directly to all the important quantities related to the retardation or retention of zones in linear chromatography, as emphasized above.

10.3   MIGRATION AND ZONE SHAPE IN NONLINEAR CHROMATOGRAPHY

Factors leading to non-Gaussian zones in separation systems were described generally in Section 5.9. One source of zone asymmetry identified was the variation of local solute velocity W with solute concentration, described as overloading. The way in which overloading causes zone asymmetry in chromatography is explained below.

Once migration commences, each narrow zone of linear chromatography moves along with little change in shape or form except for the gradual broadening effect of statistical (diffusion-like) processes. This uniform migration is altered drastically, however, for systems that are strongly nonlinear. The disturbed migration can be illustrated by assuming that nonlinearity stems from Langmuir-type adsorption. (Related distortions occur for any nonlinear isotherm or, generally, any sorption phenomenon that shifts the ratio of solute in the two phases with increasing concentration.)

The Langmuir isotherm accounts for the fact that at high concentrations the surface sites become almost saturated with solute and will adsorb very little more, even when additional amounts are added to the mobile phase. For this reason the mobile phase solute fraction R (or R') increases with solute concentration. Thus the local velocity ($W = Rv$) of a given component can no longer be counted on as a constant characteristic of that component; the system is overloaded. Since R and Rv increase with concentration, the solute migrates fastest at the center of the zone where the overall concentration is greatest. The center of the zone consequently overtakes the

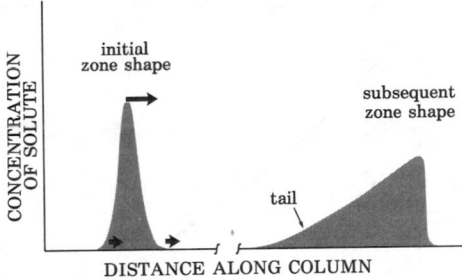

Figure 10.4. The formation of an asymmetric zone with a tail due to highly nonlinear adsorption. Arrows show relative solute migration velocities (highest at the center) at different points.

leading edge while leaving the trailing edge behind. The extreme result of such an internal velocity variation is shown in Figure 10.4: the zone profile loses its symmetry, forming a sharp rapidly moving front and a gradually descending rear [16]. This *tailing* phenomenon has undesirable effects because the "tail" often reaches back far enough to mix with and obscure slower components. The change in the "peak" velocity with sample size is another undesirable effect since components can no longer be described by a constant well-defined retention time.

If the sample size and thus the nonlinearity is not extreme, then neither is the tailing. Modest tailing is common in chromatographic peaks. The reduction of solute load is frequently a cure for tailing.

Tailing sometimes has a kinetic (rather than equilibrium) origin, stemming from slow desorption processes [28]. Thus the observation of tailing does not prove that the column is overloaded and consequently behaving nonlinearly (see Section 5.9). When kinetic factors are involved, the reduction of solute load will *not* eliminate tailing.

10.4 EFFECT OF GAS COMPRESSIBILITY ON RETENTION IN GC

The calculation of retention time t_r is simple when the flow velocity in the column is uniform throughout the column length and has a known value v which can be substituted into Eq. 10.2. However in gas chromatography, where the mobile phase is highly compressible, v varies substantially with the level of compression, which changes through the column. This requires special consideration.

In Section 4.6 we treated the flow of an ideal gas through a column, showing that the flow velocity changes with the position in the column due to gas compressibility. The velocity is greatest near the outlet because there the pressure is least and the gas is sweeping through the column in its most expanded condition. This nonuniform flow was illustrated in Figure 4.5; the

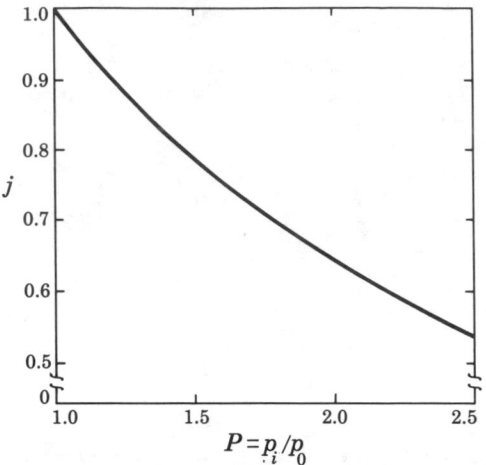

Figure 10.5. Plot of James–Martin pressure-gradient correction factor j as a function of the compression ratio $P = p_i/p_o$ for a GC column.

nonuniformity was mathematically expressed by Eq. 4.30b, repeated here for convenience

$$\frac{v}{v_i} = \left(\frac{P^2}{P^2 - (P^2 - 1)x/L} \right)^{1/2} \tag{10.13}$$

where v is the cross-sectional average flow velocity at distance x from the inlet, v_i is the value of v at the inlet, L is the column length, and $P = p_i/p_o$ is the compression ratio, the ratio of inlet to outlet pressure.

The retention time t_r can no longer be calculated as L/Rv (Eq. 10.2) because v is not constant. Below we write v as $v(x)$ to show that it varies with distance x down the column. To get t_r we must sum up (integrate) all the small time elements dt needed for a component to pass through each thin slice dx of column. We have

$$t_r = \int_0^{t_r} dt = \int_0^L \frac{dx}{Rv(x)} \tag{10.14}$$

If $v \equiv v(x)$ is substituted into this from Eq. 10.13, we get

$$t_r = \frac{1}{Rv_i P} \int_0^L \left[P^2 - (P^2 - 1)\frac{x}{L} \right]^{1/2} dx \tag{10.15}$$

This integrates readily if we change our variable of integration to y

$$y = P^2 - (P^2 - 1)\frac{x}{L} \tag{10.16}$$

from which

$$dy = -dx \frac{(P^2-1)}{L} \tag{10.17}$$

which leads to

$$t_r = \frac{-L}{(P^2-1)Rv_iP} \int_{P^2}^{1} y^{1/2} \, dy \tag{10.18}$$

When we integrate and substitute v_o for Pv_i, we get

$$t_r = \frac{L}{Rv_0} \frac{2}{3} \frac{(P^3-1)}{(P^2-1)} = \frac{L}{R(v_0 j)} \tag{10.19}$$

Where j is the James–Martin *pressure-gradient correction factor*

$$j = \frac{3}{2} = \frac{(P^2-1)}{(P^3-1)} \tag{10.20}$$

A plot of j versus P is shown in Figure 10.5. We find that $j<1$ when $P>1$, which applies to all real situations. The factor $1/j$ (which exceeds unity) represents the degree to which t_r exceeds the hypothetical retention time L/Rv_o; the latter is calculated on the basis that the peak is migrating throughout the column at its outlet velocity Rv_o. Time t_r, of course, exceeds L/Rv_o because the peak travels more slowly than Rv_o near the inlet and thus takes longer than L/Rv_o to emerge.

Insight about the meaning of j can be gained by writing t_r in terms of the average velocity $\tilde{v}$ of carrier gas in the column

$$t_r = \frac{L}{R\tilde{v}} \tag{10.21}$$

or, for a nonretained peak

$$t^0 = \frac{L}{\tilde{v}} \tag{10.22}$$

Comparison of Eq. 10.21 with Eq. 10.19 shows that

$$\tilde{v} = v_o j \tag{10.23}$$

that is, j is simply a correction factor that converts outlet velocity v_o to the average velocity $\tilde{v}$. (Technically, $\tilde{v}$ must be interpreted as a time-average rather than a distance-average velocity in the column [29].) The physical meaning of $\tilde{v}$ is made clear from Eq. 10.22; it is simply the total length L traversed divided by the time t^0 spent in traversing that length.

The retention volume, which is the volume of gas collected at the outlet before the peak center emerges, is much like retention time t_r. Without compression we have $V_r = V_m + KV_s$ but with compression we have

$$V_r = (V_m + KV_s)j^{-1} \tag{10.24}$$

10.5   EFFECT OF GRADIENTS ON MIGRATION IN TLC AND PC

In the last section we noted that zone migration (and thus retention) is complicated in GC because of the flow gradients in GC columns. Migration in thin-layer chromatography (TLC) and paper chromatography (PC) is even more complicated by gradients. As in GC, a gradient in velocity exists, but superimposed on this is a gradient in the saturation of the bed by the mobile phase, which leads also to a gradient in R and k' [19, 30].

The gradients affecting these planar methods originate with the unusual nature of capillary flow, which was discussed in Section 4.7. It was explained that capillary phenomena cause the level of liquid saturation of the porous bed to decrease on proceeding form the rear to the front of the advancing train of liquid [30]. Because the liquid *is* the mobile phase, the local phase ratio V_m/V_s also decreases toward the front. The capacity factor, $k' = KV_s/V_m$ (Eq. 10.7), correspondingly increases in response to the same trend, whereas R (Eq. 10.9) decreases upon approaching the front. The latter can be seen by writing Eq. 10.9 in the simplified form

$$R = \frac{V_m/V_s}{(V_m/V_s) + K} \tag{10.25}$$

Migration in TLC and PC is not measured by the time of elution since component zones are generally not eluted from the chromatographic bed; migration is measured instead by the R_f value. The factor R_f is the ratio of the distance $X - X_0$ migrated by the zone to the distance $X_f - X_0$ advanced by the liquid front in the same time interval

$$R_f = \frac{X - X_0}{X_f - X_0} \tag{10.26}$$

where X_0 is the initial zone position as illustrated in Figure 10.6. When X_0 is small, it can be shown that the ratio of the zone velocity $\mathcal{V} = Rv$ to the front velocity v_f becomes constant and can be equated to R_f

$$R_f = \frac{\mathcal{V}}{v_f} = \frac{Rv}{v_f} \tag{10.27}$$

Using the R from Eq. 10.25, we have

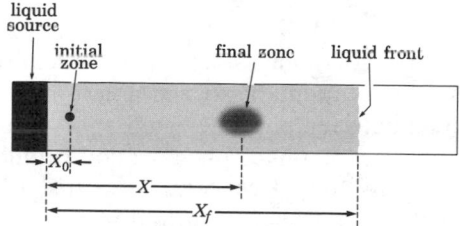

Figure 10.6. Migration of zone in thin-layer or paper chromatography. The zone appears as a spot, ideally of elliptical shape (see Section 6.3).

$$R_f = \frac{v}{v_f} \frac{V_m/V_s}{(V_m/V_s) + K} \tag{10.28}$$

This expression highlights both factors whose gradients affect R_f: the liquid velocity ratio v/v_f and the phase ratio V_m/V_s. Analysis shows that v/v_f is always less than unity, decreasing as one moves back from the liquid front [31]. In paper strips, this ratio can vary from unity at the front to somewhere around 0.6 near the liquid source. Factor V_m/V_s varies even more, increasing twofold upon retreating from 90% of the distance to the front to 10%.

The sequence of zones in TLC and PC is determined by the sequence of K. However, as Eq. 10.28 shows, the numerical spacing depends upon variations in v/v_f and V_m/V_s as well. Erroneously, the latter correction factors are often ignored in the interpretation of TLC and PC data.

It is unfortunate that the above corrections are not expressible by simple equations. The numerical procedures for dealing with these factors have, however, been worked out [31].

10.6   NONEQUILIBRIUM MIGRATION

All the cases of zone migration and retention discussed above are based, if not explicitly then implicitly, on the assumption that the solute distribution between phases is dictated by equilibrium. Our repeated reference to the zone velocity Rv equaling $R'v$ or $v/(1 + k')$—which provides our only link between zone migration rates and underlying physicochemical parameters— is based on the equilibrium assumption. If, then, we were to find some systematic departure from equilibrium, we would expect our migration and retention equations to be faulty. When we examine carefully the question of nonequilibrium (below), we do indeed find a systematic departure from equilibrium. Surprisingly, however, migration rates are little affected by the nonequilibrium; instead zone sharpness is the victim of nonequilibrium departures. This problem was discussed in Section 9.5 for $F(+)$ systems generally. The consequences for chromatography are deduced below.

Source of Nonequilibrium

We examine a solute zone migrating under linear conditions. For the moment we confine our attention to the solute in the mobile phase. If complete equilibrium between phases existed at all points, the mobile-phase solute would form a Gaussian-like concentration profile as indicated by the shaded profile in Figure 10.7. However the actual profile for the mobile phase, shown by the dashed line, is shifted ahead of the shaded (equilibrium) profile due to solute migration. The basis of the profile shift is explained as follows.

Each region of the column occupied by the zone has mobile phase flowing through it, the mobile phase bringing solute from upstream. A small volume element at point A, ahead of the zone center, will have mobile phase coming in from the high concentration regions toward the zone center. The solute-enriched mobile phase thus entering will cause a continuous increase in solute concentration. Some of the excess will be quickly passed along to the stationary phase in the drive to reach a new equilibrium condition. However, the adjustment to equilibrium is never instantaneous; it requires a finite time to be realized. In the finite time interval before equilibrium can be achieved, more enriched fluid enters with its excess solute to upset the balance once again. The continual unbalancing effect of incoming highly concentrated solute prevents equilibrium from being fully realized ahead of the zone center. The actual concentration at point A thus remains higher than the equilibrium concentration, as shown in the figure.

The situation at the rear of the zone, point B, is just the opposite. The incoming mobile phase is entering from the solute-impoverished tail of the zone. As mobile phase enters the region with its concentration deficiency, solute immediately desorbs from the stationary phase in pursuit of equilibrium. The continued entrance of impoverished fluid keeps the mobile-phase

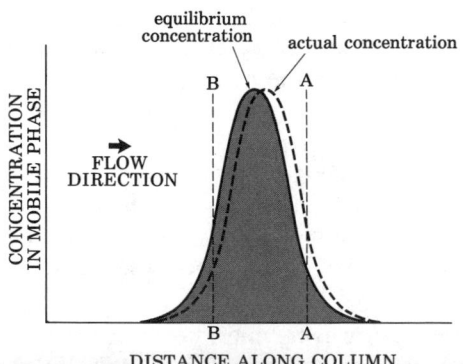

Figure 10.7. Difference between equilibrium and actual concentration profiles in the mobile phase.

concentration once again out of equilibrium, in this case somewhat below the equilibrium value. This fact is also illustrated by Figure 10.7.

The only place where full equilibrium is attained is where the equilibrium and actual profiles coincide, at a point very near the zone center. Here the concentration profile is momentarily flat, creating a situation exactly midway between the concentration gains in the front and losses in the rear.

Nonequilibrium and Zone Spreading

Although the departure from equilibrium is usually small, it has an important effect as noted earlier: it causes zones to broaden out as migration proceeds, thus negating the separation. Below we explain why.

It was shown in the previous section that solute migrates at a velocity proportional to its fraction in the mobile phase. At equilibrium this fraction is R and the migration velocity is Rv. Since equilibrium is attained at a point very near the zone center, the center's velocity is Rv. Thus the zone center, which is the reference point for retention measurement, migrates at a rate governed by equilibrium, even though this point is surrounded by regions where solute is migrating in a nonequilibrium condition.

Ahead of the center, as just noted, a fraction of solute greater than the equilibrium value R occupies the mobile phase. Hence the solute in that region moves at a velocity $> Rv$. To the rear of center the fraction is less than R, leading to a solute migration velocity $< Rv$. Thus the zone is one in which the leading parts are moving faster and the rear is advancing more slowly than the center, that is, the zone is spreading out in both directions from its own center—literally tearing itself apart—as it migrates.

Factors Governing Nonequilibrium Zone Spreading

The above phenomena have been incorporated into a theoretical approach which explicitly associates zone spreading with flow velocity and the rates of various equilibration processes. This is the *generalized nonequilibrium theory* developed by the author [5]. While the theoretical details are too lengthy to develop here, some semi-quantitative reasoning can be used to understand the nature of nonequilibrium-induced zone spreading and the parameters that control it.

It was seen above that the slight departure from equilibrium in different regions is due to solute surpluses and shortages in the incoming mobile phase and to the finite time needed to equilibrate these concentration changes. The excesses and deficiencies in solute originate in concentration gradients, that is, in the fact that the regions just upstream have more, or less, solute than the region of interest. Clearly, the degree of solute excess or deficiency in the entering mobile phase is directly proportional to the concentration gradient. The final degree of nonequilibrium, proportional to the excesses and deficiencies, may therefore be written as

$$\Delta c_m = \text{const} \times \frac{dc}{dx} \tag{10.29}$$

where

$$\Delta c_m = c_m - c_m^* \tag{10.30}$$

that is, Δc_m is the solute concentration in the mobile phase in excess of its equilibrium value c_m^*. The term dc/dx is simply the gradient in the total zone concentration (amount of solute per unit volume of the chromatographic bed), x being the distance along the column. As stated previously, the extent of nonequilibrium is generally small, $\Delta c_m \ll c_m$, even though its consequences are significant.

Three other observations are important in establishing the nonequilibrium concentration Δc_m. First, where concentration gradients are positive, as at the rear of the zone, Δc_m is made negative by the fact that the mobile phase coming up from behind is deficient in solute. Thus a negative sign belongs to the proportionality expressed by Eq. 10.29. Second, the departure from equilibrium is also proportional to the equilibration time t_{eq}, the time necessary to bring a certain initial degree of nonequilibrium to within a certain fraction of equilibrium (t_{eq} is technically the *relaxation time* for equilibration). This is logical because the slower the equilibration process, the greater the nonequilibrium. Third, Δc_m is proportional to flow velocity v, since solute surpluses and deficits enter in amounts proportional to the flow which transports them. These three deductions lead to the following modification of the previous equation

$$\Delta c_m = -\alpha \left(v t_{eq} \frac{dc}{dx} \right) \tag{10.31}$$

where α is a positive constant of proportionality. This constant is shown by detailed theory to equal

$$\alpha = \frac{R(1-R)}{\epsilon} \tag{10.32}$$

where ϵ, the porosity, is the fraction of column volume occupied by mobile phase.

In accord with the earlier discussion, zone broadening occurs because solute velocity at the front and back of the zone is such as to pull these parts further and further from the center. An instructive way to look at this is to note that when Δc_m is positive, as it is ahead of center, solute in amounts greater than the equilibrium value are being transported through each unit area of column in each second. That is, with respect to equilibrium, extra solute in proportion to Δc_m is transported forward; thus solute at the front of the zone is out-pacing all other solute. At the rear, with Δc_m negative, solute transport falls behind. Thus the zone spreads out.

The nature of this spreading is best seen by returning to the fundamentals of transport developed in Chapter 3. The solute flux contributed by flow is given generally by Eq. 3.22, that is

$$J = vc \tag{10.33}$$

However, only the solute in the mobile phase is in motion and contributing to J.

Thus concentration c must be replaced by ϵc_m, where porosity ϵ, as noted, is the fraction of total volume occupied by the mobile phase and thus the fractional volume serving as conduit space for the solute transport. The last equation thus becomes

$$J = v\epsilon c_m \tag{10.34}$$

using Eq. 10.30 to get c_m leads to

$$J = v\epsilon(c_m^* + \Delta c_m) \tag{10.35}$$

the substitution of Eq. 10.31 (using also Eq. 10.32) for Δc_m gives

$$J = v\epsilon c_m^* - R(1-R)v^2 t_{eq} \frac{dc}{dx} \tag{10.36}$$

The term $v\epsilon c_m^*$ represents the translational motion of the zone under equilibrium conditions, which is equivalent to the zone migration discussed in Section 10.2. The second term on the right, which originates with the finite nonequilibrium concentration Δc_m, is proportional to the concentration gradient, a strong reminder of classical diffusion as expressed by Fick's first law, Chapter 3. This law, it may be recalled, states that substances will diffuse from regions of high to low concentration (e.g., from the zone interior to its edges) at a rate proportional to the concentration gradient

$$J = -D \frac{dc}{dx} \tag{10.37}$$

where D is the *diffusion coefficient*. This equation shows that the nonequilibrium spreading of chromatographic zones obeys the laws of diffusion, even though the process itself is different from normal diffusion. This verifies the conclusion drawn in Chapter 5 that diffusion-like spreading (yielding Gaussian peaks) is a general consequence of random processes in separation systems.

The *effective diffusion coefficient* for nonequilibrium zone spreading, obtained by equating the last term of Eq. 10.36 to Eq. 10.37, is

$$D_n = R(1-R)v^2 t_{eq} \tag{10.38}$$

Recall that the plate height corresponding to an effective diffusion process is given by Eq. 5.38, $H = 2D_T/W$. The nonequilibrium contribution to H is obtained by replacing the total diffusion coefficient D_T by the nonequilibrium contribution D_n and, of course, general displacement velocity W by zone velocity Rv. Thus with the help of Eq. 10.38, we get

$$H_n = 2(1-R)v t_{eq} \tag{10.39}$$

The last two equations show that the nonequilibrium plate height (corresponding to the rate of outward diffusion of the zone) is directly proportional to the time of equilibration and increases also with the flow velocity. It is clear, then, that narrow zones can best be obtained if equilibration is rapid and flow velocity reasonably slow. The implications for practical chromatography will be discussed at length in Chapter 14.

The time constant t_{eq} can generally be divided into stationary and mobile phase contributions, t_{es} and t_{em}, giving H_n the additive form

$$H_n = H_s + H_m \qquad (10.40)$$

Extensive development of the generalized nonequilibrium theory has led to mathematical expressions for H_s and H_m for many different models, allowing for various microscopic configurations of the stationary phase and flow profiles for the mobile phase. Details, too lengthy to repeat here, can be found in our main reference [5].

REFERENCES

[1] L. S. Ettre and A. Zlatkis, Eds., 75 *Years of Chromatography*, Elsevier, Amsterdam, 1979.

[2] Commission on Analytical Nomenclature, *Pure and Appl. Chem.*, **37**, 447 (1974).

[3] R. C. Denney, *A Dictionary of Chromatography*, 2nd ed., Wiley-Interscience, New York, 1982.

[4] R. E. Majors, *LC · GC*, **6**(2), 1988.

[5] J. C. Giddings, *Dynamics of Chromatography*, *Part 1*, *Principles and Theory*, Marcel Dekker, New York, 1965.

[6] P. J. Schoenmakers and L. G. M. Uunk, in J. C. Giddings, E. Grushka, and P. R. Brown, Eds., *Advances in Chromatography*, Vol. 30, Marcel Dekker, New York, 1989, Chapter 1.

[7] V. G. Berezkin, in J. C. Giddings, E. Grushka, and P. R. Brown, Eds., *Advances in Chromatography*, Vol. 27, Marcel Dekker, New York, 1987, Chapter 1.

[8] K. B. Sentell and J. G. Dorsey, *Anal. Chem.*, **61**, 930 (1989).

[9] J. S. Fritz, D. T. Gjerde, and C. Pohlandt, *Ion Chromatography*, Huethig, New York, 1982.

[10] H. F. Walton, Ed., *Ion Exchange Chromatography*, Dowden, Hutchinson & Ross, New York, 1976.

[11] J. C. Giddings, E. Kucera, C. P. Russell, and M. N. Myers, *J. Phys. Chem.*, **72**, 4397 (1968).

[12] P. L. Dubin, in J. C. Giddings, E. Grushka, and P. R. Brown, Eds., *Advances in Chromatography*, Vol. 31, Marcel Dekker, New York, in press.

[13] J. Å. Jönsson, Ed., *Chromatographic Theory and Basic Principles*, Marcel Dekker, New York, 1987.

[14] P. R. Brown and R. A. Hartwick, Eds., *High Performance Liquid Chromatography*, Wiley, New York, 1989.

[15] A. Katti and G. Guiochon, *Anal. Chem.*, **61**, 982 (1989).

[16] P. Rouchon, M. Schonauer, P. Valentin, and G. Guiochon, *Sep. Sci. Technol.*, **22**, 1791 (1987).

[17] A. Tiselius, *Arkiv Kemi, Mineral. Geol.*, **16A**, No. 18, 11 pp (1943).

[18] C. Horvath, in F. Bruner, Ed., *The Science of Chromatography*, Elsevier, Amsterdam, 1985, p. 179.

[19] F. Geiss, *Fundamentals of Thin Layer Chromatography* (*Planar Chromatography*), Huethig, Heidelberg, 1987.

[20] M. L. Lee, F. J. Yang, and K. D. Bartle, *Open Tubular Column Gas Chromatography: Theory and Practice*, Wiley, New York, 1984.

[21] M. Novotny, S. R. Springston, P. A. Peaden, J. C. Fjeldsted, and M. L. Lee, *Anal. Chem.*, **53**, 407A (1981).

[22] M. Novotny, *J. Microcolumn Separations*, **2**, 7 (1990).

[23] F. J. Yang, Ed., *Microbore Column Chromatography: A Unified Approach*, Marcel Dekker, New York, 1989.

[24] P. Jendera and J. Churacek, *Gradient Elution in Column Liquid Chromatography: Theory and Practice*, Elsevier, Amsterdam, 1985.

[25] G. Guiochon, M.-F. Gonnord, and M. Zakaria, *Chromatographia*, **17**, 121 (1983).

[26] J. C. Giddings, in H. J. Cortes, Ed., *Multidimensional Chromatography: Techniques and Applications*, Marcel Dekker, New York, 1990, Chapter 1.

[27] M. Verzele, *Anal. Chem.*, **62**, 265A (1990).

[28] J. C. Giddings, *Anal. Chem.*, **35**, 1999 (1963).

[29] J. C. Giddings, *Anal. Chem.*, **36**, 741 (1964).

[30] A. L. Ruoff, D. L. Prince, J. C. Giddings, and G. H. Stewart, *Kolloid-Zeitschrift*, **166**, 144 (1959).

[31] J. C. Giddings, G. H. Stewart, and A. L. Ruoff, *J. Chromatogr.*, **3**, 239 (1960).

EXERCISES

10.1(**) Table 10.1 lists many variants of chromatography that can be combined with one another in an enormous number of ways. Some of the combinations are practical and some not. From the following list, pick out the two combinations which are most likely to fail. Explain why.

a. liquid–adsorption
b. partition–displacement
c. gas–noncolumn
d. frontal–column
e. gas–elution

10.2(*) Four very prominent combinations in chromatography are *gas solid*, *gas liquid*, *liquid solid*, and *liquid liquid* chromatography. Would you expect (a) solid solid chromatography and (b) gas gas chromatography to work also? Explain.

10.3(*)   A mixture of two solutes, whose R values are 0.10 and 0.13, respectively, is started as a narrow zone at the head of a column in which the mean flow velocity v is 0.15 cm/s. How far down the column are the centers of the two solute zones located 6 min after they are started?

10.4(*)   A given solute peak migrates at one-fourth the velocity of the mobile phase, $R = 1/4$. What is the value of the capacity factor k'?

10.5(**)   Suppose there are two different stationary phases in a column (an example of a mixed stationary phase), phase A holding twice as many molecules at equilibrium as phase B. If $R = 0.1$, what fraction of all molecules does B hold?

10.6(**)   Prove that the James–Martin pressure-gradient correction factor j approaches zero as $P \to \infty$ and approaches unity as $P \to 1$. Deduce these limits on a physical basis using the result $j = \tilde{v}/v_0$.

10.7(**)   The time-average velocity in a column can be defined by

$$\tilde{v} = \frac{\int_0^{t^0} v \, dt}{\int_0^{t^0} dt}$$

Show that $\tilde{v}$ for a nonretained peak in a GC column equals $v_0 j$.

10.8(**)   The distance-average velocity in a column can be expressed by

$$\bar{v} = \frac{\int_0^L v \, dx}{\int_0^L dx}$$

Show that $\bar{v}$ for a nonretained peak in a GC column equals $2v_0/(P + 1)$.

10.9(*)   Assuming that zone broadening is caused only by nonequilibrium, calculate the effective diffusion coefficient for a zone in gas chromatography using the typical parameters $t_{eq} = 10^{-2}$ s, $v = 10$ cm/s and $R = 0.20$. After 1 min calculate how far the zone has migrated and its width 4σ.

10.10(*)   Write Eq. 10.38 in terms of capacity factor k' instead of retention ratio R.

10.11(**)   A GC column containing 2.30 mL of void space occupied by gas and 0.125 mL of liquid stationary phase is operated at a compression ratio of 4.00. For a component having a distribution coefficient $K = 105$, calculate the volume of gas emerging from the outlet prior to the appearance of the peak center for that component. Calculate also the volume of gas entering the inlet prior to the emergence of the peak center.

11

CHROMATOGRAPHY FROM A MOLECULAR VIEWPOINT

Up to this point we have described the migration and broadening of chromatographic zones in terms of bulk transport. This approach, by following the average behavior of vast numbers of molecules, ignores the unique displacements of single molecules. Yet much can be learned by considering the chaotic pathway followed by a lone molecule during its migration. An understanding of this pathway is important because the single molecule not only typifies its numerous companions but it exhibits statistical fluctuations from the average that lead to zone broadening. Consequently, the molecular viewpoint becomes a source of both insights and models for understanding the dynamical basis of chromatography. This theme was developed in an earlier book by the author [1] from which we borrow many of the concepts below.

11.1 MOLECULAR BASIS OF MIGRATION

The molecular migration pathway in chromatography is a statistical pathway in which each molecule randomly executes the steps that constitute its particular migration path. If it were possible to outline the "choices" faced by a molecule (such as whether or not to adsorb in a given time interval or which streampath to follow around a particle), and the probability of each option, we would in theory be able to calculate the probability that the molecule would follow a particular path. Since all identical molecules are identically inclined, the probability of a certain path for a single molecule is equal to the fraction of all molecules taking that path. If, for instance, the probability is known for two paths, one displacing the molecule 20 mm and

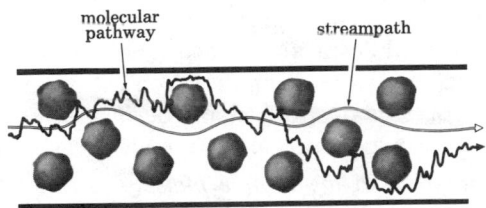

Figure 11.1. Diagram contrasting a smoothly (but randomly) varying streampath with an erratically varying molecular pathway starting at the same point. The molecular pathway is subject to rapid Brownian fluctuations. Dots represent sorption events on stationary particles, each of which delays the migration by a random time interval.

the other 25 mm in a given time, then we know the fraction of molecules at each distance; with more such data we can plot a concentration profile for the zone. The concentration profile, of course, is identical to the probability profile within a constant factor. Zone spreading, consequently, can be viewed as the broadening of a probability profile.

The two contrasting approaches, the macroscopic viewpoint which describes the bulk concentration behavior (last chapter) versus the microscopic viewpoint dealing with molecular statistics (this chapter), are not unique to chromatography. Both approaches offer their own special insights in the study of reaction rates, diffusion (Brownian motion), adsorption, entropy, and other physicochemical phenomena [2].

In chromatography, both approaches are useful [1]. The macroscopic approach, following the procedures described briefly in Section 10.6, has been most fruitful in yielding rigorous and general theoretical expressions. The molecular-statistical (or stochastic) approach, however, gives the clearest physical insight into zone migration and spreading, and provides the most direct route to the useful equations of chromatography.

Nature of Molecular Migration

By all outward appearances, the migration of a zone is a smooth continuous process. Down at the molecular level, where migration originates, the picture is different (see Figure 11.1). Each molecule is moving along with an erratic, jerk-like motion. Its irregular path is determined by several independent statistical processes which, as we shall see shortly, control zone spreading. Some of these processes were mentioned in a general context in Chapter 5; here we note only the processes relevant to chromatography:

1. Longitudinal molecular diffusion. Solute molecules are engaged in ceaseless Brownian motion, which is responsible for diffusion. The component of this erratic motion along the column axis, superimposed on the downstream displacement caused by flow, is one source of zone broadening.

2. Sorption-desorption. An important irregularity in molecular migration is caused by sorption and desorption. Each time a molecule is sorbed by the stationary phase, its downstream motion ceases. When desorbed, it proceeds again. The processes of sorption and desorption occur randomly, thus making this stop-and-go sequence an extremely erratic one.

3. Flow and diffusion in the mobile phase. As noted in Chapter 4, the microscopic flow process in a packed bed is very tortuous, each streampath suffering frequent changes in direction and velocity. A solute molecule carried in this flow will trace out a very uneven path. The randomness of the molecular path will be amplified by Brownian displacements of the molecule from one streampath to another (Figure 11.1).

The molecular picture of chromatography, based on these underlying mechanisms, is one of a complex motion having numerous variations and discontinuities. The motion of the zone as a whole appears smooth, however, because the individual molecules cannot be perceived. Instead one observes (visually or instrumentally) only the bulk motion of solute, which is a smoothed average of the motion of a great number of erratically behaving molecules.

The latter point is especially well illustrated by the sorption-desorption process. When a molecule is sorbed, its velocity is zero; when desorbed, it is on average equal to the fluid velocity v. Since a great number of molecules occupy a zone, at any given time many will be stationary and many in downstream motion. The zone as a whole, moving along as the blurred sum of these two parts, advances smoothly, assuming the extremes of neither the mobile nor stationary populations.

Mechanism of Retardation

We recognize that all molecules in chromatography occupy one of two general velocity states: one moving freely at velocity v and one with no velocity at all. The "strength" of sorption does not change either component velocity. What it does change is the relative fraction of time each molecule spends in these two extreme conditions. If a typical molecule spends one-half its time moving at velocity v and the rest of its time at zero velocity, this molecule will obviously outdistance a different kind of molecule that spends only one-tenth of its time moving at velocity v. Separation is based on just this mechanism—unlike molecules spending different fractions of their time in free motion.

The above reasoning can be made more quantitative. The molecule spending half its time in free motion at velocity v will clearly move at an average velocity of $\frac{1}{2}v$. The molecule spending one-tenth time in free motion will assume an overall average velocity of $\frac{1}{10}v$. Since each zone migrates as a composite of many such typical molecular displacements, the observed velocities of the two zones will be $\frac{1}{2}v$ and $\frac{1}{10}v$, respectively. Since retention ratio R is defined as the ratio of the zone velocity to mobile phase velocity v, we conclude that $R = \frac{1}{2}$ and $\frac{1}{10}$, respectively, in the two cases above. This shows that R is equivalent to the average fraction of time a molecule of a given kind spends desorbed, a concept first explained by LeRosen in 1945 (3). In Section 10.2, by contrast, R was shown to be equivalent to the (equilibrium) fraction of molecules in the mobile phase. These two ways of viewing R are equivalent.

11.2 MOLECULAR BASIS OF ZONE SPREADING

The molecular view of migration clarifies another point. From a bulk concentration viewpoint, a fraction R of the solute is racing ahead at velocity v while the fraction $1 - R$ remains stationary. One might expect this to lead to two distinct portions of the zone which become separated because of the difference in velocity. It is now apparent that the molecules in the moving and nonmoving portions are continually exchanging positions through sorption and desorption. The solute thus remains in a single compact zone because each molecule frequently changes its status through recurrent sorptions and desorptions, long before it gets far enough ahead of or behind its companions to be observably separated. This, of course, is closely related to the fact that the solute approaches equilibrium (a macroscopic viewpoint) quite closely due to rapid sorption and desorption.

Fluctuations from Mean Behavior

One cannot expect a molecule which follows a random migration path, full of frivolous excursions, to arrive after a fixed time at exactly the same point as its equally frivolous companions. There will be a mean distance of migration X, but the individual molecules will exhibit fluctuations about this mean due to the peculiarities of their own migration. These statistical fluctuations will lead to zone broadening. The statistical (stochastic) theory of zone broadening was first developed by Giddings and Eyring [4] and has been expanded subsequently by a number of authors [5–8].

It is well known in statistics, as explained in Section 5.3, that a great number of small random perturbations will lead to a Gaussian profile. Molecular migration ordinarily leads to a Gaussian (or to a close approximation thereof) because each migration path is made up of many small and independent random displacements. There are some exceptions, as with nonlinear chromatography, where the molecules interfere with one another's migration (Section 10.3). There is also an exception, to be discussed later, in which a single random event may lead to a relatively large (instead of small) displacement. Despite these exceptions, the Gaussian is the natural reference profile in chromatography; most zones can be considered as Gaussians to a first approximation. This view is consistent with the nonequilibrium picture of the last chapter, Section 10.6.

The Random-Walk Model

The simplest mechanisms leading to the dispersion (spreading) of a zone's molecules can be described by the classical random-walk model [9], as noted in Section 5.3. However this model does not fully account for the complexities of migration. It gives, instead, a simple approximation which inherits the most essential and important properties (foremost of all the randomness) of the real migration process. The random-walk model has been used in a similar first-approximation role in many fields (chemical kinetics, diffusion, polymer chain configuration, etc.) and is thus important in its own right.

Here we review the basic rules governing random walks and their relationship to plate height. First, the concentration (probability) profile generated by a random walk is a Gaussian with a variance σ^2 given by (see Eq. 5.24)

$$\sigma^2 = l^2 n \tag{11.1}$$

where l is the step length and n the total number of steps. Second, if two or more random walks are being executed independently, very often with different step lengths and numbers, the final cumulative result of all of these steps will be a Gaussian profile with a variance equal to the sum of component variances, one for each random walk (see Eq. 5.27)

$$\sigma^2 = \sigma_1^2 + \sigma_2^2 + \cdots = l_1^2 n_1 + l_2^2 n_2 + \cdots \tag{11.2}$$

Finally, the plate height H, given by σ^2/L (see Eq. 5.47), equals the following for a single random walk

$$H = \frac{\sigma^2}{L} = \frac{l^2 n}{L} \tag{11.3}$$

and for several independent random walks it becomes a sum of terms

$$H = \frac{\sigma^2}{L} = \frac{l_1^2 n_1}{L} + \frac{l_2^2 n_2}{L} + \cdots = H_1 + H_2 + \cdots \tag{11.4}$$

Chromatography and the Random Walk

The erratic motion of a chromatographically migrating molecule resembles a random-walk process. In order to apply random-walk ideas to chromatography, we must identify the effective step lengths and step numbers associated with the molecular migration. This is the main task to follow.

It must be kept in mind that the actual migration process is more complex than the random walk. For one thing, the steps or displacements are not of constant length. An "average" length must be used to bring migration realistically within the framework of the random-walk model. Second, the steps are not always discrete: a molecule engaged in taking a forward step may pass gradually and continuously through regions of decreasing velocity in the mobile phase until it sorbs and thereby takes a reverse step.

Despite such complications of detail, *the random-walk model describes the essence of chromatographic zone spreading.* It properly accounts for the way in which all major experimental parameters influence the broadening process.

The symmetric random walk used here is one in which forward and backward steps are equally probable. All positive and negative steps are considered as displacements with respect to the zone center. Therefore in chromatography where the zone as a whole is in a state of motion, we must consider all displacements with respect to a coordinate system moving with the zone center at velocity $\mathcal{V} = Rv$.

We will return now to the three independent random processes that underlie migration and induce zone spreading: molecular diffusion, sorption-desorption, and flow-diffusion processes in the mobile phase.

11.3 LONGITUDINAL MOLECULAR DIFFUSION

The ceaseless thermal motion of molecules will cause each of them to move in a series of random steps up and down the flow axis. It is possible to characterize this motion in terms of step length and step number, but since in this case the process is nothing but molecular diffusion, a more universal reference parameter is the diffusion coefficient D. Following this line of attack the variance can be obtained directly from Eq. 5.35 in which overall diffusion coefficient D_T is replaced by D

$$\sigma^2 = 2Dt \tag{11.5}$$

Diffusion in the mobile phase, particularly in GC, dominates the longitudinal diffusion process. In this case the molecular diffusion coefficient D will be replaced by the specific mobile phase value D_m. The diffusion time t will be replaced by the time spent in the mobile phase in the process of migrating through the column, traveling distance L. Since the average velocity while in the mobile phase is v, this time is equal to L/v. With these substitutions, the variance becomes

$$\sigma^2 = \frac{2D_m L}{v} \tag{11.6}$$

For packed columns one must account for the numerous obstructions in the flow path that prevent diffusion along a straight line. In skirting the obstructions created by the packing the diffusion process is retarded as measured by an *obstructive factor* γ (≈ 0.6), making it necessary to replace D_m by γD_m (1)

$$\sigma^2 = 2\gamma D_m L/v \tag{11.7}$$

For capillary columns, of course, $\gamma = 1$. The plate height, $H = \sigma^2/L$, becomes

$$H_D = \frac{2\gamma D_m}{v} \tag{11.8}$$

For those cases in which longitudinal diffusion in the stationary phase is also significant, a plate height term of the following form [1] should be added

$$H_D = \frac{2\gamma_s D_s}{v} \frac{1-R}{R} \tag{11.9}$$

where D_s is the stationary phase diffusion coefficient and γ_s is an obstruction factor for the stationary phase.

Both of the H_D terms given above are inversely proportional to flow velocity v. For purposes of brevity we can consequently summarize longitudinal diffusion by

$$H_D = \frac{B}{v} \tag{11.10}$$

where B is a constant. This equation is the same form as Eq. 9.14, reminding us that many perpendicular flow ($F(+)$) methods share common features. Further details on longitudinal diffusion are found in a paper by Knox and Scott [10].

11.4 STATIONARY PHASE SORPTION-DESORPTION

Once it adheres to (e.g., partitions into) the stationary phase, a solute molecule takes a step backward with respect to the onmoving zone. When desorbed, it moves faster than the partially retarded zone, thus taking a step forward. Sorption and desorption occur randomly, making this process resemble a random walk.

Consider a sorbed molecule. The desorption of the molecule takes a finite time, as do all kinetic processes. The mean lifetime before desorption occurs may be denoted by t_d. This means that on average a sorbed molecule will stay with the stationary phase a time t_d, but in individual cases the time might be much greater or smaller than this. In terms of the random-walk model, this suggests that a new random step occurs after a residence time of order of magnitude t_d in the stationary phase. A positive or forward step taken at that time is equivalent to desorption while a negative step signifies continued sorption.

During the time t_d, the zone center moves forward at velocity Rv and thus travels a distance of Rvt_d. Since the zone center is the frame of reference, this means that a molecule sorbed for a time t_d is taking a step backwards of length

$$l = Rvt_d \qquad (11.11)$$

The total number of sorptions plus desorptions may be taken to approximate the number of steps, n, in the random walk. Since each sorption process is followed by a desorption, the two occur in equal numbers and the total number of steps may be estimated as twice the number of desorptions, a quantity evaluated below.

Any given zone requires a time L/Rv to migrate distance L to the end of the column. This time is the retention time t_r (see Eq. 10.2). Molecules spend a fraction $1 - R$ of t_r, thus a time of

$$t_s = (1 - R) \frac{L}{Rv} \qquad (11.12)$$

immobilized by the stationary phase. The number of desorptions, $n/2$, occurring in this entire period of residence in the stationary phase is equal to time t_s divided by the mean desorption time t_d, that is, $n/2 = t_s/t_d$. Consequently

$$n = \frac{2}{t_d} \left(\frac{1-R}{R} \right) \frac{L}{v} \qquad (11.13)$$

Upon substituting Eqs. 11.11 and 11.13 into the random-walk expression, Eq. 11.3, we get

$$H = 2R(1 - R)t_d v \qquad (11.14)$$

If we replace t_d by the time t_{ds} necessary to escape only from the stationary phase, without consideration of the time necessary to get through a film of mobile phase into a nearby flowstream, then the H term we get will be strictly a stationary-phase term which we designate as H_s

$$H_s = 2R(1 - R)t_{ds} v = C_s v \qquad (11.15)$$

The H_s term is called a *mass-transfer term* because it accounts for the speed of transport of a molecule through one phase to reach another. It is also called a *nonequilibrium term* because it describes (although obtained here from a molecular perspective) the macroscopic nonequilibrium phenomena of the last chapter. In fact, the H_s given by Eq. 11.15 is the same as the H_s of Eqs. 10.39 and 10.40. This equivalence may not be apparent because the two Hs are expressed in terms of somewhat different time constants (t_{ds} and t_{eq}). While the relationship between time constants will not be discussed here, it is important to note the consistency of the two approaches in showing that H_s is proportional to both flow velocity v and to a time-based term (either t_{ds} or t_{eq}) describing the finite time of transfer through the stationary phase.

Equation 11.15 is applicable to most forms of adsorption and partition chromatography. In the case of adsorption chromatography, the term t_{ds} simply represents the average time needed for molecular detachment from the adsorbing surface. Detachment is a straightforward rate process involving the breaking of the ties between surface sites $-S$ and solute molecules M. If this process is represented by

$$-S-M \xrightarrow{k_d} -S + M \qquad (11.16)$$

occurring with first order rate constant k_d, then

$$t_{ds} = \frac{1}{k_d} \qquad (11.17)$$

An expression for t_d is also available for detachment from nonuniform surface sites [1].

With partition chromatography, desorption involves getting the solute molecules from the interior of the small units of absorbing liquid to the surface where they can escape. This is a diffusion process. It was seen earlier, Eqs. 5.21 and 11.5, that in time t the "average" diffusion distance σ is given by $\sigma^2 = 2Dt$. If the units of stationary phase are of depth d, the time, equal to t_d, needed to diffuse out of them will be roughly $d^2/2D_s$, where D_s is the diffusion coefficient for solute in the stationary phase (see Eq. 9.16). If this time is written as $t_d = qd^2/2D_s$, where q is a *configuration factor* with a value (near unity) accounting for the shape of the stationary phase unit, Eq. 11.15 becomes

$$H_s = qR(1-R)\frac{d^2v}{D_s} \qquad (11.18)$$

For a uniform film, $q = 2/3$; values of q for other geometries have been compiled [1].

Equation 11.18 shows that the plate height in partition chromatography is reduced by dispersing the stationary phase into extremely fine units or as a very thin film so that the average depth d is small. Other aspects of optimization will be discussed in the next chapter.

We should like to be able to obtain simple expressions like Eqs. 11.15 or 11.18 for the mobile phase. Unfortunately, mobile phase processes are complicated by flow. The consequences are discussed below.

11.5 FLOW AND DIFFUSION IN THE MOBILE PHASE

The complexity of flow in a packed column was discussed in Section 4.5. The local velocity varies wildly from point to point, depending on the size of the interstitial channel, proximity to a particle surface, obstacles blocking the flow path, and so on. A molecule's migration through such a matrix is chaotic, clearly subject to modeling by statistical methods, most simply by a random walk. However the terrible complexity of the interstitial pore space and the flow going through it makes realistic modeling more difficult than before. Nonetheless, by carefully formulating the random-walk model in accordance with appropriate scaling laws, we can arrive at a general relationship showing how plate height depends upon particle diameter, flow velocity, and diffusivity. This relationship, combined later with empirical data, will provide criteria for the inherent "goodness" of the packing in virtually any column.

We have already dealt with stationary phase processes and have noted that they can be treated with some success by either macroscopic (bulk transport) or microscopic (molecular-statistical) models. For the mobile phase, the molecular-statistical model has little competition from bulk transport theory. This is because of the difficulty in formulating mass transport in complex pore space with erratic flow. (One treatment based on bulk transport has been developed but not yet worked out in detail for realistic models of packed beds [11, 12].) Recent progress in this area has been summarized by Weber and Carr [13].

The random-walk model for the mobile phase will be based on some simple scaling concepts. These concepts require that flow behave similarly in different well-packed beds of particles (of the same shape) irrespective of particle diameter and type of mobile phase. Flow similarity, in turn, requires that different packings have almost the same statistical mix of large and small channels, with the same relative size distribution with respect to the mean. This geometrical similarity is expected to hold whether one is using 5 μm silica particles in LC or 100 μm diatomaceous earth particles in GC, providing the particles have a narrow size distribution and are equally well packed. With geometrical similarity, the velocity spectrum representing relative flow velocities in different channels will be the same irrespective of particle diameter and irrespective of the mobile phase (gas or liquid) at a given overall flowrate.

We can now express our scaling ideas in concrete form. First of all, most significant distances, such as those between adjacent channels or across channels, are scaled in proportion to mean particle diameter d_p. For example if d_p doubles, the mean distance across channels or between the larger voids found between particles also doubles. Second, all velocities in the velocity spectrum will be scaled in proportion to the mean cross-sectional flow velocity v in the packed medium as long as the flow remains laminar. Thus, if v doubles, every single velocity vector in the packed bed doubles. (Recall Eq. 4.11a, which shows that all velocities in a capillary tube are represented in a fixed constant proportion to one another, all scaled to v (or $\langle v \rangle$) irrespective of the capillary radius r_c. This notion extends to capillary bundles as well as packed beds.)

We now recall that the spreading of chromatographic peaks is due to different velocity states that molecules can occupy, allowing one molecule to get ahead of or behind another. Due to the complicated structure of a bed of packed particles, there are a number of ways in which these critical velocity increments or biases can arise [1]. These are:

1. Molecules travel faster at the center of narrow flow channels (e.g., those winding between adjacent particles) than at the outside edge.
2. Molecules travel faster in some channels than in others close by because of differences in shape, openness, obstructions, and so on.
3. Molecules travel faster in some regions (incorporating perhaps 10–100 particles) than in other regions because of fluctuations from place to place in void geometry due to particles occasionally bridging across open spaces and otherwise not fitting well together. (This class of effects resembles category 2 but is longer range.)
4. Molecules travel faster in the flow channels between particles than in the narrow pores within porous support particles because flow is inhibited (sometimes brought almost to a standstill) by the small pore size within the particles.
5. Molecules generally travel faster near the column wall than at the column center.

All the above velocity increments contribute to zone spreading. For simplicity, we will focus on number 2 above, which has been called the short-range interchannel effect [1]; the other kinds of increments are treated in about the same way.

When a molecule gets into a fast flow channel, its higher-than-average velocity will persist for some average distance S along the flow axis. After it has gone distance S, the molecule finds itself in a "new" channel and it therefore assumes a new velocity uncorrelated with its initial velocity. This change in velocity represents a random step.

Figure 11.2 shows a column schematically divided into segments of length S. There are L/S segments in the column, each corresponding to a random step. Thus the number of steps is

Figure 11.2. The division of column length L into segments of length S, where S is the distance a given velocity bias will persist.

$$n = \frac{L}{S} \tag{11.19}$$

The length of step is the distance by which a fast-track molecule surges ahead of (or a slow-track molecule falls behind) the average molecule in segment length S. If the velocity v_f of the typical fast molecule is Δv greater than the average velocity v, then

$$l = \frac{\Delta v}{v_f} S = \omega_\beta S \tag{11.20}$$

in which $\Delta v / v_f$ can be written as a constant ω_β because all velocities (including Δv and v_f) are scaled to one another and to v irrespective of particle diameter, mobile phase, and so on. Constant ω_β is dimensionless with a probable value between zero and one for most kinds of velocity bias [1].

The substitution of Eqs. 11.19 and 11.20 into 11.3 gives the plate height

$$H_c = \omega_\beta^2 S \tag{11.21}$$

indicating that we prefer ω_β values to be as small as possible to minimize plate height. Segment length S is also preferably small.

To evaluate S, and ultimately to reduce S, we must determine how a molecule is removed from a velocity bias. There are two mechanisms:

1. *Flow.* A molecule carried in a fast streampath will eventually slow down because the flow along that path encounters constrictions or obstructions that hinder fluid motion (see Figure 11.3a).
2. *Diffusion.* A molecule can diffuse from a fast streampath to a slow one, and vice versa (see Figure 11.3b).

If flow is the dominant mechanism bringing a velocity bias to an end, then S will be the average length of the channel having a consistently higher or lower than average permeability, ultimately responsible for that bias. Channel lengths, like other distances in the packed bed, are scaled to particle diameter d_p. Thus the S value based on a flow mechanism is proportional to d_p

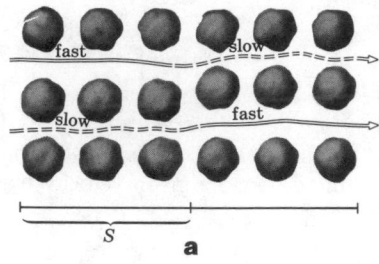

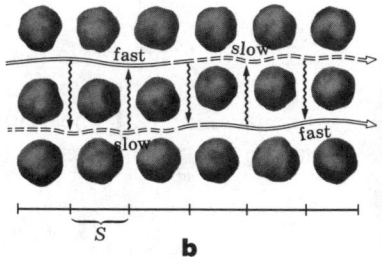

Figure 11.3. A molecule may be removed from a high or low velocity condition by (a) following the same streampath until the streampath changes velocity or (b) diffusing to a nearby streampath. The two mechanisms are termed "flow" and "diffusion," respectively.

$$S_f = 2\omega_\lambda d_p \tag{11.22}$$

When substituted into Eq. 11.21, this gives a plate height

$$H_f = 2(\omega_\beta^2 \omega_\lambda)d_p = 2\lambda d_p = A \tag{11.23}$$

where ω_λ is a packing-structure constant and λ is a constant equal to the product $\omega_\beta^2 \omega_\lambda$. The term H_f, as expressed above, is sometimes called the *eddy diffusion* term.

If diffusion terminates the velocity bias, then S is the distance a molecule is carried downstream in the average time t_D necessary to diffuse into a nearby flow channel having a different velocity. Specifically

$$S_D = vt_D \tag{11.24}$$

If the nearby flow channel is a distance d' away, diffusion time t_D can be approximated as

$$t_D = \frac{d'^2}{2D_m} \tag{11.25}$$

Since distance d', like other distances, is scaled to d_p, we can write $d' = \omega_\alpha d_p$, where ω_α is a constant. Then the last equation yields

$$t_D = \frac{\omega_\alpha^2 d_p^2}{2D_m} \tag{11.26}$$

and Eq. 11.24 becomes

$$S_D = vt_D = \frac{\omega_\alpha^2 d_p^2 v}{2D_m} \tag{11.27}$$

From Eq. 11.21 the plate height is

$$H_D = \left(\frac{\omega_\alpha^2 \omega_\beta^2}{2}\right) \frac{d_p^2 v}{D_m} = \omega \frac{d_p^2 v}{D_m} = C_m v \tag{11.28}$$

where ω is the combined constant $\omega_\alpha^2 \omega_\beta^2/2$.

We see, then, two distinct kinds of plate height terms for the mobile phase: a term H_D proportional to flow velocity v, valid when diffusion terminates a molecule's velocity bias, and a velocity-independent term H_f, valid when flow terminates the bias. The question yet unanswered is how H_D and H_f combine in contributing to the overall experimental plate height. Plate heights are usually additive because variances (for independent processes) are additive; it is tempting to apply the additive rule here. However, *additivity does not apply to H_D and H_f*. We see this in simple physical terms by looking at two extremes of flowrate.

At extremely high flow velocities (but with turbulence remaining negligible), molecules are swept rapidly into and out of different velocity biases. The exchange between velocity states by this mechanism becomes, in the limit, so rapid that diffusional exchanges are negligible and have no influence on the random walk or the resulting plate height. In this case, the overall plate height H_c due to the velocity variations becomes

$$\lim_{v \to \infty} H_c = H_f \tag{11.29}$$

The additivity rule, $H_c = H_D + H_f$, clearly does not apply because as $v \to \infty$, $H_D \to \infty$, and H_c would go to infinity rather than to its actual limit of H_f as required by Eq. 11.29.

In the low velocity extreme, flow exchanges become so slow that all velocity exchanges are caused by diffusion. In this limit we find

$$\lim_{v \to 0} H_c = H_D \tag{11.30}$$

Again additivity does not apply because, if it did, H_c would approach the much larger H_f rather than H_D.

Why has additivity failed? Additivity is valid only for independent random processes in which the step length and number arising in one process are independent of those for the steps of other processes. However the flow and diffusion processes do not have independent random walks. First of all, both participate in the same steps based on the same velocity biases. Second, a random step undertaken by the flow process can be cut short by rapid diffusional exchange, and a normal diffusion-controlled step can be cut short by flow exchange. Diffusion and flow clearly do not have independent random walks.

In actual fact, flow and diffusion are both acting independently in only one sense: to cut short the length of steps. Both terminate steps; with each termination a new random step begins. Therefore, the only way in which diffusion and flow are additive is in the number of random steps they initiate

$$n = n_f + n_D \tag{11.31}$$

This additivity has the following consequences. First by increasing the number of steps taken, it reduces segment length S, which, according to Eq. 11.19 becomes

$$S = \frac{L}{n_f + n_D} \tag{11.32}$$

Additivity also decreases plate height, Eq. 11.21, which, with Eq. 11.32, assumes the form

$$H = \omega_\beta^2 S = \frac{\omega_\beta^2 L}{n_f + n_D} \tag{11.33}$$

Rearrangement gives

$$H = \frac{1}{(n_f/\omega_\beta^2 L) + (n_D/\omega_\beta^2 L)} \tag{11.34}$$

However, since $L/n = S$ (see Eq. 11.19), L/n_f is simply the segment length S_f resulting if flow acts alone, and $L/n_D = S_D$, the diffusion-controlled segment length. Thus

$$H = H_c = \frac{1}{(1/\omega_\beta^2 S_f) + (1/\omega_\beta^2 S_D)} = \frac{1}{(1/H_f) + (1/H_D)} \tag{11.35}$$

which shows how H_f and H_D combine (which is anything but additive) and verifies our earlier conclusions that H approaches the smallest of the two participant terms H_f and H_D. In fact, we see that H is always smaller than either H_f or H_D.

If we substitute specific H_f and H_D forms into Eq. 11.33 from Eqs. 11.23 and 11.28, we get

$$H_c = \frac{1}{(1/2\lambda d_p) + (D_m/\omega d_p^2 v)} \tag{11.36}$$

which shows how the mobile phase H depends on particle diameter d_p, diffusivity D_m, and flow velocity v. We see that the last two equations are consistent with the limits expressed in Eqs. 11.29 and 11.30. Figure 11.4 illustrates the dependence of H_c, H_f, and H_D on v. Note that the difference between H_c and the value of H_c predicted by additivity $H_f + H_D$ is extreme; additivity is a very poor approximation.

The last two equations are two forms of the *coupling equation* for mobile phase plate height [1]. *Coupling* refers to the way in which flow and diffusion act in concert to reduce step length and thus plate height.

The coupling equations shown above are applicable to one kind of velocity bias, such as that between adjacent channels. However, the theory is more or less the same for the other kinds of velocity biases listed previously, but the constants (ω_β, ω_λ, ω_α) are different, yielding different ω and λ in the foregoing coupling expressions. Each kind of velocity bias generates its own independent random walk, leading to an additive plate height term H_i. The overall mobile phase term is therefore the sum

$$H_c = \sum H_i = \sum \left(\frac{1}{2\lambda_i d_p} + \frac{D}{\omega_i d_p^2 v} \right)^{-1} \tag{11.37}$$

This is a rather complete expression for the mobile phase random walk, giving, as desired, H_c as a function of d_p, D_m, and v. While the w_i and λ_i constants are unspecified, they should be the same for equally well packed columns, for all mobile phases (gases or liquids), and for all solutes. The universality of this equation will be put to good use in the following chapter.

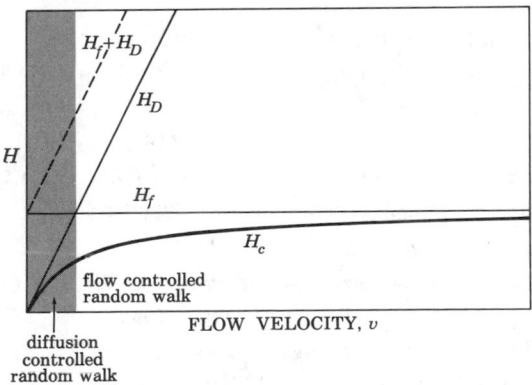

Figure 11.4. Mobile phase plate height H_c compared with the underlying terms for flow (H_f), diffusion (H_D), and their sum ($H_f + H_D$).

11.6 OVERALL PLATE HEIGHT EXPRESSION

If we add to Eq. 11.37 the longitudinal diffusion and stationary phase terms, Eqs. 11.8 and 11.18, we get the overall plate height expression

$$H = \frac{2\gamma D_m}{v} + \sum \left(\frac{1}{2\lambda_i d_p} + \frac{D_m}{\omega_i d_p^2 v} \right)^{-1} + qR(1 - R) \frac{d^2 v}{D_s} \qquad (11.38)$$

This can be condensed to highlight only the velocity dependence

$$H = \frac{B}{v} + \sum \left(\frac{1}{A_i} + \frac{1}{c_{mi}v} \right)^{-1} + C_s v \qquad (11.39)$$

The last two equations provide a good summary of how the various plate height terms are assembled into an overall plate height equation for chromatography. Whether or not all the constants (γ, q, ω_i, λ_i) of the equations are known, these expressions show the nature of the dependence of H on the controllable system parameters v, d_p, d, R, D_m, and D_s. The expressions consequently have profound consequences for guiding practical chromatography. These consequences will be discussed in the next chapter.

REFERENCES

[1] J. C. Giddings, *Dynamics of Chromatography*, Part 1, *Principles and Theory*, Marcel Dekker, New York, 1965.

[2] R. S. Berry, S. A. Rice, and J. Ross, *Physical Chemistry*, Wiley, New York, 1980.

[3] A. L. LeRosen, *J. Am. Chem. Soc.*, **67**, 1683 (1945).

[4] J. C. Giddings and H. Eyring, *J. Phys. Chem.*, **59**, 416 (1955).

[5] J. C. Giddings, *J. Chem. Phys.*, **26**, 1755 (1957).

[6] D. A. McQuarrie, *J. Chem. Phys.*, **38**, 437 (1963).

[7] D. M. Scott and J. S. Fritz, *Anal. Chem.*, **56**. 1561 (1984).

[8] F. Dondi and M. Remelli, *J. Phys. Chem.*, **90**, 1885 (1986).

[9] J. C. Giddings, *J. Chem. Ed.*, **35**, 588 (1958).

[10] J. H. Knox and R. P. W. Scott, *J. Chromatogr.*, **282**, 297 (1983).

[11] J. C. Giddings and P. D. Schettler, *J. Phys. Chem.*, **73**, 2577 (1969).

[12] P. D. Schettler and J. C. Giddings, *J. Phys. Chem.*, **73**, 2582 (1969).

[13] S. G. Weber and P. W. Carr, in P. R. Brown and R. A. Hartwick, Eds., *High Performance Liquid Chromatography*, Wiley, New York, 1989, Chapter 1.

EXERCISES

11.1(*)   A solute has a capacity factor of 9. What is the probability that a molecule of that solute will be found in the mobile phase at any given instant of time? In the stationary phase?

11.2(*)   During a period of 3 min a liquid mobile phase traveling through a silica gel column at a velocity of 0.04 cm/s carries a component 3 cm. What is the equilibrium fraction of solute in the mobile phase? What fraction of time does an average molecule spend attached to the stationary phase? At any instant what is the probability that a particular molecule is in a desorbed state?

11.3(**)   In a chromatographic peak (assume $R = 0.1$) narrow enough for good analytical separations, not many molecules migrate more than 1% ahead of the peak center. What is the probability that a molecule leading the peak center by 1% will be found in the mobile phase?

11.4(*)   What is the value of R if it takes seven times as long for a molecule to desorb as it does to sorb? What is k'?

11.5(*)   Calculate the plate height contributed by longitudinal diffusion in the mobile phase of a column for which $\gamma = 0.60$ and in which the mean flow velocity is 2.0 cm/s. First assume that the column is a GC column with a typical solute diffusivity of $D_m = 0.10$ cm^2/s; second, assume a LC column with $D_m = 1.0 \times 10^{-5}$ cm^2/s.

11.6(**)   The random-walk model assumes steps of equal length forward and backward. Prove that the desorbed molecule takes a step forward of length equal to the length Rvt_d of the backward step discussed in the text. (Hint: Assume that a desorbed molecule takes a step after each mean sorption time t_a.)

11.7(*)   Calculate the plate height contributed by sorption-desorption mass transfer (nonequilibrium) through a uniform liquid layer (configuration factor $q = 2/3$) of thickness 1.0×10^{-3} cm coated on the inside of an open tubular (capillary) column. The gas velocity v is 10 cm/s. The solute retention ratio is 0.10 and its diffusion coefficient D_s through the stationary liquid is 1.0×10^{-5} cm^2/s.

11.8(*)   An isobutylene peak is observed after passage at 50°C through a GC column packed with Chromosorb W having a 15% by weight loading of dinonyl phthalate. The measured R value is 0.603; the measured C_s value is 0.0079 s. From porosimetry data the depth of DNP in the Chromosorb W pores is estimated as 3.50 ×

10^{-4} cm. The diffusion coefficient for the penetration of iso-butylene into dinonyl phthalate is 2.5×10^{-6} cm^2/s. Assuming the pores to have a uniform cross section (equivalent to $q = 2/3$), give a theoretical estimation of C_s. (Since the pore model is obviously oversimplified, close agreement with the experimental value will ordinarily be lacking.)

11.9(**)   Let us represent the sorption/desorption process in chromatography by $A_1 \underset{k_a}{\overset{k_d}{\rightleftarrows}} A_2$ where A_1 represents the sorbed state of the molecule and A_2 represents the desorbed state. We assume for the moment that the rate constants for desorption and sorption, k_d and k_a, are simple first-order kinetic rate constants. In physical chemistry textbooks you will find that the relaxation time (equivalent to the equilibration time of Section 10.6) for the above kinetic process is governed by the expression

$$t_{eq} = \frac{1}{k_d + k_a}$$

Show that the use of this expression in Eq. 10.39, which was based on nonequilibrium considerations, gives results identical to those obtained on the basis of the random-walk model, which are found when Eq. 11.17 is substituted into Eq. 11.15.

11.10(**)   If we write the coupled plate height term (Eq. 11.36) as

$$H_c = \frac{1}{(1/A) + (1/C_m v)}$$

Then it is easy to show that the initial slope (as $v \to 0$) of an H_c versus v plot (Figure 11.4) equals C_m. Assume that $C_m = 0.0036$ s and $A = 81$ μm for a GC column. At what velocity (in cm/s) does the slope of the plot fall to $\frac{1}{2}C_m$, or 0.0018 s?

12

PLATE HEIGHT AND OPTIMIZATION IN CHROMATOGRAPHY

Chapters 10 and 11 dealt with the basic mechanisms of chromatography and with the retention and plate height equations that emerge from those mechanisms. In this chapter we will focus on the applications of plate height theory and show how it can be (i) utilized to describe different experimental systems including packed columns, capillary columns, and GC columns having large gas compression effects, (ii) reduced to a universal form called the reduced plate height equation, and (iii) used to optimize chromatographic separations.

12.1 PLATE HEIGHT EQUATIONS

In Eq. 11.39 we saw that plate height can be described by [1]

$$H = \frac{B}{v} + \sum \left[\frac{1}{(1/A_i) + (1/C_{mi}v)} \right] + C_s v \qquad (12.1)$$

where the terms are given by

$$B = 2\gamma D_m \qquad (12.2)$$

$$C_s = qR(1-R)\frac{d^2}{D_s} \qquad (12.3)$$

$$A_i = 2\lambda_i d_p \qquad (12.4)$$

$$C_{mi} = \omega_i \frac{d_p^2}{D_m} \tag{12.5}$$

in which all symbols retain the meaning given in Chapter 11. We have written C_s in the form used for partition chromatography.

Because of the complex central term representing mobile phase band broadening, Eq. 12.1 has a rather awkward form, not simple to use. The awkwardness reflects the complexity of packed-bed processes, specifically the complexity caused by (i) the many types of velocity states and (ii) the competition (coupling) between diffusion and flow in controlling random displacements.

We will use two approaches to skirt the complexity of Eq. 12.1. First, we will show how various simplified expressions, each applicable over a limited range, emerge as special cases of Eq. 12.1. Second, we will find that by using reduced variables, important conclusions can be drawn from Eq. 12.1 without regard to its awkward form.

Packed Columns

Several approximate forms of Eq. 12.1 can be used for packed columns. Over a limited low-velocity range where diffusion controls the mobile phase random walk, the terms containing A_i drop out and Eq. 12.1 reduces to

$$H = \frac{B}{v} + C_m v + C_s v \tag{12.6}$$

where $C_m = \Sigma\, C_{mi} = \omega d_p^2/D_m$; ω is now the constant $\Sigma\, \omega_i$. This equation is suitable for some descriptive applications at low and medium velocities.

At higher velocities, flow dominates diffusion in the central (coupling) term on the right of Eq. 12.1, giving the limiting form

$$H = A + \frac{B}{v} + C_s v \tag{12.7}$$

where $A = \Sigma\, A_i$ and $\lambda = \Sigma\, \lambda_i$. This high-velocity form is equivalent to the well-known *van Deemter equation* [2], once thought to be valid at all flowrates. While the van Deemter equation can be empirically fit to most H versus v data over a limited range, it does not accurately reflect column dynamics and must be used with caution in interpreting experimental results.

Another alternative to the complete coupled expression of Eq. 12.1 is the somewhat simplified coupling equation found by replacing the summation by a single term. This can be written in several forms, including

$$H = \frac{B}{v} + \frac{AC_m v}{A + C_m v} + C_s v \tag{12.8}$$

This equation also works effectively over a limited range, but removal of the summation from the coupling term gives a somewhat too abrupt transition from the low velocity limit ($C_m v$) to the high velocity limit (A). This abruptness can be moderated by replacing $C_m v$ with $C_m v^{0.3}$ [3].

A still simpler form developed by Knox et al. [4] to approximate the complex coupling phenomena is based on a power law, $A'v^n$, in which the exponent n is commonly assigned a value of $1/3$ to best fit typical experimental conditions. When this term is substituted for the entire coupling term (not just $C_m v$) of Eq. 12.1, we have

$$H = \frac{B}{v} + A'v^n + C_s v \tag{12.9}$$

This equation is more accurate over a wider practical velocity range than Eq. 12.6, 12.7, or 12.8. The equation is compact, but it is more difficult to manipulate mathematically than Eqs. 12.6 and 12.7 because of the fractional value of exponent n.

Capillary Columns

The situation is simpler for capillary columns, now used widely in GC [5] and on an experimental basis in LC [6, 7, 8]. These columns are also being used with increasing frequency in supercritical fluid chromatography [9] following the initial development of this approach by Novotny and co-workers [10].

The ideal capillary column is a tube of circular cross section coated with a uniform stationary film of thickness d (see Figure 12.1). Its uniformity and symmetry greatly simplify its theoretical description. Accordingly, Eq. 12.1 reduces to a much more tractable form and the parameters (γ, q, λ, and ω) become calculable.

First of all, because the stationary phase is assumed to be a uniform film, we have $q = 2/3$ as noted in Section 11.4. This makes C_s rigorously calculable by means of Eq. 12.3. Second, because there are no obstacles in

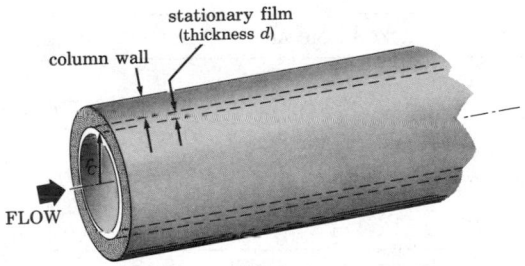

Figure 12.1. Ideal capillary column.

the flow path, the obstruction factor γ in Eq. 12.2 is unity, which fixes the coefficient B. Third, because there is only one flow conduit in the tube, the entire velocity bias, expressed by a summation of terms in a packed column, reduces to a single term. Thus the summation of terms appearing in Eq. 12.1 becomes a single mobile phase term. Fourth, in view of the fact that all flow streamlines maintain a constant velocity through the length of the tube, a random step cannot be terminated by a flow-change mechanism. Thus the term containing A_i in Eq. 12.1 is eliminated. (Technically the single A term in Eq. 12.1 is very large, effectively equal to the column length L, because of the uniform flow over L. Consequently its reciprocal, $1/A$, is vanishingly small.) Finally, by virtue of the fact that the flow profile in a round tube is a known mathematical form, a parabola (Section 4.4), the remaining C_m term can be rigorously derived using nonequilibrium theory [1] or other forms of transport theory [11].

All the above elements combined lead to the conversion of Eq. 12.1 into the expression

$$H = \frac{2D_m}{v} + \frac{1}{24} (6R^2 - 16R + 11) \frac{r_c^2 v}{D_m} + \frac{2}{3} R(1 - R) \frac{d^2 v}{D_s} \quad (12.10)$$

or in simpler form

$$H = \frac{B}{v} + C_m v + C_s v \quad (12.11)$$

For real (as opposed to ideal) capillaries, the stationary phase film tends to distribute nonuniformly on the wall unless fixed by cross-linking or other mechanisms; this will cause some deviation from Eq. 12.10 [12, 13].

For both packed and ideal capillary columns, we have simplified the plate height expressions by using coefficients A, A', B, C_m, and C_s to replace groups of more basic parameters. In Table 12.1 we summarize the expressions for these coefficients.

TABLE 12.1 Summary of Plate Height Coefficients for Packed and Ideal Capillary Columns [a]

	Packed Column	Capillary Column
A	$2\lambda d_p$	—
A'	$\alpha d_p^{n+1}/D_m^n$	—
B	$2\gamma D_m$	$2D_m$
C_m	$\omega d_p^2/D_m$	$\frac{1}{24}(6R^2 - 16R + 11)r_c^2/D_m$
C_s (partition)	$qR(1 - R)d^2/D_s$	$qR(1 - R)d^2/D_s$
C_s (adsorption)	$2R(1 - R)t_{ds}$	$2R(1 - R)t_{ds}$

[a] For uniform stationary films, $q = 2/3$. The dimensionless constants λ, α, γ, and ω are generally of order unity. Other parameters are defined in text.

12.2   PLATE HEIGHT AND GAS COMPRESSIBILITY

The foregoing expressions are valid for either gas or liquid mobile phases in uniform columns. In gas chromatography, as we have noted, compression effects often create significant velocity gradients in association with the pressure gradients driving flow through the column. It is important to understand how these nonuniformities affect plate height.

Most of the plate height terms we have discussed are velocity sensitive. For gases, any term containing the mobile (gas) phase diffusion coefficient D_m (D_g for gases) is also pressure dependent, since D_g varies in inverse proportion to pressure p. For an ideal gas

$$D_g = \frac{D_{go}p_o}{p} \tag{12.12}$$

where D_{go} is the D_g value at outlet pressure p_o. Because $pv = $ constant, we can write, in analogy to Eq. 4.26

$$pv = p_o v_o \tag{12.13}$$

When we substitute Eq. 12.13 into 12.12, we get D_g in a velocity-dependent form

$$D_g = \frac{D_{go}v}{v_o} \tag{12.14}$$

or

$$\frac{D_g}{v} = \frac{D_{go}}{v_o} = \text{a constant} \tag{12.15}$$

which shows that the ratio D_g/v is constant down the length of the column, a consequence of the fact that both D_g and v are increasing at the same rate with decompression. Inspection of Table 12.1 shows that each of the two terms B/v and $C_m v$ are constant throughout the column (based on Eq. 12.15) because both vary with D_g/v. The A term has no velocity-dependent terms and is thus also constant. All of these terms involve gas-phase processes; their collective contribution to plate height can be labeled H_g. We conclude that throughout the column

$$H_g = \text{constant} \tag{12.16}$$

Note that this conclusion is independent of how the various mobile phase terms combine; thus it is a general conclusion valid for any of the H equations of Section 12.1, whether in a "coupled" form (Eq. 12.1) or one of the subsequent simplified forms.

Unlike mobile phase terms, the stationary phase term in both packed and capillary columns, $C_s v$, varies throughout the column, increasing continuously from inlet to outlet as v increases. Its "average" value is given by

$$H_s = C_s \tilde{v} \tag{12.17}$$

where $\bar{v} = v_o j$ is the time-average velocity defined by Eq. 10.23.

We can approximate the *observed plate height* $\hat{H}$ (which is measured experimentally as $L\tau^2/t_r^2$, Eq. 5.48) as the sum of the last two expressions

$$\hat{H} \simeq H_g + C_s \tilde{v} \tag{12.18}$$

We have given a special symbol $\hat{H}$ to the observed plate height because it assumes a single measurable value in contrast to the variability of the local plate height; the latter changes from point to point because of column nonuniformities.

A more careful analysis of the problem shows that some rather subtle effects exist in all nonuniform columns, which, when carefully analyzed for gas chromatography, lead to the following fully corrected equation [14, 15]

$$\hat{H} = H_g f_1 + C_s \tilde{v} \tag{12.19}$$

where

$$f_1 = \frac{9(P^4 - 1)(P^2 - 1)}{8(P^3 - 1)^2} \tag{12.20}$$

While the correction term f_1 looks imposing, it varies only within a narrow range, from unity at $P = 1$ to 1.125 for $P \to \infty$.

Equations 12.18 and 12.19 show that gas compression does not cause undue complications. Although the plate height varies along the column, the variation is confined to only one term. Furthermore, to a good approximation, the plate height can be represented by an average of its value within the column. Thus our upcoming discussions of optimization will, by and large, apply to GC as well as LC if we keep in mind the need to deal in averages in GC. For brevity, we will not explore the precise and quantitative implications of compressibility for these applications.

12.3 REDUCED PLATE HEIGHT

Equation 12.1 provides a good description of H in packed columns but its application is complicated by its complex form and by its dependence on unknown constants λ_i and ω_i. Furthermore, as with the other equations of Section 12.1, a plot of the equation as H versus v yields curves that differ enormously for different forms of GC and LC because of wide variations in D_m, d_p, and so on. This is shown schematically in Figure 12.2.

Despite the apparent difficulties in application, these equations, when examined more closely, are found to possess global features essential to understanding and correlating chromatographic behavior. These global characteristics can be uncovered by transforming the H equation and related experimental data to reduced (or dimensionless) variables. In this way the dependence of H on v can be approximated by a single universal curve applicable to all types of GC and LC. This universal reference curve (or series of close-lying curves) provides a means for comparing and evaluating specific columns drawn from widely different chromatographic families. The concept of a universal curve based on reduced variables was first proposed by this author [16].

The situation is similar to that for nonideal gases. Under nonideal conditions (e.g., at high pressures), each gas at a given temperature has its own pressure-volume curve. However, by introducing the right set of reduced coordinates (reduced pressure, reduced volume, and reduced temperature) and plotting pressure-volume data in terms of these dimensionless coordinates instead of the normal pressure-volume coordinates, one gets universal curves to which most gases closely adhere. (This universality is described by the principle of corresponding states [17].) The common curves can be described by a single reduced equation of state, whereas in normal pressure and volume units every gas has its own specific equation of state. Thus the use of properly reduced coordinates brings order out of chaos for gases.

The key step in developing a reduced coordinate system lies in identifying characteristic scaling parameters. In the case of gases, these characteristic parameters are the critical pressure, the critical volume, and the critical temperature. We seek similar scaling parameters for chromatography.

We start by noting that the plate height curves in Figure 12.2 are of similar form (as they should be since they are described by the same equations) but are strung out widely along the velocity scale. Consequently,

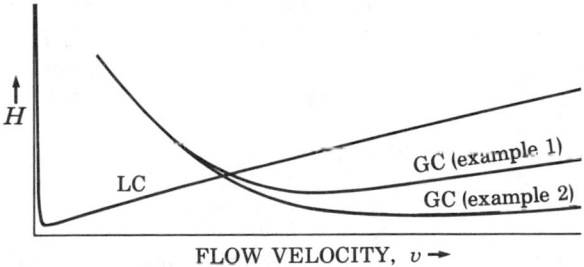

Figure 12.2. Comparison of plate height versus flow velocity plots for one LC column and two different GC columns.

we need a new velocity scale to bring these curves into concordance. We require, for this purpose, a characteristic velocity that will serve as a scaling base for the flow velocity axis.

Two basic components of plate height in packed columns are H_f and H_D, given by Eqs. 11.23 and 11.28, respectively. When these two terms are equal—which occurs at the specific velocity $v'_c = A/C_m$—we are at the transition point between a flow-controlled and diffusion-controlled random walk. (This transition, as it turns out, occurs near the minimum of each of the plate height curves shown in Figure 12.2.) The velocity v'_c is a fundamental parameter characterizing every packed column system. Its value can be expressed by (see Eqs. 11.23 and 11.28)

$$v'_c = \frac{A}{C_m} = \left(\frac{2\lambda}{\omega}\right)\frac{D_m}{d_p} \tag{12.21}$$

However, since λ and ω are constants of order unity whose exact values are difficult to determine, we can replace v'_c by a simpler expression that still characterizes (i.e., provides a scale for) the transition, namely

$$v_c = \frac{D_m}{d_p} \tag{12.22}$$

The state of the dynamic balance between mobile phase processes in every chromatographic column is governed by the flow velocity *relative to* the fundamental velocity v_c, that is, by [16]

$$\nu = \frac{v}{v_c} = \frac{d_p v}{D_m} \tag{12.23}$$

where ν is termed the *reduced velocity*. This dimensionless velocity provides a more basic measure of the state of flow than the flow velocity itself.

We must now find a scale factor with length dimensions that can be used to reduce plate height to dimensionless form. In Section 11.5 we identified particle diameter d_p as a fundamental unit of length to which most distances in the column are scaled. Scaling H in units of d_p seems, then, a logical choice. This choice is supported by noting that when v equals the fundamental velocity v_c, all mobile phase H terms are close to d_p in value. Thus when $v = v_c$, $C_m v$ becomes $C_m v_c$, which equals

$$C_m v_c = \frac{\omega d_p^2 v_c}{D_m} = \omega d_p \tag{12.24}$$

which shows more concretely that plate height is scaled to particle diameter d_p. For a universal representation, then, plate height is expressed in multiples of d_p, that is, in terms of the *reduced plate height* defined by

$$h = \frac{H}{d_p} \tag{12.25}$$

If we have reasoned correctly, we now have what amounts to a principle of corresponding states for chromatography. Given a reduced velocity v, the reduced plate height should be approximately the same for virtually all well-packed columns. In other words, a single h versus v curve should come close to representing all such chromatographic systems.

To obtain the universal h versus v curve we replace H by h (using $H = d_p h$) and v by v ($v = D_m v / d_p$) in our basic plate height expression, Eq. 12.1. We get after rearrangement

$$h = \frac{2\gamma}{v} + \sum \frac{1}{(1/2\lambda_i) + (1/\omega_i v)} \quad + \quad \Omega v \tag{12.26}$$

mobile phase   stationary phase

terms   term

which shows the desired h versus v relationship in the form of a *reduced plate height equation*. The coefficient Ω is

$$\Omega = \frac{C_s D_m}{d_p^2} \tag{12.27}$$

Equation 12.26 is indeed remarkable in that, as far as the mobile phase is concerned, h is a function only of v and the packing-structure constants γ, λ_i, and ω_i. All dependence on d_p and D_m has dropped out. Therefore for similarly packed columns with similar packing constants, the h versus v curves representing mobile phase effects will be about the same. This will be true in comparing GC and LC columns as well, despite the fact that diffusivity D_m typically differs by 10^4 between gases and liquids and d_p varies by a factor of ten or so.

While theory does not provide the exact values for λ_i and ω_i needed for plotting h versus v curves, empirical plots serve just as well. If we plot h versus v for a series of well-packed columns (GC as well as LC) for which stationary effects are negligible, we find that they nearly superimpose in a narrow band representing the norm for high performance columns (see Figure 12.3). Poorly packed columns yield plots that fall significantly above this band. Any column, with any d_p or D_m, can be plotted on such a graph to determine its relative efficiency. Thus without specific theoretical values, our theoretical treatment has shown that using the reduced variables h and v provides a means for the comparison of unlike columns—a method for picking the good from the bad.

If the stationary phase term in Eq. 12.26 is significant, it will increase h values somewhat, especially at high v. However another goal of good

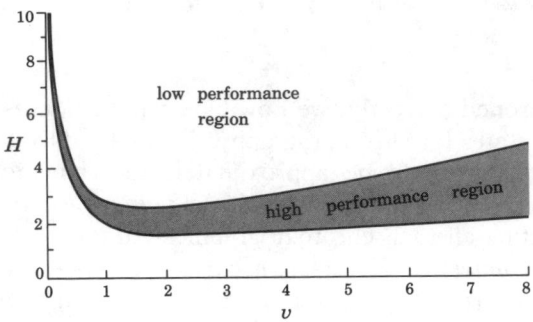

Figure 12.3. Reduced plate height plot showing the band in which h versus v curves lie for high performance columns.

column building is to distribute the stationary phase in a manner that will keep this term relatively small.

In cases where we need a tractable mathematical expression of h versus v it has been found empirically that the coupling term in Eq. 12.26 can be replaced by αv^n, where α is a constant and n is an exponent of order 1/3. This is the same term, shown here in reduced form, as was shown in Eq. 12.9. With the use of this term, Eq. 12.26 becomes

$$h = \frac{2\gamma}{v} + \alpha v^n + \Omega v \qquad (12.28)$$

which is known as the Knox equation.

12.4   OPTIMIZATION: HIGHER RESOLUTION

For many chromatographic runs, especially with complex samples, resolution or peak capacity is inadequate for the desired separation. In these cases steps can be taken to improve resolution. These steps are described in this section. If resolution is already adequate, other steps can be taken to do the separation faster without losing resolution. Methods for increasing speed will be described in Section 12.5.

In Chapter 5 we obtained the following general expression for resolution (see Eq. 5.58)

$$R_s = \left(\frac{N}{16}\right)^{1/2} \frac{\Delta W}{W} \qquad (12.29)$$

The velocity W along the separation axis in chromatography is simply Rv (or $v/(1 + k')$); the difference ΔW in the velocity of two zones is therefore $(\Delta R)v$. Thus Eq. 12.29 becomes

$$R_s = \left(\frac{N}{16}\right)^{1/2} \frac{\Delta R}{R} \tag{12.30}$$

For difficult separation problems, ΔR is small and can be reasonably treated (for the moment) as a differential term dR. We note that

$$d\left(\frac{1}{R}\right) = \frac{-dR}{R^2} \tag{12.31}$$

Thus approximately

$$\Delta\left(\frac{1}{R}\right) = \frac{-\Delta R}{R^2} \tag{12.32}$$

or

$$\frac{-\Delta R}{R} = R\Delta\left(\frac{1}{R}\right) \tag{12.33}$$

We choose to ignore the negative sign in these equations because we are interested only in the absolute values of ΔR, $\Delta(1/R)$, and so on. (A change in the sign of ΔR would occur if the peaks exchanged positions; this would not alter resolution and is thus unimportant.)

If we substitute the last equation (without the minus sign) into Eq. 12.30, we get

$$R_s = \left(\frac{N}{16}\right)^{1/2} R\Delta\left(\frac{1}{R}\right) \tag{12.34}$$

If now we write for R (see Eq. 10.9)

$$R = \frac{V_m}{V_m + KV_s} \tag{12.35}$$

and its reciprocal

$$\frac{1}{R} = \frac{V_m + KV_s}{V_m} = 1 + K\frac{V_s}{V_m} \tag{12.36}$$

we find

$$\Delta\left(\frac{1}{R}\right) = \Delta K \frac{V_s}{V_m} \tag{12.37}$$

When we combine Eqs. 12.35 and 12.37 we obtain

$$R\Delta\left(\frac{1}{R}\right) = \frac{\Delta K}{K}(1 - R) \tag{12.38}$$

This expression can go into Eq. 12.34 to yield [1]

$$R_s = \left(\frac{N}{16}\right)^{1/2} \frac{\Delta K}{K} (1 - R) = \left(\frac{N}{16}\right)^{1/2} \frac{\Delta K}{K} \left(\frac{k'}{1 + k'}\right) \qquad (12.39)$$

This equation contains a wealth of information on how to improve resolution.

Some paths to improvement appear explicitly in Eq. 12.39. It is clear, for example, that R_s increases with the relative selectivity $\Delta K / K$. The latter can be altered by changing various chemical (e.g., stationary phase) and physical (e.g., temperature) parameters. However, Eq. 12.39 shows that these changes must not be made in such a manner as to drive the components excessively into the mobile phase or else R will approach unity and $1 - R$ (and k') will approach zero and thus erode resolution. It is desirable therefore to keep R small, at least below 0.5 (equivalent to $k' > 1$) if possible. This simply means that one must maintain a reasonable level of retention, that is, peaks should not emerge until well after the void peak.

Clearly, $N = L/H$ should be large. This can be realized by decreasing (as we will discuss shortly) plate height H or increasing L (mile long capillary and packed columns have been constructed to yield about 10^6 theoretical plates), but many practical problems of excess pressure drop, difficulty of construction and operation, and excessive separation time are encountered. The challenge of reducing separation time will be addressed in the subsequent section; for now we will look at some of the more subtle factors affecting resolution.

Resolution and Thermodynamics

Distribution coefficient K can be expressed as (see Eq. 2.19)

$$K = \exp\left(-\frac{\Delta\mu^0}{\mathscr{R}T}\right) \qquad (12.40)$$

By differentiation we find that a small increment in $\Delta\mu^0$, $d(\Delta\mu^0)$, leads to a shift in K of

$$dK = K\left(\frac{-d(\Delta\mu^0)}{\mathscr{R}T}\right) \qquad (12.41)$$

which gives

$$\frac{dK}{K} = \frac{-d(\Delta\mu^0)}{\mathscr{R}T} \qquad (12.42)$$

If the increments are finite rather than infinitesimal, the relationship is still approximately valid. The finite increments represent the difference in K (i.e., ΔK) and the difference in $\Delta\mu^0$ (i.e., $\Delta(\Delta\mu^0)$) found upon going from one peak to a neighboring peak that one is trying to resolve. We have

$$\frac{\Delta K}{K} = \frac{\Delta(\Delta\mu^0)}{\mathscr{R}T} \qquad (12.43)$$

where the minus sign has again been dropped because we are interested only in absolute differences between the parameters for the two peaks.

If we substitute Eq. 12.43 into Eq. 12.39, we have

$$R_s = \left(\frac{N}{16}\right)^{1/2} \frac{\Delta(\Delta\mu^0)}{\mathscr{R}T}(1 - R) \qquad (12.44)$$

which shows explicitly how R_s depends on the difference in thermodynamic parameter $\Delta\mu^0$ between two peaks [1].

Of equal interest, Eq. 12.44 can be rearranged to show the number of theoretical plates N needed to achieve a desired resolution level R_s with a given increment $\Delta(\Delta\mu^0)$

$$N = \frac{16R_s^2}{(1 - R)^2}\left(\frac{\mathscr{R}T}{\Delta(\Delta\mu^0)}\right)^2 \qquad (12.45)$$

Table 12.2 shows the number of plates needed to achieve unit resolution at 300 K for different increments of $\Delta\mu^0$. For simplicity we assume $R = 0.2$.

We note that a difference in $\Delta\mu^0$ values of only 3 cal/mol will resolve two peaks in a powerful column of 10^6 plates; 15 cal/mol requires only 40,000 plates. Differences of the order of 10 cal/mol (requiring $N \sim 10^5$) serving to split two peaks is very small compared to the absolute value of $\Delta\mu^0$ for either peak, typically in the thousands of calories/mole. Hence we can understand how small differences in intermolecular interactions resulting from small structural differences between solute species, such as between isomers, can lead to chromatographic separation in high N columns.

We note that N increases with R_s^2. Thus if we can find some way for extracting the necessary analytical information from less fully resolved peaks, the reduction in N and the associated reduction in column length and analysis time (see next section) can be substantial. Mathematical methods

TABLE 12.2 Number of Plates Required to Obtain Unit Resolution for Different $\Delta(\Delta\mu^0)$ Values at 300 K and $R = 0.2$

	$\Delta(\Delta\mu^0)$	
N	cal/mol	J/mol
400	150	620
2500	60	250
10,000	30	125
40,000	15	62
250,000	6	25
1,000,000	3	12

for the deconvolution (breaking down into component peaks) of overlapping peaks, and multiwavelength detection for sensing individual components within overlapping groups of peaks, constitute promising approaches [18, 19].

Complex Multicomponent Separations

For complex samples that contain a large number (10^2 or more) of components, the resolution of a given pair is not enough; the repetitive resolution of successive pairs is generally needed. The ability to separate successive pairs is best measured by peak capacity n_c. As explained in Section 5.8, n_c is a measure of the maximum number of peaks that can be accommodated in a chromatogram; n_c rather than resolution best measures the capability for handling multicomponent samples. According to Eq. 5.62, the peak capacity for chromatographic elution systems (in the absence of programming) is [20]

$$n_c = 1 + \left(\frac{N}{4}\right)^{1/2} \ln \frac{V_{max}}{V_{min}} \qquad (12.46)$$

We recall from Section 6.7 that to cleanly separate the majority of components in a complex sample, there must be much more space along the axis of separation than that required to hold the peaks if they lined up side by side. As in separations generally, peaks in chromatography are rarely so accommodating and orderly as to assume even spacing. In the random jumble of peaks commonly found in chromatograms, many component peaks overlap by chance even if there is space for them (see Figure 6.11). The only way to assure that most (rarely all) component peaks separate cleanly is to provide considerably more space for peaks, measured by n_c, than there are components m to fill that space. The statistical arguments of Section 6.7 show that the number of components successfully isolated is disappointingly small, approximately [21]

$$s = m \exp\left(-\frac{2m}{n_c}\right) \qquad (12.47)$$

Thus in a high-powered column with $n_c = 200$, 100 components ($m = 100$) scrambled randomly will overlap so much that only 37 component peaks will be fully isolated. Hence n_c values should be much higher than m for effective separation. This requires a wide elution volume (or time) range and a large plate number N.

Equations 12.39 and 12.46 agree that whether we seek high resolution or a large peak capacity, plate count N should be large. This involves using a long column and/or achieving a small plate height H, as just noted. Below we discuss ways of reducing H.

Minimizing Plate Height

As shown by the equations in Section 12.1, column plate height is affected by many parameters, including flow velocity, particle diameter, packing nonuniformity, diffusivities, degree of retention, stationary phase structure, temperature, pressure drop, and pressure. Some of these parameters are interdependent, such as diffusivity and temperature; also velocity and pressure drop. Finding a minimum with respect to all of these parameters is an extended task we shall not attempt here. However, we can readily uncover some simple rules for optimizing a few of the major parameters. First we choose flow velocity.

Following Eq. 12.6, which we choose for simplicity, H can be approximated by

$$H = \frac{B}{v} + C_m v + C_s v = \frac{B}{v} + Cv \tag{12.48}$$

Because of the longitudinal diffusion (B/v) term, H becomes very large at low v, and because of the nonequilibrium (Cv) term, H is large at high v (see Figure 12.2). Clearly there is a minimum H at some intermediate v. We find the minimum by differentiating Eq. 12.48 with respect to v and writing $dH/dv = 0$. Thus

$$\frac{dH}{dv} = -\frac{B}{v^2} + C = 0 \tag{12.49}$$

which, when solved for v, yields

$$v_{\text{opt}} = \left(\frac{B}{C}\right)^{1/2} \tag{12.50}$$

When this optimum v is substituted back into Eq. 12.48 the minimum H emerges

$$H_{\text{min}} = 2(BC)^{1/2} \tag{12.51}$$

The numerical values obtained for v_{opt} and H_{min} depend strongly on the relative roles of mobile phase and stationary phase nonequilibrium terms, represented by C_m and C_s, respectively. In the most favorable case for chromatographic separation, C_s (which can be more readily modified than C_m) is reduced to negligible proportions, $C_s \ll C_m$. From Table 12.1 we find $B = 2\gamma D_m$ and $C_m = \omega d_p^2/D_m$; the substitution of these coefficients for B and C in Eqs. 12.50 and 12.51 yields

$$v_{\text{opt}} = \frac{D_m}{d_p}\left(\frac{2\gamma}{\omega}\right) \tag{12.52}$$

and

$$H_{\min} = d_p(8\gamma\omega) \tag{12.53}$$

Since γ and ω are of order unity, we have

$$v_{\mathrm{opt}} \sim \frac{D_m}{d_p} \tag{12.54}$$

and

$$H_{\min} \sim d_p \tag{12.55}$$

These equations show the order of magnitude of the optimum point for the case in which stationary phase effects are negligible. We observe that the minimum H differs only by a numerical constant from the particle diameter d_p.

We note that v_{opt} is of similar magnitude to the transition velocity $v_c = D_m/d_p$ (see Eq. 12.22). Since the reduced velocity v is defined as v/v_c and the reduced plate height h is H/d_p (Eq. 12.25), the above optimum conditions (v near v_c and H near d_p) correspond roughly to unit values of h and v, a conclusion consistent with Figure 12.3.

When stationary phase nonequilibrium effects are so large that $C_s \gg C_m$ and therefore $C \simeq C_s$, the increased magnitude of C shifts v_{opt} to lower values and $H_{\min}$ to higher levels than found above, as shown by Eqs. 12.50, 12.51, and Figure 12.4. If for B and C we use the values of B and C_s defined by Table 12.1, $H_{\min}$ from Eq. 12.51 becomes

$$H_{\min} = d[8\gamma q R(1 - R)(D_m/D_s)]^{1/2} \tag{12.56}$$

which applies when stationary phase nonequilibrium is dominant.

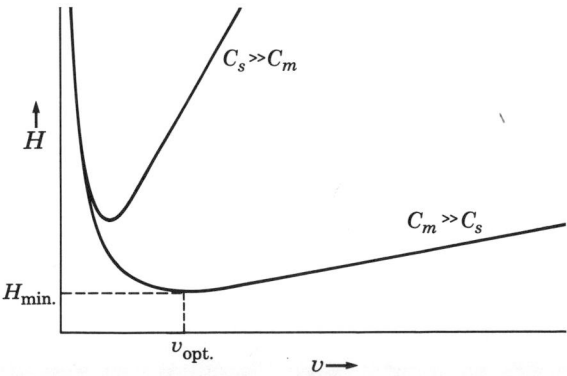

Figure 12.4. Plate height versus flow velocity plot for negligible stationary phase nonequilibrium effects (bottom curve) and for dominant stationary phase effects (top).

In the most general case, in which we use both C_m and C_s to get the sum $C = C_m + C_s$ for use in Eq. 12.51, we get

$$H_{min} = [8\gamma(\omega d_p^2 + qR(1 - R)(D_m/D_s)d^2)]^{1/2} \qquad (12.57)$$

Many chromatographic systems are run at or above the optimum flow velocity v_{opt} to achieve faster separation. In this case Eq. 12.57 is no longer valid for plate height. However, for purposes of the present discussion, in which we are seeking ways to reduce plate height but not time, we will assume that the first step, velocity optimization, has been taken, and that Eq. 12.57 is applicable.

Equation 12.57 suggests that adjustments in a number of parameters other than v will reduce plate height. Where stationary phase effects are important, reducing the stationary phase film thickness d is clearly profitable. Increasing diffusivity D_s in the stationary film is also helpful; this can be done by increasing temperature or choosing a stationary phase of lower viscosity, providing these changes are consistent with other requirements of the separation (such as maintaining high selectivity). It is not recommended that mobile phase diffusivity D_m be reduced, as suggested by Eq. 12.56, because the gain in resolution will generally be offset by a reduction in separation speed (see next section).

To reduce mobile phase effects, we should use uniformly packed columns, for which ω is minimal. Furthermore, the particle diameter d_p of the packing should be as small as possible. The reduction of d_p to a minimum is an important aspect of modern chromatography with advantages for both low plate height and high separation speed. However, decreasing d_p has several other implications, especially related to the need for increased pressure gradients to force fluid through beds of finer particles. Clearly, if the pressure drop applicable to a column is constrained to some maximum value established by mechanical limitations, there must follow a limit on the fineness of particles to be used. This point is discussed in the next section.

Finally, peak spreading can originate outside a column as well as within; both forms of peak spreading degrade resolution. Any peak spreading of external origin is called an *extracolumn effect*. Such extracolumn effects are caused by finite sample volumes along with spreading within the injection chamber, connecting tubing, and detector. Extracolumn effects are best reduced by using small sample sizes and by keeping all external volumes small compared to the volume of the chromatographic column. The external volumes should be narrow and streamlined in configuration [22].

12.5 OPTIMIZATION: FASTER SEPARATION

In simple theory, as noted above, we can make columns of such great length that the most difficult separation can be achieved. However, as separation difficulty and column length increase, the time needed to achieve separation rises dramatically, soon reaching impractical levels. The time of separation then becomes an overriding consideration; steps must be taken to reduce it to acceptable levels. In some applications, such as process control, high-speed separations (often a few minutes or less) are critical.

We start with the premise that any given separation problem can be solved if we have enough theoretical plates N. The required value of N can be obtained from Eq. 12.39

$$N = \frac{16R_s^2}{(\Delta K/K)^2(1 - R)^2} \tag{12.58}$$

Thus if we know $\Delta K/K$ and the average R for a solute pair, and we require a specific value of R_s, we can calculate how many plates we need. Or, if we need to develop a certain peak capacity n_c to achieve our goals, we can calculate the necessary N from Eq. 4.46.

The length of column needed to get the required N plates when the plate height is H equals the product of N and H

$$L = NH \tag{12.59}$$

The time of separation (specifically, the time required to get through the N plates) is simply the time t needed for the components in question to travel distance L at average velocity Rv

$$t = \frac{L}{Rv} = \frac{N}{R}\frac{H}{v} \tag{12.60}$$

With this equation, we have a fundamental starting point for time optimization.

We note, first of all, that t increases with N. Consequently, any step that will reduce the required plate number N will reduce t. For example, the use of a column that has high relative selectivity $\Delta K/K$ for a critical solute pair will reduce N (see Eq. 12.58) and thus t. Also, special detection techniques (such as multiwavelength UV detection) that make it possible to extract analytical information from partially overlapping peaks reduce the required R_s and thus, as shown by Eq. 12.58, the required N.

The dependence of t on the N/R ratio means that R must not approach zero, but also must not approach unity because then N goes to infinity (see Eq. 12.58). It is easy to show that the minimum N/R ratio with respect to R variations occurs at $R = 1/3$.

For a given column type, mobile phase, temperature, and detection system, both N and R are constant and t depends only on the H/v ratio. We have looked at H versus v curves before (Figures 12.2 and 12.4); we note here that H/v is simply the slope of the straight line on such a plot connecting the origin with the point (the specific H and v values) representing a particular column operation. For a column with a plate height curve as shown in Figure 12.5, this slope is shown for the dashed lines that correspond to the two specific operating velocities v_1 and v_2.

We see from Figure 12.5 that the slope H/v becomes continuously smaller as v increases. We can surmise that the minimum H/v value is found only as $v \to \infty$. This result can be readily shown for all the plate height equations of Section 12.1. In the simple case $H = (B/v) + Cv$ (Eq. 12.48) we have

$$\frac{H}{v} = \frac{B}{v^2} + C \tag{12.61}$$

which clearly approaches its smallest possible value, C, when $v \to \infty$.

More specifically, if we substitute Eq. 12.61 into Eq. 12.60, we get for time

$$t = \frac{N}{R}\left(\frac{B}{v^2} + C\right) \tag{12.62}$$

Now suppose we have achieved our N plates and thus the desired separation at a certain velocity (e.g., v_1 in Figure 12.5). The last equation and Figure 12.5 show us that we can always do the separation faster if we increase v. Unfortunately, increasing v sets off a chain reaction of necessary adjustments.

First, as we increase v, we increase the pressure drop Δp proportionately. We see this by repeating Eq. 4.24a, using v in place of $\langle v \rangle$

$$\Delta p = \frac{\phi \eta v L}{d_p^2} \tag{12.63}$$

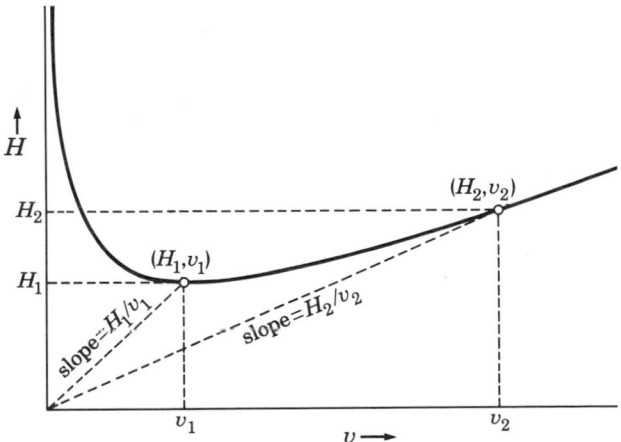

Figure 12.5. The H/v ratio shown as a slope for the two operating velocities v_1 and v_2.

In addition to the direct increase of Δp with v shown by this equation, the plate height also increases with v (for $v > v_{\mathrm{opt}}$), requiring that the column length be increased to keep the needed N plates. The increased L means a further increase in pressure drop Δp. We see then that an increase in v leads to a greater Δp for two reasons: Δp increases directly with v, and with increasing velocity the value of L must be greater, requiring another jump in Δp. Thus our attempt to increase speed leads invariably to a much larger Δp. As we continue to increase flow in pursuit of speed, we reach a practical limit beyond which Δp cannot be increased because of pump limitations, column bursting strength, and so on. We thus conclude that maximum separation speed requires, and is limited by, the maximum pressure drop available. This key conclusion was first stated by Knox for GC in 1961 [23].

Given the critical role of Δp, we next question if the best way to fully utilize the available Δp is to push v to the limit as just discussed. We constrained ourself above to a single column type with a specific H versus v curve (like that in Figure 12.5); for that column we increased v, H, and L to increase speed. However a specific column has a given packing and, of course, a fixed d_p. But there is no *a priori* reason to hold d_p constant while other parameters are allowed to vary. The key question is whether significant additional gains in speed would arise by varying particle diameter d_p.

To determine this, we write Eq. 12.60 in reduced form, which requires that H must be replaced by h and v by ν using Eqs. 12.25 and 12.23, respectively. We get

$$t = \frac{N}{R} \frac{h}{\nu} \frac{d_p^2}{D_m} \tag{12.64}$$

which shows the dramatic result that for a given position on the reduced plate height curve (i.e., a given h and ν), t is quadratically linked to d_p. Thus a powerful approach to increased speed appears to be the decrease of d_p. Clearly, however, in reducing d_p, an increasing Δp is required (see Eq. 12.63). Thus, rather than seeking speed (and expending our limited Δp) by increasing v (and ν) at a fixed particle size, we can boost speed more effectively (because of the quadratic dependence) by reducing d_p at constant ν. There is no need to push flow velocity beyond the point of minimum plate height providing we can effectively pack a column with smaller particles. However, the reduction in particle diameter is clearly limited by the Δp available to maintain adequate flow through the column. To optimize, we must find out how small d_p can become with a fixed Δp.

Theory shows that the smallest workable particle diameter is inversely related to the available pressure drop Δp. This is shown by rearranging Eq. 12.63 and substituting NH for L, which gives the d_p-Δp relationship as

$$d_p^2 = \frac{\phi \eta N H v}{\Delta p} \tag{12.65}$$

The product $H\upsilon$ can be replaced by $h\upsilon D_m$ using reduced variables, giving

$$d_p^2 = \frac{\phi \eta N D_m}{\Delta p} h\upsilon \tag{12.66}$$

When this expression for d_p^2 is substituted into the reduced time expression, Eq. 12.64, we obtain

$$t = \frac{\phi \eta N^2}{R \Delta p} h^2 \tag{12.67}$$

which shows that, generally, the smallest possible reduced plate height consistent with the upper operating limit of Δp should be sought. In practice, this means that we should use only well-packed columns with a minimal stationary phase contribution to the plate height so that the h versus υ curves are as low-lying as possible. We then do best if we specifically select a column with a d_p value calculated from Eq. 12.66, using h and υ values that fall at the minimum of the h versus υ curves. (The means for finding minimum h values are much like those used to get minimum H, discussed in Section 12.4.)

The optimization procedure outlined above is independent of the specific plate height equations chosen to represent chromatography. It depends only on the validity of the scaling relationships used to define reduced plate height and reduced velocity, and the emergence of universal curves from the scaling relationships.

The foregoing considerations explain the basis of the change in direction taken in the late 1960s in the development of high performance liquid chromatography (HPLC). Despite the low D_m values (relative to gaseous mobile phases) that tend to increase separation time (see Eq. 12.64), the replacement of large particles by small particles (3–10 μm) and low pressure drops by high pressure drops (hundreds of atmospheres), in accordance with the above theory, has led to the development of an efficient, high-speed separation tool of almost universal applicability.

It has been noted frequently in this book that separation speed for diverse methods is increased by reducing viscosity. Equation 12.67 shows that this rather universal conclusion is valid for chromatography as well.

For more details on time optimization, the reader is referred to treatments by Guiochon [24], Knox and Saleem [25], and Hartwick and Dezaro [26].

REFERENCES

[1] J. C. Giddings, *Dynamics of Chromatography*, Part 1, *Principles and Theory*, Marcel Dekker, New York, 1965.

[2] J. J. van Deemter, F. J. Zuiderweg, and A. Klinkenberg, *Chem. Eng. Sci.*, **5**, 271 (1956).

[3] J. H. Knox and J. F. Parcher, *Anal. Chem.*, **41**, 1599 (1969).

[4] J. N. Done, G. J. Kennedy, and J. H. Knox, in E. S. Perry, Ed., *Gas Chromatography* 1972, Applied Science, London, 1973.

[5] M. L. Lee, F. J. Yang, and K. D. Bartle, *Open Tubular Column Gas Chromatography: Theory and Practice*, Wiley, New York, 1984.

[6] T. Tsuda and M. Novotny, *Anal. Chem.*, **50**, 632 (1978).

[7] D. Ishii and T. Takeuchi, in *Advances in Chromatography*, Vol. 21, Marcel Dekker, New York, 1983, Chapter 4.

[8] J. W. Jorgenson, D. J. Rose, and R. T. Kennedy, *American Laboratory*, April 1988, p. 32.

[9] J. C. Fjeldsted and M. L. Lee, *Anal. Chem.*, **56**, 619A (1984).

[10] M. Novotny, S. R. Springston, P. A. Peaden, J. C. Fjeldsted, and M. L. Lee, *Anal. Chem.*, **53**, 407A (1981).

[11] M. J. E. Golay, in D. H. Desty, Ed., *Gas Chromatography* 1958, Academic Press, New York, 1958, pp. 36–53.

[12] J. C. Giddings, *Anal. Chem.*, **34**, 458 (1962).

[13] K. D. Bartle, C. L. Woolley, K. E. Markides, M. L. Lee, and R. S. Hansen, *J. High Resolut. Chromatogr. Chromatogr. Commun.*, **10**, 128 (1987).

[14] J. C. Giddings, *Anal. Chem.*, **35**, 353 (1963).

[15] J. C. Giddings, *Anal. Chem.*, **36**, 741 (1964).

[16] J. C. Giddings, *J. Chromatogr.*, **13**, 301 (1964).

[17] R. C. Reid, J. M. Prausnitz, and T. K. Sherwood, *The Properties of Gases and Liquids*, 3rd ed., McGraw-Hill, New York, 1977.

[18] F. J. Knorr, H. R. Thorsheim, and J. M. Harris, *Anal. Chem.*, **53**, 821 (1981).

[19] D. W. Osten and B. R. Kowalski, *Anal. Chem.*, **56**, 99 (1984).

[20] J. C. Giddings, *Anal. Chem.*, **39**, 1927 (1967).

[21] J. M. Davis and J. C. Giddings, *Anal. Chem.*, **55**, 418 (1983).

[22] J. C. Giddings, *J. Gas Chromatogr.*, **1**, 12 (1963).

[23] J. H. Knox, *J. Chem. Soc.*, 433 (1961).

[24] G. Guiochon, *Anal. Chem.*, **52**, 2002 (1980).

[25] J. H. Knox and M. Saleem, *J. Chromatogr. Soc.*, **7**, 614 (1969).

[26] R. A. Hartwick and D. D. Dezaro, in P. Kucera, Ed., *Microcolumn High-Performance Liquid Chromatography*, Elsevier, Amsterdam, 1984.

EXERCISES

12.1(*)   Using the van Deemter equation, estimate the plate height and number of theoretical plates in a 1-m long column using particles of diameter 0.014 cm and a helium flow velocity of 10 cm/s. Ignore gas compression effects. Previous work with the column has shown that the parameters λ and γ are 1.0 and 0.70, respectively; the depth of the pools of stationary liquid is $d = 4.0 \ \mu\text{m}$; $q = 2/3$. For the particular solute being considered assume $D_m = 0.25 \ \text{cm}^2/\text{s}$, $D_s = 3.2 \times 10^{-6} \ \text{cm}^2/\text{s}$, and $R = 0.25$.

12.2.(*) Plot H against v to find the velocity that gives the smallest value of plate height in the following typical cases. Use Eq. 12.8. The parameters that may be assumed equal in the two systems are $\lambda = 2.0$, $\gamma = 0.60$, $\omega = 1.5$, $d = 2.0 \ \mu\text{m}$, $q = 2/3$, and $D_s = 8.0 \times 10^{-6} \ \text{cm}^2/\text{s}$.

a. Gas chromatography. Assume $d_p = 0.020$ cm, $R = 0.10$, $D_m = 0.40 \ \text{cm}^2/\text{s}$.

b. Liquid chromatgraphy. Assume $d_p = 0.0050$ cm, $R = 0.30$, $D_m = 1.0 \times 10^{-5}$.

You will notice that the "optimum" velocity is considerably different in the two kinds of methods. Which parameter is mainly responsible for this?

12.3.(**) What is the apparent plate height $\hat{H} = L\tau^2/t_r^2$ for a toluene peak in a 20-m capillary with a uniform coating which yields the following component plate heights? $H_g = 2.00 \times 10^{-2}$ cm; $C_s v_i(\text{inlet}) = 3.00 \times 10^{-2}$ cm; $C_s v_0(\text{outlet}) = 7.50 \times 10^{-2}$ cm.

12.4.(**) Prove that the pressure-gradient correction term f_1 from Eq. 12.20 approaches unity as $P \rightarrow 1$ and approaches $9/8 \rightarrow \infty$.

12.5.(*) In Section 12.3 we found that when the two mobile phase terms A and $C_m v$ are equal, the flow velocity is approximately equal to the fundamental value $v_c = D_m/d_p$, which is used as a scale factor to get reduced velocity. Derive expressions for the velocities at which $A = B/v$ and $C_m v = B/v$, showing that these velocities are also of the same order of magnitude as fundamental velocity v_c.

12.6.(**) What length column is needed to separate at unit resolution components A and B where the distribution coefficients are $K_A = 110$, $K_B = 120$, and where the ratio $\beta = V_m/V_s = 20$? The plate height H for both components is approximately 0.050 cm.

12.7.(*) A gas chromatographic column with a benzene solute peak operates with the following parameters: $\gamma = 0.60$, $D_m = 0.50$ cm^2/s, $D_s = 5.0 \times 10^{-6} \ \text{cm}^2/\text{s}$, $d = 1.00 \times 10^{-3}$ cm, $q = 2/3$, and $R = 0.50$. Assuming that the plate height in this case is described to a good approximation by the equation $H = B/v + C_s v$, determine the optimum velocity (the velocity at which H is smallest). What is the plate height at the optimum velocity?

12.8.(*) Derive expressions for v_{opt} and H_{min} from the van Deemter equation, $H = A + (B/v) + Cv$.

12.9(**)   Use the form of the Knox equation given by Eq. 12.9, along with values from Table 12.1, to show that

$$v_{\text{opt}} = \frac{D_m}{d_p} \left(\frac{2\gamma}{\eta\alpha} \right)^{(1/n)+1}$$

when $C_s = 0$. (Note the difficulty of finding an analytical expression for v_{opt} when $C_s \neq 0$.)

12.10(***)   Starting with the Knox equation for the reduced plate height due to mobile-phase effects alone, $h = (2\gamma/v) + av^n$, show that h reaches the minimum value

$$h_{\min} = \frac{n+1}{n} (2\gamma)^{n/(n+1)} (na)^{1/(n+1)}$$

at a reduced velocity of

$$v_{\text{opt}} = \left(\frac{2\gamma}{na} \right)^{1/(n+1)}$$

12.11(**)   Transform the low-velocity form of the plate height equation

$$H = (B/v) + C_m v + C_s v$$

into reduced variables, expressing h as a function of v and of the parameters γ, ω, and so on.

12.12(**)   Verify the conclusion in the text that minimum time is found by adjusting R to equal 1/3, given constant values of H, v, R_s, and $\Delta K/K$.

12.13(***)   A chromatographer, fed up with waiting 36 min to accomplish a separation requiring 8000 plates, packs a new column to reduce separation time. He boldly slashes particle diameter by a factor of three. He adjusts the flow velocity to work at the same reduced velocity as before (the h versus v curves of the two columns are the same) and chooses a column length to get the same number of plates. Comparing the new column to the old, by what factor will the following parameters change: (a) plate height, (b) column length, (c) flow velocity, (d) pressure drop, and (e) separation time? What is his new separation time?

12.14(**)   The difficult separation of two closely related polypeptides by LC requires 120,000 plates in a system in which the slowest peak has $R = 0.160$. The LC column can be packed in a way that will produce a minimum h of 2.85 at $v = 4.60$. The maximum pressure drop for the system is 400 atm while $\phi = 620$. The viscosity equals 0.0100 poises; $D_m = 0.314 \times 10^{-5}$ cm^2/s. Calculate the shortest time necessary to achieve the separation and the particle diameter of the packing material to be used.

APPENDIX I

ANSWERS TO EXERCISES

1.1	$(a + b + c + d + e) \rightarrow (a + d) + (c + e) + (b)$
1.2	Proof
1.3	(a) 3.2 L, (b) 2592 L
1.4	$\Delta S_{sep} - -11.02$ cal/deg; not spontaneous
1.5	$\alpha' = 2.0 \times 10^4$
1.6	0.811
2.1	$\Delta \mu^0 = -1150$ cal/mol
2.2	7.2×10^7 spheres/cm^3
2.3	Proof
2.4	$K = 1 - as$
2.5	(a) $K = 0.12$, (b) $K = 0.35$
3.1	$A = Fm/f^2$, $B = F/f$, $C = -Fm/f^2$, $k = f/m$
3.2	(a) $\tau_r = 2.76 \times 10^{-14}$ s, (b) $\tau_r = 6.80 \times 10^{-11}$ s
3.3	Proof
3.4	$J = 5 \times 10^{-11}$ mol/cm^2 s
3.5	$\bar{U} = -5 \times 10^{-4}$ cm/s
3.6	(a) $f = 4.61 \times 10^{15}$ g s^{-1} mol^{-1}, (b) $2r = 7.12 \times 10^{-8}$ cm = 7.12 Å
3.7	(a) $f = 2.91 \times 10^{15}$ g s^{-1} mol^{-1}, (b) $D = 0.88 \times 10^{-5}$ cm^2 s^{-1}
3.8	MW $= 2.22 \times 10^7$ g/mol
3.9	(a) $r = 31.1$ Å, (b) $r = 27.3$ Å, (c) $t = 3.8$ Å; MW (total) = 93,000
3.10	(a) $r = 311$ Å, (b) $r = 45.6$ Å, (c) $t = 265$ Å; MW (total) = 75,800,000
3.11	(a) $U = 4.3 \times 10^{-7}$ cm/s, (b) 46,000 teams
3.12	$x_s = 5.20 \times 10^{-4}$ cm, $dc/dt = -0.0237$ mol/L per s
3.13	7.42×10^{-12} s
4.1	0.5 psi
4.2	$9Q_1$, $81Q_2$

4.3   (a) 1.9×10^{-2} cm/s, (b) same

4.4   (a) $Re = 5.2 \times 10^{-3}$, (b) no

4.5   45.8 cm

4.6   2.236 atm

4.7   Proof

4.8   Proof

4.9 $\langle v \rangle = 122$ cm/s

4.10 (a) $\langle v \rangle = 0.328$ cm/s, (b) $v_{max} = 0.492$ cm/s, (c) $\Delta p = 2.41 \times 10^{-3}$ atm

4.11 (a) $K_0 = 1.33 \times 10^{-10}$ cm^2, (b) $d_c = 1.92 \ \mu$m

4.12 $D = 2.04 \times 10^{-8}$ cm^2/s

4.13 (Table 4.1 values in parentheses): (a) 0.47 (0.33), (b) 0.44 (0.58), (c) 0.52 (0.46), (d) 0.74 (0.59), (e) 0.52 (0.97)

5.1 $2Dt$

5.2   (a) $\sigma^2 = w^2/12$, (b) $\sigma = 0.289$ cm

5.3   (a) $\sigma^2 = w^2/24$, (b) $\sigma = 0.204$ cm

5.4   Proof

5.5   (a) 0.0152 cm, (b) 0.118 cm, (c) 4.46 cm

5.6   (a) 4.65 cm, (b) 36.1 cm, (c) 1368 cm

5.7 $y = 10.5 \ \sigma$

5.8 $y = 54.8 \ l$

5.9   Proof

5.10 4.5×10^{-6} cm^2/s

5.11 6.2×10^{-4} cm

5.12 (a) 5.0×10^{-5} cm^2/s, (b) 0.001 cm, (c) 30,000 plates

5.13 3.5 s

5.14 100 cm

5.15 Proof

5.16 Proof

5.17 Proof

5.18 σ (theor.) = 3.2 steps

5.19 $30^{1/2}$ mm

5.20 $3^{1/2}$ mm

5.21 $H = \sigma^2/L$, $\sigma^2 = HL$
(a) $\sigma^2 = 0.01$ cm $\times$ 25 cm = 0.25 cm^2; $\sigma = 0.5$ cm, $4\sigma = 2$ cm
(b) $\sigma^2 = 0.01$ cm $\times$ 100 cm = 1 cm^2; $\sigma = 1$ cm, $4\sigma = 4$ cm

5.22 (a) $\tau = \sigma/V = \sigma t_r/L = 2$ cm $\times$ 200 s/100 cm = 4 s
(b) $H = \sigma^2/L = 4$ cm^2/100 cm = 0.04 cm
(c) $H = L\tau^2/t_r^2 = 100$ cm $\times$ 16 s^2/40,000 s^2 = 0.04 cm

5.23 Proof

5.24 0.0167

5.25 Proof

6.1 8.2×10^{-2} cm

6.2 1.43×10^{-4} cm/s

6.3 $4\sigma = 0.079$ cm

6.4	17 μm
6.5	5.8 μm
6.6	(a) $c_0 = 2.9$ mol/L, (b) precipitation and membrane fouling
6.7	(a) $\sigma_x = 0.180$ cm, (b) $\sigma_y = 0.072$ cm, (c) $C = 1.14$ cm^2
6.8	Proof
6.9	0.1005σ
6.10	0.0367σ
6.11	42
6.12	(a) 27 and 30, (b) 15 and 18
6.13	$a = 0.804$ cm, $b = 0.463$ cm
6.14	$n_c = 62.5$
7.1	μ^* sequence: high, very high, high, med. high, low, very high
7.2	Sd
7.3	Table required
8.1	Proof
8.2	Proof
8.3	(a) 6.38×10^{-3} cm/s, (b) 1.27×10^{-4} cm^2/V s
8.4	(a) 2.6×10^{-5} cm/s, (b) 3.6×10^{-13} s
8.5	(a) 61,000, (b) not substantially affected by hydration shell
8.6	(a) 0.727 cm, (b) 244 s, (c) 288 plates
8.7	$V = 625$ V, $t = 6.92$ min, $t = 173$ min
8.8	(a) $H_{min} = 0.665$ cm, (b) $L = 266$ m
8.9	$R_s = 2.1$
8.10	$R_s = 3.54$
8.11	9.2×10^{-3} pH units
8.12	Proof
9.1	$R = 1$
9.2	(a) $H = 0.25$ cm, $N/t = 0.64$ plates/s,
	(b) $H = 0.0025$ cm, $N/t = 64$ plates/s
9.3	Proof
9.4	Proof
9.5	Proof
9.6	$M = 10^5$: $\ell = 74.6$ μm, $\lambda = 0.294$, $R = 0.848$
	$M = 10^6$: $\ell = 7.46$ μm, $\lambda = 0.0294$, $R = 0.166$
	$M = 10^7$: $\ell = 0.746$ μm, $\lambda = 0.00294$, $R = 0.0175$
9.7	$M = 10^6$: $H = 1.47$ cm, $N = 34$ plates
	$M = 10^7$: $H = 0.00245$ cm, $N = 20,407$ plates
9.8	$R(A) = 0.338$, $R(B) = 1.50$
9.9	2.3 mm/s
10.1	b and c
10.2	(a) no, (b) no
10.3	(a) 5.4 cm, (b) 7.0 cm
10.4	$k' = 3$
10.5	0.3
10.6	Proof

10.7 Proof

10.8 Proof

10.9 (a) $D_n = 0.16$ cm^2/s, (b) $L = 120$ cm, (c) $4\sigma = 17.4$ cm

10.10 $D_n = k'/(1 + k')^2 \, vt_{eq}^2$

10.11 (a) $V_r = 43.2$ mL, (b) V_r (inlet) $= 10.8$ mL

11.1 (a) $R = 0.1$, (b) $1 - R = 0.9$

11.2 (a) $R = 0.42$, (b) $1 - R = 0.58$, (c) $R = 0.42$

11.3 0.101

11.4 (a) $R = 0.125$, (b) $k' = 7$

11.5 (a) $H_D = 0.06$ cm, (b) $H_D = 6 \times 10^{-6}$ cm

11.6 Proof

11.7 $H_n = 0.06$ cm

11.8 $C_s = 0.0078$ s

11.9 Proof

11.10 $v = 0.93$ cm/s

12.1 $H = 0.126$

12.2 (a) $v_{opt} = 21.2$ cm/s ($H_{min} = 0.0508$ cm); (b) $v_{opt} = 0.00269$ cm/s ($H_{min} = 0.0112$ cm); (c) D_m

12.3 $\hat{H} = 6.14 \times 10^{-2}$ cm

12.4 Proof

12.5 (a) $v = (\gamma/\lambda)v_c$, (b) $v = (2\gamma/\omega)^{1/2} \, v_c$

12.6 $L = 146$ cm

12.7 (a) $v_{opt} = 4.2$ cm/s, (b) $H_{min} = 0.28$ cm

12.8 Proof

12.9 Proof

12.10 Proof

12.11 $$H = \frac{2\gamma}{v} + \omega v + qR(1 - R) \frac{d^2 D_m}{d_p^2 D_s} v$$

12.12 Proof

12.13 (a) 1/3, (b) 1/3, (c) 3, (d) 9, (e) 1/9, (f) $t = 4$ min

12.14 (a) $t = 187$ min, (b) $d_p = 2.75 \, \mu$m

AUTHOR'S RELATED PUBLICATIONS

The titles of publications by the author and his co-workers cited in this book are given in list A. Other publications of general and theoretical interest in separations appear in list B. Many specialized publications are omitted.

A. CITED PUBLICATIONS

A Molecular-Dynamic Theory of Chromatography, J. C. Giddings and H. Eyring, *J. Phys. Chem.*, **59**, 416 (1955).

Kinetic Model for Chromatographic Dispersion and Electrodiffusion, J. C. Giddings, *J. Chem. Phys.*, **26**, 1755 (1957).

The Random Downstream Migration of Molecules in Chromatography, J. C. Giddings, *J. Chem. Phys.*, **35**, 588 (1958).

Spot Distribution and Size in Paper Chromatography, J. C. Giddings and R. A. Keller, *J. Chromatogr.*, **2**, 626 (1959).

Diffusion Analogy for Solvent Flow in Paper, A. L. Ruoff, D. L. Prince, and J. C. Giddings, *Kolloid-Zeitschrift*, **166**, 144 (1959).

Nonequilibrium and Diffusion: A Common Basis for Theories of Chromatography, J. C. Giddings, *J. Chromatogr.*, **2**, 44 (1959).

Paper Geometry and Flow Velocity in Paper Chromatography, A. L. Ruoff and J. C. Giddings, *J. Chromatogr.*, **3**, 438 (1960).

Zone Migration in Paper Chromatography, J. C. Giddings, G. H. Stewart, and A. L. Ruoff, *J. Chromatogr.*, **3**, 239 (1960).

Diffusion of Liquids in Unsaturated Paper, A. L. Ruoff, G. H. Stewart, H. K. Shin and J. C. Giddings, *Kolloid-Zeitschrift*, **173**, 14 (1960).

Liquid Distribution on Gas Chromatographic Support: Relationship to Plate Height, J. C. Giddings, *Anal. Chem.*, **34**, 458 (1962).

Theoretical Basis for a Continuous Large-Capacity Gas Chromatographic Apparatus, J. C. Giddings, *Anal. Chem.*, **34**, 37 (1962).

Kinetic Origin of Tailing Chromatography, J. C. Giddings, *Anal. Chem.*, **35**, 1999 (1963).

Principles of Column Performance in Large Scale Gas Chromatography, J. C. Giddings, *J. Gas Chromatogr.*, **1**, 12 (1963).

Plate Height of Nonuniform Chromatographic Columns. Gas Compression Effects, Coupled Columns, and Analogous Systems, J. C. Giddings, *Anal. Chem.*, **35**, 353 (1963).

The Theoretical Plate as a Measure of Column Efficiency, J. C. Giddings, *J. Gas Chromatogr.*, **2**, 167 (1964).

Role of Column Pressure Drop in Gas Chromatographic Resolution, J. C. Giddings, *Anal. Chem.*, **36**, 741 (1964).

Reduced Plate Height Equation: A Common Link Between Chromatographic Methods, J. C. Giddings, *J. Chromatogr.*, **13**, 301 (1964).

J. C. Giddings, *Dynamics of Chromatography, Part I: Principles and Theory*, Marcel Dekker, New York, 1965.

Pressure Profiles, Fluctuations, and Pulses in Gas Chromatographic Columns, P. D. Schettler, Jr. and J. C. Giddings, *Anal. Chem.*, **37**, 835 (1965).

A New Separation Concept Based on a Coupling of Concentration and Flow Nonuniformities, J. C. Giddings, *Sep. Sci.*, **1**, 123 (1966).

Turbulent Gas Chromatography, J. C. Giddings, W. A. Manwaring, and M. N. Myers, *Science*, **154**, 146 (1966).

Physico-Chemical Basis of Chromatography, J. C. Giddings, *J. Chem. Ed.*, **44**, 704 (1967).

Maximum Number of Components Resolvable by Gel Filtration and Other Elution Chromatographic Methods, J. C. Giddings, *Anal. Chem.*, **39**, 1927 (1967).

Statistical Theory for the Equilibrium Distribution of Rigid Molecules in Inert Porous Networks. Exclusion Chromatography, J. C. Giddings, E. Kucera, C. P. Russell, and M. N. Myers, *J. Phys. Chem.*, **72**, 4397 (1968).

Resolution and Optimization in Gel Filtration and Permeation Chromatography, J. C. Giddings, *Anal. Chem.*, **40**, 2143 (1968).

High Pressure Gas Chromatography of Nonvolatile Species, J. C. Giddings, M. N. Myers, L. McLaren, and R. A. Keller, *Science*, **162**, 67 (1968).

Application of the Nonequilibrium Theory of Chromatography to a Variable Flow Correlation Model of Complex Flow and Coupling, P. D. Schettler and J. C. Giddings, *J. Phys. Chem.*, **73**, 2582 (1969).

General Nonequilibrium Theory of Chromatography with Complex Flow Transport, J. C. Giddings and P. D. Schettler, *J. Phys. Chem.*, **73**, 2577 (1969).

Dense Gas Chromatography at Pressures to 2000 Atmospheres, J. C. Giddings, M. N. Myers, and J. W. King, *J. Chromatogr. Sci.*, **7**, 276 (1969).

Generation of Variance, "Theoretical Plates," Resolution, and Peak Capacity in Electrophoresis and Sedimentation, J. C. Giddings, *Sep. Sci.*, **4**, 181 (1969).

Computer Characterization of Chromatographic Peaks by Plate Height and Higher Central Moments, E. Grushka, M. N. Myers, P. D. Schettler, and J. C. Giddings, *Anal. Chem.*, **41**, 889 (1969).

Column Parameters in Thermal Field-Flow Fractionation, M. E. Hovingh, G. E. Thompson, and J. C. Giddings, *Anal. Chem.*, **42**, 195 (1970).

Moments Analysis for the Discernment of Overlapping Chromatographic Peaks, E. Grushka, M. N. Myers, and J. C. Giddings, **42**, 21 (1970).

Resolution and Peak Capacity in Equilibrium-Gradient Methods of Separation, J. C. Giddings and K. Dahlgren, *Sep. Sci.*, **6**, 345 (1971).

Electrical Field-Flow Fractionation of Proteins, K. D. Caldwell, L. F. Kesner, M. N. Myers, and J. C. Giddings, *Science*, **176**, 296 (1972).

The Conceptual Basis of Field-Flow Fractionation, J. C. Giddings, *J. Chem. Educ.*, **50**, 667 (1973).

Thermal Field-Flow Fractionation: Extension to Lower Molecular Weight Separations by Increasing the Liquid Temperature Range Using a Pressurized System, J. C. Giddings, L. K. Smith, and M. N. Myers, *Anal. Chem.*, **47**, 2389 (1975).

Flow Field-Flow Fractionation: A Versatile New Separation Method, J. C. Giddings, F. J. Yang, and M. N. Myers, *Science*, **193**, 1244 (1976).

Mass Analysis of Particles and Macromolecules by Field-Flow Fractionation, J. C. Giddings, M. N. Myers, F. J. F. Yang, and L. K. Smith, in M. Kerker, Ed., *Colloid and Interface Science*, Vol. IV, Academic Press, New York, 1976, pp. 381–398.

Exclusion Chromatography in Dense Gases: An Approach to Viscosity Optimization, J. C. Giddings, L. M. Bowman, Jr., and M. N. Myers, *Anal. Chem.*, **49**, 243 (1977).

Basic Approaches to Separation. Analysis and Classification of Methods According to Underlying Transport Characteristics, J. C. Giddings, *Sep. Sci. Technol.*, **13**, 3 (1978).

Steric Field-Flow Fractionation: A New Method for Separating 1–100 μm Particles, J. C. Giddings and M. N. Myers, *Sep. Sci. Technol.*, **13**, 637 (1978).

Basic Approaches to Separation: Steady-State Zones and Layers, J. C. Giddings, *Sep. Sci. Technol.*, **14**, 871 (1979).

Thermogravitational Field-Flow Fractionation: An Elution Thermogravitational Column, J. C. Giddings, M. Martin, and M. N. Myers, *Sep. Sci. Technol.*, **14**, 611 (1979).

Observations on Anomalous Retention in Steric Field-Flow Fractionation, K. D. Caldwell, T. T. Nguyen, M. N. Myers, and J. C. Giddings, *Sep. Sci. Technol.*, **14**, 935 (1979).

Analysis of Biological Macromolecules and Particles by Field-Flow Fractionation, J. C. Giddings, M. N. Myers, K. D. Caldwell, and S. R. Fisher, in D. Glick, Ed., *Methods of Biochemical Analysis*, Vol. 26, Wiley, New York, 1980, pp. 79–136.

Future Pathways for Analytical Separations, J. C. Giddings, *Anal. Chem.*, **53**, 945A (1981).

Field-Flow Fractionation: Methodological and Historical Perspectives, J. C. Giddings, M. N. Myers, and K. D. Caldwell, *Sep. Sci. Technol.*, **16**, 549 (1981).

Field-Flow Fractionation: A Versatile Method for the Characterization of Macromolecular and Particulate Materials, J. C. Giddings, *Anal. Chem.*, **53**, 1170A (1981).

Principles of Chemical Separations, J. C. Giddings, in I. M. Kolthoff and P. J. Elving, Eds., *Treatise on Analytical Chemistry*, Part I, Vol. 5, Wiley, New York, 1982, Chapter 3.

Statistical Theory of Component Overlap in Multicomponent Chromatograms, J. M. Davis and J. C. Giddings, *Anal. Chem.*, **55**, 418 (1983).

Colloid Characterization by Sedimentation Field-Flow Fractionation. I. Monodisperse Populations, J. C. Giddings, G. Karaiskakis, K. D. Caldwell, and M. N. Myers, *J. Colloid Interface Sci.*, **92**, 66 (1983).

Field-Flow Fractionation, J. C. Giddings, *Sep. Sci. Technol.*, **19**, 831 (1984).

Field-Flow Fractionation: Choices in Programmed and Nonprogrammed Operation, J. C. Giddings and K. D. Caldwell, *Anal. Chem.*, **56**, 2093 (1984).

Test of the Statistical Model of Component Overlap by Computer-Generated Chromatograms, J. C. Giddings, J. M. Davis, and M. R. Schure, in S. Ahuja, Ed., *Ultrahigh Resolution Chromatography*, ACS Symposium Series No. 250, American Chemical Society, Washington, DC, 1984, pp. 9–26.

Two-Dimensional Separations: Concept and Promise, J. C. Giddings, *Anal. Chem.*, **56**, 1258A (1984).

Statistical Method for Estimation of Number of Components from Single Complex Chromatograms: Application to Experimental Chromatograms, J. M. Davis and J. C. Giddings, *Anal. Chem.*, **57**, 2178 (1985).

Crossflow Gradients in Thin Channels for Separation by Hyperlayer FFF, SPLITT Cells, Elutriation, and Related Methods, J. C. Giddings, *Sep. Sci. Technol.*, **21**, 831 (1986).

High-Speed Separation of Large ($>1\ \mu$m) Particles by Steric Field-Flow Fractionation, T. Koch and J. C. Giddings, *Anal. Chem.*, **58**, 994 (1986).

Fast Particle Separation by Flow/Steric Field-Flow Fractionation, J. C. Giddings, X. Chen, K.-G. Wahlund, and M. N. Myers, *Anal. Chem.*, **59**, 1957 (1987).

Fractionating Power in Sedimentation Field-Flow Fractionation with Linear and Parabolic Field Decay Programming, P. S. Williams, J. C. Giddings, and R. Beckett, *J. Liq. Chromatogr.*, **10**, 1961 (1987).

Measuring Particle Size Distribution of Simple and Complex Colloids Using Sedimentation Field-Flow Fractionation, J. C. Giddings, K. D. Caldwell, and H. K. Jones, in T. Provder, Ed., *Particle Size Distribution: Assessment and Characterization*, ACS Symposium Series No. 332, American Chemical Society, Washington, DC, 1987, Chapter 15.

Concepts and Comparisons in Multidimensional Separation, J. C. Giddings, *J. High Res. Chromatogr. Chromatogr. Commun.*, **10**, 319 (1987).

Gravity Augmented High-Speed Flow/Steric Field-Flow Fractionation: Simultaneous Use of Two Fields, X. Chen, K.-G. Wahlund, and J. C. Giddings, *Anal. Chem.*, **60**, 362 (1988).

Field-Flow Fractionation, J. C. Giddings, *C&E News*, **66**, 34 (1988).

Separation and Size Characterization of Colloidal Particles in River Water by Sedimentation Field-Flow Fractionation, R. Beckett, G. Nicholson, B. T. Hart, M. Hansen, and J. C. Giddings, *Wat. Res.*, **22**, 1535 (1988).

Rapid Particle Size Analysis of Ground Minerals by Flow/Hyperlayer Field-Flow Fractionation, B. N. Barman, M. N. Myers and J. C. Giddings, *Powder Technol.*, **59**, 53 (1989).

High-Speed Size Characterization of Chromatographic Silica by Flow/Hyperlayer Field-Flow Fractionation, S. K. Ratanathanawongs and J. C. Giddings, *J. Chromatogr.*, **467**, 341 (1989).

Colloid Characterization by Sedimentation Field-Flow Fractionation. VII. Colloidal Aggregates, J. C. Giddings, B. N. Barman and H. Li, *J. Colloid Interface Sci.*, **132**, 554 (1989).

Retention Perturbations Due to Particle-Wall Interactions in Sedimentation Field-Flow Fractionation, M. E. Hansen and J. C. Giddings, *Anal. Chem.*, **61**, 811 (1989).

Field-Flow Fractionation: An HPLC Analogue, L. F. Kesner and J. C. Giddings, in P. R. Brown and R. A. Hartwick, Eds., *High Performance Liquid Chromatography*, Wiley, New York, 1989, Chapter 15.

Separation and Characterizing of Colloidal Materials of Variable Particle Size and Composition by Coupled Column Sedimentation Field-Flow Fractionation, H. K. Jones and J. C. Giddings, *Anal. Chem.*, **61**, 741 (1989).

Field-Flow Fractionation: An Alternative to Size Exclusion Chromatography, J. C. Giddings, in B. J. Hunt and S. Holding, Eds., *Size Exclusion Chromatography*, Blackie and Son, Glasgow, 1989, Chapter 8.

Harnessing Electrical Forces for Separation: CZE, IEF, FFF, SPLITT, and Other Techniques, J. C. Giddings, *J. Chromatogr.*, **480**, 21 (1989).

Use of Multiple Dimensions in Analytical Separations, J. C. Giddings, in H. J. Cortes, Ed., *Multidimensional Chromatography: Techniques and Applications*, Marcel Dekker, New York, 1990, Chapter 1.

Two-Dimensional Field-Flow Fractionation, J. C. Giddings, *J. Chromatogr.*, **504**, 247 (1990).

B. OTHER PUBLICATIONS IN SEPARATION SCIENCE

Stochastic Considerations on Chromatographic Dispersion, J. C. Giddings, *J. Chem. Phys.*, **26**, 169 (1957).

Eddy Diffusion in Chromatography, J. C. Giddings, *Nature*, **184**, 357 (1959).

Nonequilibrium Kinetics and Chromatography, J. C. Giddings, *J. Chem. Phys.*, **31**, 1462 (1959).

Multiple Zones and Spots in Chromatography, R. A. Keller and J. C. Giddings, *J. Chromatogr.*, **3**, 205 (1960).

Zone and Boundary Diffusion in Electrophoresis, J. R. Boyack and J. C. Giddings, *J. Biological Chem.*, **235**, 1970 (1960).

Theory of Programmed Temperature Gas Chromatography: The Prediction of Optimum Parameters, J. C. Giddings, in *Gas Chromatography*, Academic Press, New York, 1962, Chapter V.

Theory of Electrophoretic Mobility in Stabilized Media, J. R. Boyack and J. C. Giddings, *Arch. Biochem. Biophys.*, **100**, 16 (1963).

Mechanism of Electrophoretic Migration in Paper, J. C. Giddings and J. R. Boyack, *Anal. Chem.*, **36**, 1229 (1964).

Failure of the Eddy Diffusion Concept of Gas Chromatography, J. C. Giddings and R. A. Robison, *Anal. Chem.*, **34**, 885 (1964).

Elementary Theory of Programmed Temperature Gas Chromatography, J. C. Giddings, *J. Chem. Educ.*, **39**, 569 (1962).

Generalized Nonequilibrium Theory of Plate Height in Large Scale Gas Chromatography, J. C. Giddings, *J. Gas Chromatogr.*, **1**, 38 (1963).

General Combination Law for C_1 Terms in Gas Chromatography, J. C. Giddings, *J. Phys. Chem.*, **68**, 184 (1964).

Evidence on the Nature of Eddy Diffusion in Gas Chromatography from Inert (Non-sorbing) Column Data, J. C. Giddings, *Anal. Chem.*, **35**, 1338 (1963).

Liquid Chromatography with Operating Conditions Analogous to Those of Gas Chromatography, J. C. Giddings, *Anal. Chem.*, **35**, 2215 (1963).

Theory of Gas-Solid Chromatography: Potential for Analytical Use and the Study of Surface Kinetics, J. C. Giddings, *Anal. Chem.*, **36**, 1170 (1964).

Contribution of Interfacial Resistance to Theoretical Plate Height in Gas Chromatography, M. R. James, J. C. Giddings, and H. Eyring, *J. Phys. Chem.*, **68**, 1725 (1964).

Measurement and Interpretation of the C Terms of Gas Chromatography, J. C. Giddings and P. D. Schettler, *Anal. Chem.*, **36**, 1483 (1964).

Comparison of the Theoretical Limit of Separating Speed in Gas and Liquid Chromatography, J. C. Giddings, *Anal. Chem.*, **37**, 60 (1965).

Thermodynamic Pitfalls in Gas Chromatography, M. R. James, J. C. Giddings, and R. A. Keller, *J. Gas Chromatogr.*, **3**, 57 (1965).

A Critical Evaluation of the Theory of Gas Chromatography, J. C. Giddings, in A. Goldup, Ed., *Gas Chromatography* 1964, Elsevier, Amsterdam, 1964.

Surface Active Agents and Interfacial Transfer in Gas Liquid Chromatography: A New Tool for Measuring Interfacial Resistance, M. R. James, J. C. Giddings, and H. Eyring, *J. Phys. Chem.*, **69**, 2351 (1965).

Dynamics of Mass Transfer and the Generalized Nonequilibrium Theory of Chromatography, J. C. Giddings, *Ber. Bunsenges. Phys. Chem.*, **69**, 773 (1965).

Theory: A Practical Tool in Gas Chromatography, J. C. Giddings, *Méthodes Phys. Anal. (GAMS)*, **1**, 13 (1966).

High Column Efficiency in Gas Liquid Chromatography at Inlet Pressures to 2500 psi, M. N. Myers and J. C. Giddings, *Anal. Chem.*, **37**, 1453 (1965).

The Nature of Eddy Diffusion, J. C. Giddings, *Anal. Chem.*, **38**, 490 (1966).

High Inlet Pressure Micro Column System for Use in Gas Chromatography, M. N. Myers and J. C. Giddings, *Anal. Chem.*, **38**, 294 (1966).

Some Aspects of Pressure Induced Equilibrium Shifts in Chromatography, J. C. Giddings, *Sep. Sci.*, **1**, 73 (1966).

Theory of Gel Filtration (Permeation) Chromatography, J. C. Giddings and K. L. Mallik, *Anal. Chem.*, **38**, 997 (1966).

Nonideal Gas Corrections for Retention and Plate Height in Gas Chromatography, P. D. Schettler, M. Eikelberger, and J. C. Giddings, *Anal. Chem.*, **39**, 146 (1967).

Theoretical Basis of Macromolecular Chromatography, J. C. Giddings, *J. Gas Chromatogr.*, **5**, 143 (1967).

Chromatography—Trend to Unity, J. C. Giddings and R. A. Keller, *Ind. Res.*, May 5, 1967, p. 70.

Dense-Gas Chromatography of Nonvolatile Substances of High Molecular Weight, L. McLaren, M. N. Myers, and J. C. Giddings, *Science*, **159**, 197 (1968).

Nonequilibrium Theory of Field-Flow Fractionation, J. C. Giddings, *J. Chem. Phys.*, **49**, 81 (1968).

Evaluation of Coupling and Turbulence by the Dynamical Comparison of Gas and Liquid Chromatography, H. Kaizuma, M. N. Myers, and J. C. Giddings, *J. Chromatogr. Sci.*, **8**, 630 (1970).

Programmed Exclusion Chromatography: A Method for the Continuous Control of Retention, J. C. Giddings and K. Dahlgren, *Sep. Sci.*, **5**, 717 (1970).

Solubility Phenomena in Dense Carbon Dioxide Gas in the Range 270–1900 Atmospheres, J. J. Czubryt, M. N. Myers, and J. C. Giddings, *J. Phys. Chem.*, **74**, 4260 (1970).

Parameters for Optimum Separations in Field-Flow Fractionation, J. C. Giddings, *Sep. Sci.*, **8**, 567 (1973).

Theory of Chromatography, J. C. Giddings, in E. Heftmann, Ed., *Chromatography*, 3rd ed., Van Nostrand Reinhold, New York, 1975, Chapter 3.

Theoretical Basis of Partition Chromatography, R. A. Keller and J. C. Giddings, in E. Heftmann, Ed., *Chromatography*, 3rd ed., Van Nostrand Reinhold, New York, 1975, Chapter 6.

Programmed Sedimentation Field-Flow Fractionation, F. J. F. Yang, M. N. Myers, and J. C. Giddings, *Anal. Chem.*, **46**, 1924 (1974).

Field-Flow Fractionation: Extending the Molecular Weight Range of Liquid Chromatography to One Trillion, *J. Chromatogr.*, **125**, 3 (1976).

Isolation of Peak Broadening Factors in Exclusion (Gel) Chromatography, J. C. Giddings, L. M. Bowman, Jr., and M. N. Myers, *Macromolecules*, **10**, 443 (1977).

The Flow Field-Flow Fractionation Channel as a Versatile Pressure Dialysis and Ultrafiltration Cell, J. C. Giddings, F. J. Yang, and M. N. Myers, *Sep. Sci.*, **12**, 499 (1977).

Flow Field-Flow Fractionation as a Methodology for Protein Separation and Characterization, J. C. Giddings, F. J. Yang, and M. N. Myers, *Anal. Chem.*, **81**, 395 (1977).

Polymer Analysis and Characterization by Field-Flow Fractionation (One-Phase Chromatography), J. C. Giddings, M. N. Myers, G. C. Lin, and M. Martin, *J. Chromatogr.*, **142**, 23 (1977).

Displacement and Dispersion of Particles of Finite Size in Flow Channels with Lateral Forces. Field-Flow Fractionation and Hydrodynamic Chromatography, J. C. Giddings, *Sep. Sci. Technol.*, **13**, 241 (1978).

Field-Flow Fractionation—One Phase Chromatography for Macromolecules and Particles, J. C. Giddings, S. R. Fisher, and M. N. Myers, *Am. Lab.*, **10**, 15 (1978).

J. C. Giddings, Autobiographical Chapter, in L. S. Ettre and A. Zlatkis, Eds., 75 *Years of Chromatography—A Historical Dialogue*, Elsevier, Amsterdam, 1979, pp. 87–98.

Analysis of Fundamental Obstacles to the Size Exclusion Chromatography of Polymers of Ultra High Molecular Weight, J. C. Giddings, in J. C. Giddings, E. Grushka, J. Cazes, and P. R. Brown, Eds., *Advances in Chromatography.*, Vol. 20, Marcel Dekker, New York, 1982, Chapter 6.

Supercritical Fluid (Dense Gas) Chromatography/Extraction with Linear Density Programming, L. M. Bowman, Jr., M. N. Myers, and J. C. Giddings, *Sep. Sci. Technol.*, **17**, 271 (1982).

Capillary Liquid Chromatography in Field-Flow Fractionation-type Channels, J. C. Giddings, J. P. Chang, M. N. Myers, J. M. Davis, and K. D. Caldwell, *J. Chromatogr.*, **225**, 359 (1983).

Hyperlayer Field-Flow Fractionation, J. C. Giddings, *Sep. Sci. Technol.*, **18**, 765 (1983).

Liquid-Gas Partition Chromatography (LGC): An LC System with a Gaseous Stationary Phase for Gas Analysis, J. C. Giddings and M. N. Myers, *J. High Res. Chromatogr. Chromatogr. Commun.*, **6**, 381 (1983).

Origin and Characterization of Departures from the Statistical Model of Component-Peak Overlap in Chromatography, J. M. Davis and J. C. Giddings, *J. Chromatogr.*, **289**, 277 (1984).

Separation of Human and Animal Cells by Steric Field-Flow Fractionation, K. D. Caldwell, Z.-Q. Cheng, P. Hradecky, and J. C. Giddings, *Cell Biophys.*, **6**, 233 (1984).

Cyclical-Field Field-Flow Fractionation: A New Method Based on Transport Rates, J. C. Giddings, *Anal. Chem.*, **58**, 2052 (1986).

Statistical Method for Estimation of Number of Components from Single Complex Chromatograms: Theory, Computer-Based Testing, and Analysis of Errors, J. M. Davis and J. C. Giddings, *Anal. Chem.*, **57**, 2168 (1985).

A System Based on Split-Flow Lateral-Transport Thin (SPLITT) Separation Cells for Rapid and Continuous Particle Fractionation, J. C. Giddings, *Sep. Sci. Technol.*, **20**, 749 (1985).

Simplified Nonequilibrium Theory of Secondary Relaxation Effects in Programmed Field-Flow Fractionation, J. C. Giddings, *Anal. Chem.*, **58**, 735 (1986).

Feasibility Study of Dielectrical Field-Flow Fractionation, J. M. Davis and J. C. Giddings, *Sep. Sci. Technol.*, **21**, 969 (1986).

Calculation of Flow Properties and End Effects in Field-Flow Fractionation Channels by a Conformal Mapping Procedure, P. S. Williams, S. B. Giddings, and J. C. Giddings, *Anal. Chem.*, **58**, 2397 (1986).

Thermal Field-Flow Fractionation Using Supercritical Fluids, J. J. Gunderson and J. C. Giddings, *Anal. Chem.*, **59**, 23 (1987).

Comparison of Polymer Resolution in Thermal Field-Flow Fractionation and Size Exclusion Chromatography, J. J. Gunderson and J. C. Giddings, *Anal. Chim. Acta*, **189**, 1 (1986).

Separation in Thin Channels: Field-Flow Fractionation and Beyond, J. C. Giddings, in J. D. Navratil and C. J. King, Eds., *Chemical Separations*, Vol. 1, Litarvan, Denver, 1986, pp. 3–20.

Transport, Space, Entropy, Diffusion, and Flow: Elements Underlying Separation by Electrophoresis, Chromatography, Field-Flow Fractionation and Related Methods, J. C. Giddings, *J. Chromatogr.*, **395**, 19 (1987).

Continuous Separation in Split-Flow Thin (SPLITT) Cells: Potential Applications to Biological Materials, J. C. Giddings, *Sep. Sci. Technol.*, **23**, 931 (1988).

Separation of B and T Lymphocytes by a Hybrid Field-Flow Fractionation/Adhesion Chromatography Technique, J. C. Bigelow, J. C. Giddings, Y. Nabeshima, T. Tsuruta, K. Kataoka, T. Okano, N. Yui, and Y. Sakurai, *J. Immunol. Methods*, **117**, 289 (1989).

Use of Secondary Equilibria for the Separation of Small Solutes by Field-Flow Fractionation, A. Berthod, D. W. Armstrong, M. N. Myers, and J. C. Giddings, *Anal. Chem.*, **60**, 2138 (1988).

Field-Flow Fractionation of Macromolecules, J. C. Giddings, *J. Chromatogr.*, **470**, 327 (1989).

Continuous Separation of Particles from Macromolecules in Split-Flow Thin (SPLITT) Cells, S. Levin and J. C. Giddings, *J. Chem. Tech. Biotechnol.*, in press.

Continuous Separation of Proteins in Electrical Split-Flow Thin (SPLITT) Cell with Equilibrium Operation, S. Levin, M. N. Myers, and J. C. Giddings, *Sep. Sci. Technol.*, **24**, 1245 (1989).

AUTHOR INDEX

SUBJECT INDEX

Gaussian (*Continued*)
    point of inflection, 91
Gradient:
  density, 180
  pH, 178
  primary, 180
  secondary, 178, 180
Gravitation, *see* Sedimentation
Gravitational sedimentation, 170, 171

Hagen–Poiseuille equation, 60
Height equivalent to theoretical plate, 97.
  See also Plate height
Hydrogen bonds, 25, 27
Hydrophobic effect, 26, 31, 227

Interactions, intermolecular, 22, 25, 26, 27,
  227
Ion exchange, 146
Irreversible thermodynamics, 38, 39, 174
  reciprocity, 38, 39
Isoelectric focusing, 112, 115, 116, 146–148,
  165, 177–182
  peak capacity, 183
  separation power, 180–182
Isoelectric point, 178, 179, 182
Isopycnic point, 182
Isopycnic sedimentation, 115, 116, 145, 146,
  177, 180, 182, 183
  peak capacity, 183
  resolution, 183
  separation power, 182–184
Isotachophoresis, 146, 148, 163, 165
Isotopes, separation by thermogravitation,
  200

Kinetics, interfacial, 196
Kozeny–Carman equation, 65

Laminar flow, 74–75
Langevin equation, 42
Langmuir isotherm, 229, 236
Liquid:
  compression effects, 68
  viscosity, 57
Liquid chromatography, *see*
  Chromatography, liquid

Membrane methods, 146, 148, 152
  steady layers, 117–119
Membranes, 31–35
Micelles, partition into, 162
Moments, statistical, 90–92
Multidimensional separation, 125–126,
  231

Navier–Stokes equation, 56
Nonequilibrium, 197, 241–246
  theory, 198, 243–246
Nonlinearities, 107, 108
Nonuniformities, gas compressibility, 99,
  273, 274
Nonuniform separation systems, 99
Number of theoretical plates, 98, 101, 190,
  281

Optimization, 195–197. *See also*
  Chromatography, optimization
  of electrophoresis, 176, 177
  of $F(+)$ systems, 195–197
  of sedimentation, 176, 177
  in static field systems, 176, 177
Overloading, 107

Packed beds, flow, 94–95, 252, 259–265
Partition coefficient, 21
Peak(s). *See also* Zones
  apparent, 130, 132
  area, 128
  capacity, 9, 105, 106, 128, 129, 132,
    133–134, 136, 159, 181, 183, 282
    two-dimensional, 126–128
  doublets, 132
  evenly spaced, 129
  fused, 130–132, 136
  isolated, 133
  number of, 132–136
  overlap, 128–137, 282
    pervasive, 134
    random, 130–136
    statistics, 131–136
  randomly spaced, 130–136, 282
  resolution, numerical, 136
  saturation, 133, 134, 136
  shape differences, 136
  singlets, 132, 133–134, 136
Permeability, specific, 63, 64
Phases, relative displacement, 214
Photophoresis, 155–156
Plate height, 9, 96–101, 126, 195, 198
  effect of flow rate, 196, 245, 246
  elution systems, 99–101
  and gas compressibility, 273, 274
  local, 99
  longitudinal diffusion, 195
  nonequilibrium, 245, 246
  observed, 274
Plate model, 97
Plates, 158–159
  number of, 9, 98, 101, 158, 159, 190,
    281